A complex system (Charlene Lam, 1987).

Introduction to Nonlinear Physics

Springer
New York
Berlin
Heidelberg
Barcelona
Budapest
Hong Kong
London
Milan
Paris
Santa Clara
Singapore
Tokyo

Lui Lam
Editor

Introduction to
Nonlinear Physics

With 264 Figures

Springer

Lui Lam
Department of Physics
San Jose State University
San Jose, CA 95192-0106
USA

Library of Congress Cataloging-in-Publication Data
Introduction to nonlinear physics / Lui Lam, editor.
    p.   cm.
   Includes bibliographical references and index.
   ISBN 0-387-94758-2 (alk. paper)
   1. Nonlinear theories.   2. Mathematical physics.   I. Lam, Lui.
QC20.7.N6I67   1996
530.1′55252—dc20   96-14764

Printed on acid-free paper.

Production coordinated by Chernow Editorial Services, Inc., and managed by Bill Imbornoni;
manufacturing supervised by Jeffrey Taub.
Typeset by Asco Trade Typesetting Ltd., Hong Kong.
Printed and bound by R.R. Donnelley and Sons, Harrisonburg, VA.
Printed in the United States of America.

9 8 7 6 5 4 3 2 1

ISBN 0-387-94758-2 Springer-Verlag New York Berlin Heidelberg   SPIN 10536582

Preface

A revolution occurred quietly in the development of physics—or, more accurately, of science—in the last three decades. The revolution touches upon *every* discipline in both the natural and social sciences. We are referring to the birth of a new science—nonlinear science—which, for the sake of presentation, may be divided into six parts: fractals, chaos, pattern formation, solitons, cellular automata, and complex systems.

Yet, in spite of all the excitement about this new science, there is not a single textbook covering all these topics. To remedy this situation and in view of the diversity of the subject, a number of pioneers and experts were invited to write about their own fields of research. The result is a textbook intended for advanced undergraduates and graduate students, which is also suitable for self-study. The materials contained in this book have been test taught in classrooms in universities, and in summer and winter schools. Examples and homework problems are included in most chapters.

Emphasis is placed on fractals, chaos, pattern formation, and solitons, which form Parts I to IV in the book. Special topics, including cellular automata, turbulence, and complex systems, are grouped in Part V. Part I to Part IV can be studied independently of each other, whereas Part V can be ignored in a first reading, except that Chapter 15 is a useful supplement to Part III.

Although most of the applications in this book are taken from examples in the physical sciences, the general principles and theories expounded are definitely applicable to other branches of science. The book is thus of use to students and researchers not just in physics but also in, for example, chemistry, biology, astronomy, meteorology, geology, mathematics, computer science, engineering, medicine, economics, and ecology. The multidisciplinary nature of nonlinear science makes it the ideal course for broadening the perspective and education of students.

I am grateful to the contributors and the publisher for their professional skills and patience in making this book possible. For discussion and encouragement I want to thank numerous colleagues, in particular, Armin Bunde, David Campbell, Patricia Cladis, James Crutchfield, Herman Cummins,

John Holland, Alfred Hübler, Stuart Kauffman, Yuji Kodama, Mitsugu Matsushita, Michael Nauenberg, Ru-Pin Pan, Gene Stanley, Harry Swinney, and Wing-Yim Tam. I am indebted to Marilyn Lam and Charlene Lam for contributing to the artwork and providing the sketch on page ii, and for a lot of other things.

San Jose   Lui Lam

Contents

Preface   v

1 Introduction   1
   Lui Lam

   1.1 A Quiet Revolution   1
   1.2 Nonlinearity   2
   1.3 Nonlinear Science   4
        1.3.1 Fractals   4
        1.3.2 Chaos   5
        1.3.3 Pattern Formation   6
        1.3.4 Solitons   6
        1.3.5 Cellular Automata   7
        1.3.6 Complex Systems   8
   1.4 Remarks   9
   References   10

Part I Fractals and Multifractals

2 Fractals and Diffusive Growth   15
   Thomas C. Halsey

   2.1 Percolation   16
   2.2 Diffusion-Limited Aggregation   17
   2.3 Electrostatic Analogy   20
   2.4 Physical Applications of DLA   22
        2.4.1 Electrodeposition with Secondary Current Distribution   24
        2.4.2 Diffusive Electrodeposition   27
   Problems   28
   References   28

3 Multifractality   **30**
Thomas C. Halsey

3.1 Definition of $\tau(q)$ and $f(\alpha)$ 30
3.2 Systematic Definition of $\tau(q)$ 32
3.3 The Two-Scale Cantor Set   33
    3.3.1 Limiting Cases   35
    3.3.2 Stirling Formula and $f(\alpha)$ 36
3.4 Multifractal Correlations   37
    3.4.1 Operator Product Expansion and Multifractality 38
    3.4.2 Correlations of Iso-α Sets   39
3.5 Numerical Measurements of $f(\alpha)$ 40
3.6 Ensemble Averaging and $\tau(q)$ 41
Problems   42
References   43

4 Scaling Arguments and Diffusive Growth   **44**
Thomas C. Halsey

4.1 The Information Dimension   44
4.2 The Turkevich–Scher Scaling Relation   45
4.3 The Electrostatic Scaling Relation   47
4.4 Scaling of Negative Moments   50
4.5 Conclusions   51
Problems   52
References   52

Part II Chaos and Randomness

5 Introduction to Dynamical Systems   **55**
Stephen G. Eubank and J. Doyne Farmer

5.1 Introduction   55
5.2 Determinism Versus Random Processes   55
5.3 Scope of Part II   57
5.4 Deterministic Dynamical Systems and State Space   58
5.5 Classification   61
    5.5.1 Properties of Dynamical Systems   61
    5.5.2 A Brief Taxonomy of Dynamical Systems Models   63
    5.5.3 The Relationship Between Maps and Flows   63
5.6 Dissipative Versus Conservative Dynamical Systems   67
5.7 Stability   68
    5.7.1 Linearization   68
    5.7.2 The Spectrum of Lyapunov Exponents   70
    5.7.3 Invariant Sets   71
    5.7.4 Attractors   73

	5.7.5 Regular Attractors	75
	5.7.6 Review of Stability	82
5.8	Bifurcations	82
5.9	Chaos	89
	5.9.1 Binary Shift Map	90
	5.9.2 Chaos in Flows	92
	5.9.3 The Rössler Attractor	94
	5.9.4 The Lorenz Attractor	97
	5.9.5 Stable and Unstable Manifolds	98
5.10	Homoclinic Tangle	100
	5.10.1 Chaos in Higher Dimensions	101
	5.10.2 Bifurcations Between Chaotic Attractors	102
	Problems	103
	References	105

6 Probability, Random Processes, and the Statistical Description of Dynamics   **106**
Stephen G. Eubank and J. Doyne Farmer

6.1	Nondeterminism in Dynamics	106
6.2	Measure and Probability	107
	6.2.1 Estimating a Density Function from Data	110
6.3	Nondeterministic Dynamics	113
6.4	Averaging	115
	6.4.1 Stationarity	115
	6.4.2 Time Averages and Ensemble Averages	116
	6.4.3 Mixing	120
6.5	Characterization of Distributions	122
	6.5.1 Moments	122
	6.5.2 Entropy and Information	133
6.6	Fractals, Dimension, and the Uncertainty Exponent	138
	6.6.1 Pointwise Dimension	139
	6.6.2 Information Dimension	140
	6.6.3 Fractal Dimension	140
	6.6.4 Generalized Dimensions	141
	6.6.5 Estimating Dimension from Data	142
	6.6.6 Embedding Dimension	144
	6.6.7 Fat Fractals	144
	6.6.8 Lyapunov Dimension	145
	6.6.9 Metric Entropy	146
	6.6.10 Pesin's Identity	148
6.7	Dimensions, Lyapunov Exponents, and Metric Entropy in the Presence of Noise	148
	Problems	149
	References	150

7 Modeling Chaotic Systems   **152**
Stephen G. Eubank and J. Doyne Farmer

7.1 Chaos and Prediction   153
7.2 State Space Reconstruction   154
    7.2.1 Derivative Coordinates   155
    7.2.2 Delay Coordinates   155
    7.2.3 Broomhead and King Coordinates   157
    7.2.4 Reconstruction as Optimal Encoding   157
7.3 Modeling Chaotic Dynamics   157
    7.3.1 Choosing an Appropriate Model   157
    7.3.2 Order of Approximation   158
    7.3.3 Scaling of Errors   160
7.4 System Characterization   162
7.5 Noise Reduction   163
    7.5.1 Shadowing   164
    7.5.2 Optimal Solution of Shadowing Problem
        with Euclidean Norm   167
    7.5.3 Numerical Results   168
    7.5.4 Statistical Noise Reduction   170
    7.5.5 Limits to Noise Reduction   172
Problems   174
References   174

Part III Pattern Formation and Disorderly Growth

8 Phenomenology of Growth   **179**
Leonard M. Sander

8.1 Aggregation: Patterns and Fractals Far from Equilibrium   179
8.2 Natural Systems   181
    8.2.1 Ballistic Growth   181
    8.2.2 Diffusion-Limited Growth   183
    8.2.3 Growth of Colloids and Aerosols   190
Problems   190
References   190

9 Models and Applications   **192**
Leonard M. Sander

9.1 Ballistic Growth   192
    9.1.1 Simulations and Scaling   192
    9.1.2 Continuum Models   196
9.2 Diffusion-Limited Growth   197
    9.2.1 Simulations and Scaling   197
    9.2.2 The Mullins–Sekerka Instability   200
    9.2.3 Orderly and Disorderly Growth   201

9.2.4 Electrochemical Deposition: A Case Study   203
9.3 Cluster–Cluster Aggregation   205
Appendix: A DLA Program   206
Problems   208
References   209

Part IV Solitons

10 Integrable Systems   **213**
Lui Lam

10.1 Introduction   213
10.2 Origin and History of Solitons   215
10.3 Integrability and Conservation Laws   218
10.4 Soliton Equations and their Solutions   219
    10.4.1 Korteweg–de Vries Equation   219
    10.4.2 Nonlinear Schrödinger Equation   219
    10.4.3 Sine–Gordon Equation   219
    10.4.4 Kadomtsev–Petviashvili Equation   223
10.5 Methods of Solution   224
    10.5.1 Inverse Scattering Method   224
    10.5.2 Bäcklund Transformation   225
    10.5.3 Hirota Method   226
    10.5.4 Numerical Method   226
10.6 Physical Soliton Systems   227
    10.6.1 Shallow Water Waves   227
    10.6.2 Dislocations in Crystals   228
    10.6.3 Self-Focusing of Light   229
10.7 Conclusions   230
Problems   230
References   231

11 Nonintegrable Systems   **234**
Lui Lam

11.1 Introduction   234
11.2 Nonintegrable Soliton Equations with Hamiltonian Structures   236
    11.2.1 The θ^4 Equation   236
    11.2.2 Double Sine–Gordon Equation   237
11.3 Nonlinear Evolution Equations   237
    11.3.1 Fisher Equation   238
    11.3.2 The Damped θ^4 Equation   239
    11.3.3 The Damped Driven Sine–Gordon Equation   240
11.4 A Method of Constructing Soliton Equations   243
11.5 Formation of Solitons   244
11.6 Perturbations   245

11.7 Soliton Statistical Mechanics   247
      11.7.1 The θ^4 System   248
      11.7.2 The Sine–Gordon System   251
11.8 Solitons in Condensed Matter   252
      11.8.1 Liquid Crystals   252
      11.8.2 Polyacetylene   261
      11.8.3 Optical Fibers   264
      11.8.4 Magnetic Systems   266
11.9 Conclusions   266
Problems   267
References   268

Part V Special Topics

12 Cellular Automata and Discrete Physics   **275**
     David E. Hiebeler and Robert Tatar

12.1 Introduction   275
      12.1.1 A Well-Known Example: Life   277
      12.1.2 Cellular Automata   278
      12.1.3 The Information Mechanics Group   279
12.2 Physical Modeling   280
      12.2.1 CA Quasiparticles   280
      12.2.2 Physical Properties from CA Simulations   281
      12.2.3 Diffusion   282
      12.2.4 Sound Waves   285
      12.2.5 Optics   287
      12.2.6 Chemical Reactions   289
12.3 Hardware   290
12.4 Current Sources of Literature   291
12.5 An Outstanding Problem in CA Simulations   291
Problems   292
References   294

13 Visualization Techniques for Cellular Dynamata   **296**
     Ralph H. Abraham

13.1 Historical Introduction   296
13.2 Cellular Dynamata   297
      13.2.1 Dynamical Schemes   297
      13.2.2 Complex Dynamical Systems   297
      13.2.3 CD Definitions   297
      13.2.4 CD States   299
      13.2.5 CD Simulation   299
      13.2.6 CD Visualization   299
13.3 An Example of Zeeman's Method   300
      13.3.1 Zeeman's Heart Model: Standard Cell   300
      13.3.2 Zeeman's Heart Model: Physical Space   300

13.3.3 Zeeman's Heart Model: Beating   300
13.4 The Graph Method   300
13.4.1 The Biased Logistic Scheme   301
13.4.2 The Logistic/Diffusion Lattice   301
13.4.3 The Global State Graph   302
13.5 The Isochron Coloring Method   305
13.5.1 Isochrons of a Periodic Attractor   305
13.5.2 Coloring Strategies   305
13.6 Conclusions   306
References   306

14 From Laminar Flow to Turbulence   **308**
Geoffrey K. Vallis

14.1 Preamble and Basic Ideas   308
14.1.1 What Is Turbulence?   309
14.2 From Laminar Flow to Nonlinear Equilibration   311
14.2.1 A Linear Analysis: The Kelvin–Helmholz Instability   312
14.2.2 A Weakly Nonlinear Analysis: Landau's Equation   314
14.3 From Nonlinear Equilibration to Weak Turbulence   321
14.3.1 The Quasi-Periodic Sequence   322
14.3.2 The Period Doubling Sequence   324
14.3.3 The Intermittent Sequence   335
14.3.4 Fluid Relevance and Experimental Evidence   337
14.4 Strong Turbulence   341
14.4.1 Scaling Arguments for Inertial Ranges   341
14.4.2 Predictability of Strong Turbulence   348
14.4.3 Renormalizing the Diffusivity   352
14.5 Remarks   355
References   357

**15 Active Walks: Pattern Formation, Self-Organization, and
Complex Systems**   **359**
Lui Lam

15.1 Introduction   359
15.2 Basic Concepts   360
15.3 Continuum Description   361
15.4 Computer Models   363
15.4.1 A Single Walker   363
15.4.2 Branching   366
15.4.3 Multiwalkers and Updating Rules   366
15.4.4 Track Patterns   368
15.5 Three Applications   371
15.5.1 Dielectric Breakdown in a Thin Layer of Liquid   371
15.5.2 Ion Transport in Glasses   375
15.5.3 Ant Trails in Food Collection   376
15.6 Intrinsic Abnormal Growth   378

15.7   Landscapes and Rough Surfaces   380
         15.7.1 Groove States   382
         15.7.2 Localization–Delocalization Transition   383
         15.7.3 Scaling Properties   387
15.8   Fuzzy Walks   390
15.9   Related Developments and Open Problems   393
15.10   Conclusions   395
References   396

Appendix: Historical Remarks on Chaos   **401**
Michael Nauenberg

Contributors   407
Index   411

1

Introduction

Lui Lam

1.1 A Quiet Revolution

Quantum mechanics and relativity, the two important discoveries in physics developed at the beginning of this century, are well recognized as revolutions. These two revolutions present unexpected concepts and insights by going beyond the classical domains (Fig. 1.1). New results are obtained in quantum mechanics when one goes to the microscopic level ($< 10^{-8}$ cm) and, in the case of relativity, the speed of the object has to be close to that of light ($\sim 10^{10}$ cm/s).

Here comes a new branch of science—nonlinear science—which, like quantum mechanics and relativity, delivers a whole set of fundamentally new ideas and surprising results. Yet, unlike quantum mechanics and relativity, nonlinear science covers systems of *every* scale, and objects moving with *any* speed; that is, the whole area displayed in Fig. 1.1. Then, by the same standard, nonlinear science is more than qualified to be called a revolution. The fact that nonlinear science delivers within the conventional system sizes and speed limits should not be counted as negative toward its novelty but, on the contrary, in view of its wide applicability, makes nonlinear science more important and powerful as a true revolution. In particular, nonlinear science can be studied with daily macroscopic systems with ordinary tools, such as a camera or a copying machine, making it accessible to almost everybody.

If nonlinear science appears to be a somewhat quiet revolution, it is perhaps due to its wide scope of coverage. The important works were done by so many researchers and accumulated over such a long period of time that it was hard for a single person or a group to call a press conference. Or, following a long scientific tradition, no one bothered to call a press conference.

For pedagogical purposes, nonlinear science may be divided into six areas of study, namely, fractals, chaos, pattern formation, solitons, cellular automata, and complex systems. The common theme underlying this diversity of subjects is the nonlinearity of the systems under study.

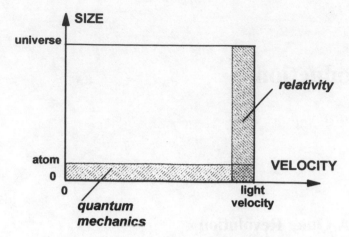

FIGURE 1.1. Sketch of the limited domains of applicability of quantum mechanics and relativity theory, the two well-recognized revolutions in physics. In contrast, nonlinear science applies to the whole domain shown here (not to scale).

1.2 Nonlinearity

A system is nonlinear if the output from the system is not proportional to the input (Fig. 1.2). For example, a dielectric crystal becomes nonlinear if the output light intensity is no longer proportional to the incident light intensity. The examination system used by a professor is nonlinear if the grade points earned by a student do not increase linearly as a function of the hours put in by the student, which is usually the case. In fact, almost all known systems in natural or social sciences are nonlinear when the input is large enough.

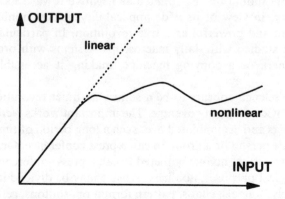

FIGURE 1.2. Definition of a nonlinear system. The broken line represents a linear system with the output proportional to the input. The solid line represents a nonlinear system.

A well-known example is the spring. When the displacement of the spring becomes large, the Hook's law breaks down and the spring becomes a non-linear oscillator. A second example is the simple pendulum. Only when the displacement angle of the pendulum is small does the pendulum behave linearly. And there are important qualitative differences between the behavior of a system in its linear and nonlinear regimes. For example, the period of the pendulum oscillation does not depend on the amplitude (the maximum displacement angle) in the linear regime, but does so in the nonlinear regime [1].

Mathematically, the signature of a nonlinear system is the breakdown of the superposition principle, which states that the sum of two solutions of the equation(s) describing the system is again a solution. There are two ways that the superposition principle may break down.

First, the equation itself is nonlinear. For example, the equation of motion for the point mass in a simple pendulum is given by

$$d^2\theta/dt^2 + (g/L)\sin\theta = 0, \tag{1.1}$$

where θ is the angle between the vertical and the pendulum, g the acceleration due to gravity, and L the length of the pendulum. It is easy to show that if $\theta_1(t)$ and $\theta_2(t)$ are each a solution of Eq. (1.1), then the sum $\theta_1(t) + \theta_2(t)$ is *not* a solution, a consequence of the simple fact that $\sin\theta_1 + \sin\theta_2 \neq \sin(\theta_1 + \theta_2)$. Consequently, Eq. (1.1) is a nonlinear equation due to the presence of the nonlinear term $\sin\theta$. [In contrast, in the linear regime where θ is small, one could replace $\sin\theta$ by θ, a linear term, in Eq. (1.1) and the superposition principle becomes valid.]

Second, the equation itself may be linear but the boundary is unknown or moving. For example, in the viscous fingering problem of pattern formation in a Hele-Shaw cell [2], one tries to determine the shape and movement of a *single*, unknown interface separating two immiscible liquids when one of them is pushed into the other. The pressure field in each type of liquid P is simply given by the Laplace equation, $\nabla^2 P = 0$, which is a linear equation. However, the superposition of two solutions of this problem (corresponding to different external conditions set at the far ends of the cell) contain two "interfaces," and obviously does not represent a solution of the original problem.

The nonlinearity of a system makes the system highly nontrivial and its analysis difficult. For example:

(i) For a nonlinear system, a small disturbance, such as a slight change of the initial conditions, can result in a big difference in the behavior of the system at a later time. This could make the behavior of a nonlinear system very complex [as happens in the case of chaos (see Part II)].

(ii) If the equations describing the nonlinear system are known, the breakdown of the superposition principle renders the Fourier transform technique—which makes the analysis of a linear problem so "easy"—inapplicable. And there is no similar systematic method in solving nonlinear equations. [For example, the celebrated inverse scattering method (see section 10.5.1) in soliton theory is applicable only to a subset of integrable systems, and there is still no way to know, a priori, which integrable system is susceptible to this method.]

(iii) In many cases, from the simple diffusion-limited aggregation model of fractal pattern growth (see section 2.2) to many examples in complex systems such as the economic system, the equations are not even known or may not exist.

All these complications make the use of computers an invaluable tool in the study of nonlinear systems because computers do not differentiate linear equations from nonlinear equations, can be used for direct simulations, and can display complex results for easy visualization. The important role played by computers is partly responsible for the fact that the rise of nonlinear science is a quite recent phenomenon, correlated with the widespread accessibility of personal computers. The other reason for the late coming of nonlinear science is that it takes time for the "easy" problems of linear systems to be exhausted first, especially because many linear problems, such as the propagation of electromagnetic waves in telecommunications, are technologically very important in our daily lives.

1.3 Nonlinear Science

An outline of nonlinear science is given here.

1.3.1 Fractals

Many spatial structures in nature result from the self-assembly of a large number of identical components. To be efficient the self-assembly process takes advantage of and occurs via some kind of simple prescription, which we call the principle of organization [3]. The two simplest principles are the principle of regularity and the principle of randomness. With the former, the components arrange themselves in a periodic or quasiperiodic regular fashion, resulting in crystals, alloys, a formation of soldiers in a parade, and so forth. Examples of structures (or nonstructures) resulting from the latter are those in gases and the distribution of animal hairs. Between these two extremes there is the principle of self-similarity, leading to self-similar structures called fractals. In a fractal, part of the system resembles the whole. A fractal usually has a fractional dimension. Many fractals are also multifractals, which can be roughly considered as a collection of fractals (see Part I). The ubiquitous existence of fractals in natural and mathematical systems became widely known to scientists in the early 1980s after the book by Benoit Mandelbrot [4] was published. Examples of fractals include aggregates and colloids, trees, rocks, mountains, clouds, galaxies, rough surfaces and interfaces, polymers, and the stock market [5].

Self-similarity, spatial power laws, and scale invariance are three equivalent ways of expressing the fact that the system lacks a characteristic length scale [6]. Similarly, the absence of a characteristic time scale in the system leads to temporal power laws (e.g., the $1/f$ noise, another ubiquitous phenomenon in nature). (Note that power laws are generally nonlinear equations.) To explain the widespread existence of fractals and scale-free behaviors in nonequilibrium systems, the hypothesis of self-organized criticality was proposed by Per Bak, Chao Tang, and Kurt Wiesenfeld in 1987, which is supposed to be applicable to sandpiles and many other natural and social systems [7].

1.3.2   Chaos

Chaos, the sensitive dependence of a nonlinear system's behavior on the initial conditions, has been investigated by Henri Poincaré at about the turn of the century and subsequently by a number of mathematicians. Recent frenzy about chaos occurred in the late 1970s, after Mitchell Feigenbaum discovered the universality properties of some simple maps, which was preceded by the important but obscure work of Edward Lorentz related to weather predictions (see Appendix). The signature of chaos in a dissipative system is the existence of strange attractor(s) in the phase space, which is a fractal. The basins of attraction could also be fractals. These linkages between chaos and fractals are not fully understood (see Part II).

Two findings of chaotic systems are particularly significant.

(i) In the chaotic regime, the behavior of a deterministic system appears random. This single finding forces every experimentalist to reexamine their data to determine whether some random behavior attributed to noise is instead due to deterministic chaos.

(ii) Nonlinear systems with only a few degrees of freedom can be chaotic and appear very complex. This finding gives hope that the complex behavior observed in many real systems may have a simple origin and may indeed be discernable.

The *apparent* unpredictability of a chaotic, deterministic, *real* system (such as the weather) arises from the system's sensitive dependence on initial conditions *and* the fact that the system's initial conditions can be measured or determined only approximately in practice, due to the finite resolution of any measuring instrument. This difficulty precludes the long-term predictability of any chaotic, real system. On the other hand, for a system that turns out to be deterministic and chaotic in nature, there is order behind its seemingly complex behavior and short-term predictability is possible. The problems are how to determine whether there is a chaotic origin behind a complicated behavior and how to do the short-term prediction (see chapter 7).

For systems such as the weather or the stock market, due to the insurmountable complexity, the complete equations describing the system, if they exist, may never be known. Or, when the equations can be written down, there may not be a computer powerful enough to solve them. Besides, for practical reasons a successful short-term prediction for these sytems is usually good enough. (For example, one only needs to know the trend of the stocks slightly ahead of time—without knowing the mechanisms of the market—to make a killing in the market.) Short-term prediction of the behavior of complex systems has become one of the two most exciting practical applications of chaos.

The other important application is controlling chaos [8]. This application is based on the fact that there are many unstable periodic orbits embedded within a strange attractor, and one of these, if desirable, can be made stable and reached by the chaotic system with a small perturbation applied to the system, without knowing the system's dynamics in advance. The technique has been applied successfully in the control of mechanical sytems, electronics, lasers, chemical systems, and heart tissues.

For other practical applications of chaos, see chapter 14 and the proceedings [9].

1.3.3 Pattern Formation

One can hardly fail to notice the striking similarity between the ramified patterns formed by rivers, trees, leaf veins, and lightning. These branching patterns are different from the compact patterns observed in clouds and algae colonies. How does nature generate these patterns? Is there a simple principle or universal mechanism behind these pattern-forming phenomena? These are the profound questions that interest lay people and experts alike. Although final answers to these questions are still lacking, tremendous progress has been made in the last 15 years.

Models for aggregation and diffusive growth have been much studied and are quite successful in mimicking many real systems, which are often fractals (see Part III and also chapter 4). A new type of model based on active walks is especially effective in dealing with filamentary patterns, even though compact patterns can also be generated by these so-called active walk models (see chapter 15).

Other types of pattern formation, such as the Rayleigh-Bénard convection and chemical reaction patterns, originating in the linear instabilities of a homogeneous state, have been successfully treated in a unified manner via the amplitude equations [10, 11]. There is also a good understanding for yet another type of pattern, which involve an interface between two materials (e.g., viscous fingering in Hele-Shaw cells [2]) or two different phases of the same material (e.g., solidification patterns [10]).

1.3.4 Solitons

In some media, such as a layer of shallow water or an optical fiber, under suitable conditions, the widening of a wave packet due to dispersion could be balanced exactly by the narrowing effect due to the nonlinearity of the medium. In these cases, it is possible to have spatially localized waves that propagate with constant velocities and undistorted shapes. These localized waves appear frequently in the bell shape or in the form of a kink. In the mathematics literature, these localized waves are called solitons if they possess further the elastic collision property that two such waves will emerge from a head-on collision with their velocities and shapes unchanged. It turns out that such an elastic collision property is related to the integrability of the system and appears only rarely in nonintegrable systems. Because most real systems are nonintegrable, in the physics literature, the elastic collision property is dropped from the definition of solitons which, under perturbations, may even have their velocities and shapes slightly distorted during propagation.

Even though solitons were first observed by John Scott Russell in 1834, it was only after 1965—the year that the word soliton was coined by Norman Zabusky and Martin Kruskal—that the significance of solitons was widely appreciated and the study of solitons as a discipline took shape. Since then solitons as nonlinear waves, and as nonlinear excitations in materials, have been intensely studied in various systems, including liquid crystals, conducting polymers, and optical fibers (see Part IV). In particular, optical solitons in glass fibers are becoming very important because of their demonstrated applicability in multigigabits optical transmissions over very long distances (going around the Earth four thousand times, say).

1.3.5   Cellular Automata

Cellular automata are discrete dynamical systems whose evolution is dictated by local rules. In practice, they are usually realized on a lattice of cells, with a finite number of discrete states associated with each cell, and with local rules specifying how the state of each cell should be updated in discrete time steps. Because of the discreteness of all the quantities involved, cellular automata calculations obtained from computers are exact (see chapter 12).

Cellular automata were introduced by John von Neumann in the late 1940s, soon after the creation of electronic computers. He used cellular automata to prove that, in principle, self-reproducing machines are possible. Then, in the early 1970s the "game of life"—a very simple two-dimensional cellular automaton capable of creating lifelike "creatures" on the lattice—invented by John Conway, became very popular with the public. But scientifically, nothing serious happened.

Then in 1986 Uriel Frisch, Brosl Hasslacher, and Yves Pomeau demonstrated that it is possible to simulate the Navier-Stokes equation of fluid flows by using a cellular automaton of gas particles on a hexagonal lattice, with extremely simple translation and collision rules governing the movement of the particles. That a microscopic system of interacting particles with oversimplified dynamics could result in the physically correct macroscopic Navier-Stokes equation, is due to the fact that the Navier-Stokes equation is the consequence of appropriate conservation laws and is quite insensitive to microscopic details, as long as the appropriate symmetries are obeyed by microscopic dynamics. The spirit of this lattice gas approach in "simulating" nature runs opposite to that of the molecular dynamics or Monte Carlo approach, where the more realistic the molecular interactions are, the better. By using very simple rules in the cellular automata, the lattice gas approach provides gains in computational speed, is error free, and can easily take care of very irregular boundaries. It is one of the rare occasions that nature can be "cheated," if only on the computer.

Presently, lattice gas automata provides the most efficient method in calculating fluid flow through real porous media, which has found application in the oil industry. Armed with the same "cheating" principle, lattice gas automata with different updating rules have been devised to simulate various partial differential equations [12].

Moreover, cellular automata are being investigated as complex systems per se, and as simple devices in simulating real processes in biological, physical, and chemical systems (see Chapters 12 and 13). Finally, it is interesting to note that the "game of life" has been credited with leading Christopher Langton to the very concept of artificial life [14] that is now established as a vigorous discipline of its own.

1.3.6 Complex Systems

The fact that there is one and only one doctoral degree, the Doctor of Philosophy (Ph.D.), but not the Doctor of Physics or Doctor of Economics, attests to the fact that not too long ago, science was considered and studied as a whole. There was no division of social and natural sciences, not to mention no fragmentation of the natural sciences into physics, chemistry, biology, and so forth. And, as suspected by some, this compartmentalization of science is due more to administrative convenience than to the nature of science itself [15].

The hope of being able to return to the appealing state of a unified science was rekindled in the 1970s and early 1980s. A major influence came from the success of the chaos theory. At that time chaos was better understood and time series obtained in almost every discipline—from both social and natural sciences—were subjected to the same analyses as inspired by the chaos theory. The importance of this development was that chaos theory seemed to offer scientists a handle, or an excuse if one is needed, to tackle problems from any field of their liking. A psychological barrier was broken; no complex system was too complex to be touched.

A secondary but crucial influence that helps to propel and sustain complex systems as a viable research discipline is the prevalence of personal computers and the availability of powerful computational tools, such as parallel computers. Although the theoretical study of complex systems is usually quite difficult and sometimes appears impossible, one can always resort to some form of computer simulations, or computer experiments, as some like to call them.

The topics studied span a large spectrum, including human languages, the origin of life, computers and DNA, evolutionary biology and spin glasses, economics, psychology, ecology, ant swarms, earthquakes, immunology, self-organization of nonequilibrium systems, and so forth [15–18]. Although a precise definition of complex systems is unavailable, as is frequently the case in the early stage of a new research field, it is safe to say that almost all the subjects covered in the various departments of a university—except for those in the conventional curriculums of the physics, chemistry, and engineering departments—are in the realm of complex systems.

Although some general concepts, such as complex adaptive systems and symmetry breaking, have been found to be useful in their descriptions, no unifying theory governing all complex systems yet exists. However, two simple ideas capable of explaining the behavior of many complex systems have emerged. One is self-organized criticality [7]; the other is the principle of active walks (see chapter 15). The former asserts that large dynamical systems tend to drive themselves to a critical state with no characteristic spatial and temporal scales. The latter describes how the elements in a complex system communicate with their environment and with each other, through the interaction with the landscape they share. The principle of active walks has been applied successfully to very different problems, such as the formation of dielectric breakdown patterns, ion transport in glasses, and the cooperation of ants in food collection.

1.4 Remarks

Nonlinear science, as outlined above, involves the interplay of order and disorder, as well as the simple and the complex [19]. But technically, what makes the fascinating outcomes possible is nonlinearity.

The basic theories and principles expounded in the following chapters are applicable to *all* branches of science, although many of the examples covered are taken from physics. Parts I to IV contain the basics; they are not complete surveys. For further reading beyond this book, the references quoted in this chapter can be consulted.

Nonlinear science is still a science in the making. Exciting results keep appearing in the leading research journals such as *Physical Review Letters*, and in special journals such as *Physica D, Chaos*, and *Complexity*. Conference proceedings are also a good source for the latest developments.

Due to the multidisciplinary nature of nonlinear science, it is not easy for newcomers or even practitioners to be knowledgeable in all the subjects covered in the applications. In this regard, the general magazine *Scientific American* is a valuable source of information. And, provided the hypes are recognized, semipopular [6, 18] and popular books [20–24] are often helpful and make for entertaining reading.

References

[1] L.D. Landau and E.M. Lifshitz, *Mechanics* (Pergamon, Oxford, 1960).

[2] D. Bensimon, L.P. Kadanoff, S. Liang, B.I. Shraiman, and C. Tang, Rev. Mod. Phys. **58**, 977 (1986); P.G. Saffman, in *Nonlinear Structures in Physical Systems: Pattern Formation, Chaos and Waves*, edited by L. Lam and H.C. Morris (Springer-Verlag, New York, 1990).

[3] L. Lam, in *Lectures on Thermodynamics and Statistical Mechanics*, edited by M. Costas, R. Rodriquez, and A.L. Benavides (World Scientfic, River Edge, 1994).

[4] B.B. Mandelbrot, *The Fractal Geometry of Nature* (Freeman, New York, 1982).

[5] J.-F. Gouyet, *Physics and Fractal Structures* (Springer-Verlag, New York, 1996); *Fractals in Science*, edited by A. Bunde and S. Halvin (Springer-Verlag, New York, 1994); A.-L. Barabási and H.E. Stanley, *Fractal Concepts in Surface Growth* (Cambridge University, Cambridge, 1995).

[6] M. Schroeder, *Fractals, Chaos, Power Laws* (Freeman, New York, 1991).

[7] P. Bak, in [16].

[8] E. Ott and M. Spano, Phys. Today **48**(5), 34 (1995).

[9] *Applied Chaos*, edited by J.H. Kim and J. Stringer (Wiley, New York, 1992); *Proceedings of the 1st Experimental Chaos Conference*, edited by S. Vohra, M. Spano, M. Schlesinger, L. Pecora, and W. Ditto (World Scientific, River Edge, 1992).

[10] M.C. Cross and P.C. Hohenberg, Rev. Mod. Phys. **65**, 851 (1993).

[11] D. Walgraef, *Spatiotemporal Pattern Formation, with Examples from Physics, Chemistry and Materials Science* (Springer-Verlag, New York, 1997); *Pattern Formation in Liquid Crystals*, edited by A. Buka and L. Kramer (Springer-Verlag, New York, 1996).

[12] *Lattice Gas Methods for Partial Differential Equations*, edited by G.D. Doolen et al. (Addison-Wesley, Menlo Park, 1990).

[13] *Cellular Automata: Theory and Experiment*, edited by H. Gutowitz (North-Holland, Amsterdam, 1990) [Physica D, **45** (1990)].

[14] *Artificial Life*, edited by C.G. Langton (Addison-Wesley, Menlo Park, 1988).

[15] *Modeling Complex Phenomena*, edited by L. Lam and V. Naroditsky (Springer-Verlag, New York, 1992).

[16] *Complexity: Metaphors, Models and Reality*, edited by G.A. Cowan, D. Pines, and D. Meltzer (Addison-Wesley, Menlo Park, 1994).

[17] *Evolution of Dynamical Structures in Complex Systems*, edited by R. Friedrich and A. Wunderlin (Springer-Verlag, New York, 1992); *Self-Organization of Complex Structures: From Individual to Collective Dynamics*, edited by F. Schweitzer (Gordon and Breach, London, 1996).

[18] G. Nicholas and I. Prigogine, *Exploring Complexity: An Introduction* (Freeman, New York, 1989).

[19] An expanded version of section 1.3, with added details and concrete examples, is given in the book by L. Lam, *Nonlinear Physics for Beginners* (World Scientific, River Edge, 1996). This elementary text is at the undergraduate level and includes a collection of student project reports, programs with source codes, and selected semipopular reprints.

[20] W. Poundstone, *The Recursive Universe: Cosmic Complexity and the Limits of Scientific Knowledge* (Contempary Books, Chicago, 1985).

[21] J. Gleick, *Chaos: Making a New Science* (Viking, New York, 1987).

[22] J. Briggs and F.D. Peat, *Turbulent Mirror: An Illustrated Guide to Chaos Theory and the Science of Wholeness* (Harper & Row, San Francisco, 1989).

[23] M.M. Waldrop, *Complexity: The Emerging Science at the Edge of Order and Chaos* (Simon & Schuster, New York, 1992).

[24] M. Gell-Mann, *The Quark and the Jaguar: Adventures in the Simple and the Complex* (Freeman, New York, 1994).

I
Fractals and Multifractals

2

Fractals and Diffusive Growth

Thomas C. Halsey

In many contexts in physics, geometrically self-similar structures appear [1]. Notable examples include:

(i) Structures in the phase space of a system, such as strange attractors of dynamical systems [2].

(ii) Energy spectra of certain types of systems. An important example here is the band structure of a two-dimensional tight binding model in a magnetic field [3].

(iii) Real-space systems. These are highly heterogeneous systems, such as appear principally in condensed matter physics. A preliminary list of such structures would include percolation clusters, diffusion-limited aggregates, and perhaps "avalanches" in self-organized critical systems [4–6].

Before we proceed further, we should at least attempt to define self-similarity. This term, although rather vague and imprecise, refers basically to a symmetry of a system under dilatation (in either the time or space coordinates.) A simple example of a self-similar system is a Cantor set.

A Cantor set is a recursively constructed set. Consider the unit interval. We construct a Cantor set by first removing the central third of the interval, then the central thirds of the remaining portions, and so on (Fig. 2.1). We define the Cantor set as what is left over if this process is pursued indefinitely [7].

The Cantor set possesses an exact dilatational symmetry. Suppose we define a probability measure $d\mu(x)_0$, which on the initial interval is constant. At the first interval, we have $d\mu(x)_1$ constant on the first and last third of the interval, and zero in the missing middle third (Fig. 2.2). At each stage, we require $\int_0^1 d\mu = 1$. Then the limiting measure $d\mu_\infty$ obeys

$$(d\mu(x))_\infty = 3^D d\mu(x/3)_\infty, \tag{2.1}$$

with $D = \log(2)/\log(3)$. D is called a fractal or Hausdorff dimension. A simple geometrical interpretation of D has been widely popularized by Mandelbrot [1]. Consider $N(l)$, the number of boxes of size l needed to cover a set. As $l \to 0$, we expect $N(l)$ to increase. If it increases as a power law in l, then

FIGURE 2.1. The Cantor set is defined by an iterative procedure: first the middle third of a unit interval is removed, then the middle thirds of the remaining intervals, and so forth. The Cantor set is made up of the points that survive in the limit that this procedure is repeated an arbitrarily large number of times.

we can define the fractal dimension D as

$$D = -\lim_{l \to 0} \frac{\log(N(l))}{\log(l)}. \tag{2.2}$$

For the Cantor set above, $N = 2^m$, with $m = \log(l)/\log(1/3)$. It follows immediately that $D = \log(2)/\log(3)$.

In this exposition, attention will be concentrated upon the application of ideas of self-similarity to real-space structures. One of the best understood such self-similar structures is the percolation cluster.

2.1 Percolation

Percolation is a simple and widely applicable statistical model of a highly heterogeneous system. The simple example of "bond percolation" in two dimensions is given.

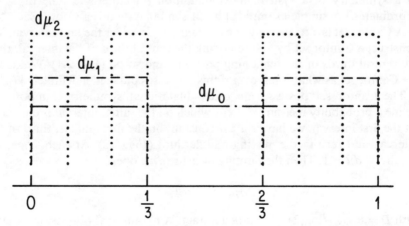

FIGURE 2.2. A normalized probability measure $d\mu(x)$ defined on the first few approximants to the Cantor set. The scaling of this measure defines the Hausdorff dimension D of the set.

Consider a two-dimensional square lattice. With probability p, any particular bond is of one type (say, metallic) and with probability $1 - p$, of a complementary type (say, insulating). We can define ensemble averaged quantities in terms of ensemble averages over all possible configurations of bonds, with their relative probabilities. Clearly, in the thermodynamic limit, the density of metallic and insulating bonds will be given by p and $1 - p$, respectively.

Now we ask what is the probability P to find an entirely metallic path spanning the entire system. The result is dependent upon p,

$$P = \begin{cases} 0, & \text{if } p < 1/2; \\ 1, & \text{if } p > 1/2. \end{cases} \tag{2.3}$$

For our purposes, the most interesting case is $p = 1/2$. Here, the cluster of metallic bonds connected to this metallic path (the "infinite cluster") is a self-similar, "fractal" object with $D = 1.89$ [4]. Of course, any particular cluster is only approximately, rather than exactly, symmetric under dilation, due to the statistical nature of the process. The correlation function on the infinite cluster $C(r)$, which is defined as the ensemble averaged probability that a point on the infinite cluster will have another infinite cluster point at a distance r, is

$$C(r) = r^{D-d}. \tag{2.4}$$

This dimension is partly important because it is universal: a wide range of two dimensional percolation problems have the same D for the infinite cluster, although the percolation threshold p_c ($p_c = 1/2$ above) will vary with lattice type or any other local detail of the particular percolation model defined. Like most critical exponents, D does depend upon the dimensionality of the problem; thus three-dimensional infinite percolation clusters have $D \approx 2.5$ [4].

2.2 Diffusion-Limited Aggregation

Our next and central physical example, and one in which multifractal analysis has proven particularly fruitful, is diffusion-limited aggregation (DLA). DLA was originally introduced as a physically motivated computer algorithm for the simulation of non-equilibrium growth. One starts with a seed particle, and introduces into the system (sufficiently far away) a randomly walking particle, which walks until it first touches the seed. At this point, it sticks permanently, and another particle may be introduced to the system (Fig. 2.3). There are numerous variants of this simple model, in which the dimensionality of space is varied, the random walker and/or the clustered particles are restricted to a lattice, the particle sticks with a probability less than one on each contact with the cluster, or other alterations are made in the

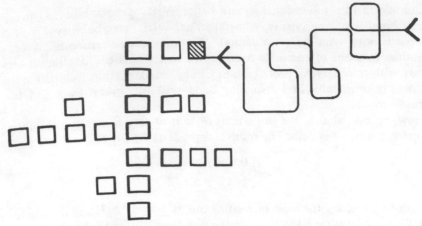

FIGURE 2.3. In diffusion-limited aggregation, a particle randomly walks until it sticks to the growing cluster. Then another particle is introduced at infinity, which walks until it sticks to the cluster.

rules [5]. In principle, arbitrarily large clusters can be simulated by simple iteration of this process. The current computational upper limit on the number of particles N is $N \sim 10^7$ [8].

The clusters generated by this method are highly ramified, presenting a strongly branched appearance (Fig. 2.4). The basic physical reason for the

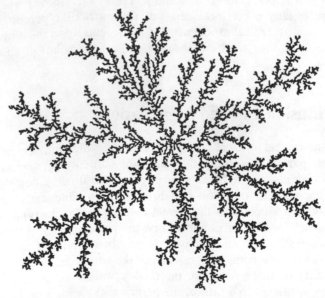

FIGURE 2.4. An off-lattice DLA cluster of 35,000 particles (grown by M. Leibig).

branching is that the inner regions of the cluster are screened by the tips. It is highly unlikely that a random walker will succeed in penetrating deep into one of the "fjords" shown in Fig. 2.4 without first striking, and adhering, at one of the outer tips of the cluster. Thus, most of the growth takes place in an "active zone" near the outer radius of the cluster. By contrast, the inner regions of the cluster quickly reach an asymptotic state, in which almost no change will take place during the further growth of the cluster.

The clusters are also "fractal." In this case, three quantitative observations are meant by this remark.

(i) The radius of gyration R_g of a cluster, which is simply the square root of the expectation value of R^2, where R is the radius of the particles from the seed, does not scale with number of particles N as $1/d$, with d the dimension of space. Instead we have,

$$R_g \equiv \sqrt{\frac{\langle \sum_{i=1}^{N} R_i^2 \rangle}{N}} = A N^{v_1}, \tag{2.5}$$

with $v_1 \neq 1/d$, as would be the case for a compact structure. We define the dimension $D = 1/v_1$; obviously $D < d$.

(ii) In the inner "dead zone" of the cluster, we have a scaling of density $\rho(r)$ about the center,

$$\rho(r) = B r^{1/v_2 - d}. \tag{2.6}$$

Simple scaling implies that $v_1 = v_2$.

Observations (i) and (ii) imply that the cluster has no asymptotic density. Defining a size dependent density by $\sigma - N/R_g^d \propto R_g^{1/v-d}$, we see that the density σ goes to zero as the cluster grows arbitrarily large. We have a final observation, which specifically involves the idea of "self-similarity,"

(iii) The scaling of density around local regions of the cluster is analogous to the scaling of the entire cluster mass about the center,

$$\int_{\Delta} d^d r \langle \rho(r) \rho(r + x) \rangle = C x^{1/v_3 - d}; \tag{2.7}$$

where the integral is over the dead zone.

Example 2.1 Suppose that the radius of gyration exponent is defined in terms of the expectation value of the radius of the nth particle,

$$(\langle r_n^2 \rangle)^{1/2} \equiv A' n^{v'}.$$

Find the relation between v' and A' and the amplitude and exponent defined in Eq. (2.5) above.

If $\langle r_n^2 \rangle = A'^2 n^{2\nu'}$, then

$$\sum_{n=1}^{N} \langle r_n^2 \rangle \approx A'^2 \int_0^N dn\, n^{2\nu'} = \frac{A'^2}{2\nu'+1} N^{2\nu'+1},$$

so that $\nu' = \nu$, $A' = \sqrt{2\nu+1}A$.

When analyzed carefully, numerical evidence is consistent with $\nu_1 = \nu_2 = \nu_3 \equiv 1/D$ (although there has been some controversy regarding this point) [9]. In two dimensions, for off-lattice (or small on-lattice) clusters, $D \approx 1.71$. In three dimensions, $D \approx 2.5$. D has been measured for clusters in up to eight dimensions. Clearly $D \leq d$, but it has also been shown that $D \geq d-1$ [10].

Actually results (ii) and (iii) above can be understood in terms of different limits of the same correlation function [11]. Define

$$f(r_1, r_2, r_1 - r_2) = \langle \rho(r_1)\rho(r_2) \rangle_0, \tag{2.8}$$

where the brackets $\langle\ \rangle_0$ denote averaging with a seed particle at the origin. Now, using Δ for the dead zone,

$$\int_\Delta dr_1 f(r_1, r_2, r_1 - r_2) = N(\Delta)\langle \rho(r_1) \rangle = N(\Delta)Br_1^{D-d}, \tag{2.9}$$

and

$$\int_\Delta dr_1 f(r_1, r_1 + x, x) = Cx^{D-d}. \tag{2.10}$$

We can define an amplitude ratio $\gamma = (N(\Delta)B/C)$, which measures the effective density around the seed by comparison with that about the typical nonseed particle. In two dimensions, $\gamma \approx 1.4$ [11].

2.3   Electrostatic Analogy

DLA can be discussed profitably in terms of an electrostatic analogy [12, 13]. Suppose that we launch our aggregating particles not from the point at infinity, but from a circle of large radius R centered about the seed particle. The rule is that the particles walk randomly until striking either the surface of the cluster (in which case they aggregate) or until they return to the circle of radius R, in which case they are destroyed. We index the number of random walk steps by t, and we start a particle at $t = 0$. For simplicity, we stay on a square lattice, and we restrict ourselves to two dimensions. We define the function $\rho(x, t)$ as the probability for the particle to be at position x at time t.

Using δ_t to indicate the finite differential with respect to t, and Δ for the lattice Laplacian, we have

$$\delta_t \rho(x, t) = \tfrac{1}{2}\Delta\rho(x, t). \qquad (2.11)$$

We must use boundary conditions that $\rho = 0$ on the trapping sites on the surface of the cluster (Fig. 2.5), and also $\rho = 0$ on the outer circle for $t \neq 0$. This is because the cluster surface and the outer circle serve as sinks of the probability density. At $t = 0$, $\rho(x, 0) = 1/2\pi R$ on the outer circle.

Finally, for the growth probability at the point w on the cluster surface, $g(w, t)$, we have

$$g(w, t) = \tfrac{1}{4}\delta_n \rho(x = w, t), \qquad (2.12)$$

with δ_n a lattice normal derivative.

Now let us define a potential function $\phi(x)$ by

$$\phi(x) = \sum_{t=0}^{\infty} \rho(x, t). \qquad (2.13)$$

We may easily show that $\phi(x)$ satisfies the lattice Laplace equation $\Delta\phi = 0$, with the boundary conditions $\phi = 1/2\pi R$ on the exterior boundary, and $\phi = 0$ on the surface of the cluster. The total probability that the particle will aggregate at the point w is proportional to $\delta_n \phi(x = w)$.

We now pass to a continuum version of this problem. We can define a continuum DLA model in the following manner. Given a cluster and an outer ring, we solve the Laplace equation

$$\nabla^2 \phi = 0 \qquad (2.14)$$

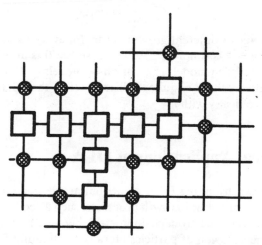

FIGURE 2.5. The probability density of a free random walker is zero on any of the "trapping sites" (filled circles) adjacent to the growing cluster (squares).

in the intervening space, with the boundary conditions

$$\phi = \begin{cases} 1, & \text{on outer ring,} \\ 0, & \text{on cluster surface.} \end{cases} \tag{2.15}$$

We then find the normal derivative of ϕ, $P(w) = \partial_n \phi$. (We use an outward pointing normal so that $P(w) > 0$.) Then we define a local growth probability on the surfaces, $G(w)$.

$$G(w) = \frac{P(w)}{\int_S dw\, P(w)}. \tag{2.16}$$

The next particle is then added to the cluster surface at a point chosen with this probability. The procedure is then repeated for the next particle: the Laplace equation is solved with the new cluster boundary, the local growth probabilities are calculated, and the next particle is added according to these probabilities.

We will assume that this "electrostatic" model is equivalent to a model in which a particle performing an isotropic random walk is added to the cluster upon contact. A slight inaccuracy arises because very close to the surface, a Boltzmann equation description of the particle motion is more appropriate than the diffusion equation description that we have been using.

This electrostatic analogy can also be used to define a class of related models, termed the η-models [14]. These models were first introduced by Niemeyer, Pietronero, and Wiesmann as models for the dielectric breakdown of insulating media. In the η-models, the boundary conditions at the cluster and at ∞ are identical to those described above, but the growth probability is proportional to some power of the electric field,

$$G(w) = \frac{P^\eta(w)}{\int_S dw\, P^\eta(w)}. \tag{2.17}$$

This relation defines a different growth model for every value of the parameter η. The case $\eta = 1$ is simply DLA. The case $\eta = 0$ is another well-known growth model, the Eden model [15]. Various η-models (with $\eta \geq 0$) have been studied; qualitatively they are quite similiar to DLA, though the dimension D decreases monotonically with η.

2.4   Physical Applications of DLA

The clusters grown using the DLA algorithm are very similiar to those observed in a variety of natural growth processes, such as colloidal aggregation, viscous fingering, and electrodeposition (see chapter 8). Only in the first of these cases is the diffusion of particles literally the limiting factor in growth. In viscous fingering the pressure field, and in electrodeposition the electric potential, approximately obey the Laplace equation, so that these problems are mathematically analogous to DLA. Of course, in none of these problems is diffusion (or its analog) the only physical effect. In the absence of a general theory of DLA, it has proven difficult to predict theoretically which physical perturbations will destroy the ramified, fractal structures characteristic of DLA, and which will not. Nevertheless, the numerous observations of DLA-like structures in nature suggest that DLA should be robust against many possible variations in the underlying microscopic physics.

To illustrate in a heuristic manner the way in which DLA might be relevant to a physical situation, let us consider the phenomenon of electrodeposition. Electrodeposition is a type of metallic solidification in which metallic cations deposit on a metallic cathode out of an electrolyte (Figure 2.6). The chemical reaction is

$$M^{n+} + ne^- \rightarrow M_{\text{solid}}. \tag{2.18}$$

Suppose that the electrodeposition is *transport limited*, so that there is effectively no chemical potential barrier to adsorption and crystallization of

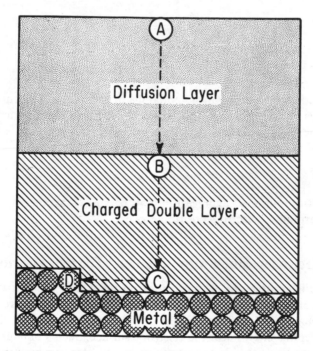

FIGURE 2.6. As an ion approaches the cathode, it passes through a diffusion layer, a double layer, and undergoes a crystallization process. In the secondary current distribution approximation, only the potential drop across the double layer is taken into account.

the ions as they arrive at the surface. Then the electrochemical potential of an ion adsorbed at the surface may be set equal to that of an ion in the solution very close to the surface, and the transport of ions through the solution to the surface will determine the local growth rate of the cathode. This transport proceeds by three mechanisms [16].

(i) Electrical migration, the generally linear current that arises in the presence of an electric field.
(ii) Diffusion of ions in response to concentration gradients.
(iii) Convection of ions with bulk flows of the solution. This convection can arise because of concentration gradients, which lead to density gradients. Also, thermal convection may occur due to heating effects at the cathode surface.

Of course, convection can destroy the very concentration gradients that would lead to a diffusive transport of ions. Thus, one natural approximation is to assume that the concentration of the depositing ions is constant, so that the only transport is due to electrical migration. Then, because there is no space charge, and the potential at the surface of the cathode is constant, we have

$$\nabla^2 \phi = 0, \tag{2.19}$$

with

$$\phi = 0 \tag{2.20}$$

at the cathode, and a constant potential Φ_0 at the anode. The current $\vec{i}_{cat} = \sigma_e \vec{E}$ will thus be determined by the normal electric field $\vec{E}$ at the surface of the cathode times a cationic conductivity σ_e. Clearly, this problem is analogous to the diffusion-limited aggregation problem, with

$$G(w) \propto \vec{i}_{cat}(w) \cdot \hat{n}. \tag{2.21}$$

Thus, diffusion-limited aggregation may be viewed as a stochastic model for this deterministic electrodeposition process.

2.4.1 Electrodeposition with Secondary Current Distribution

A more realistic model of electrodeposition takes into account the surface kinetics accompanying the deposition process [13, 16]. As electrodeposition proceeds, charge will build up in the double layer (of size λ_D, the Debye-Hückel screening length) near the electrode surface. Such a buildup of charge is necessary to drive the electrodeposition, as the strongly hydrated metallic cations have a large potential barrier (corresponding to the removal of one or more water molecules) to cross before depositing on the surface. The rate at which the metallic cations are deposited from the double layer to the surface as a function of the potential v across the double layer is determined by the

above. In the
nates over solu-

(2.27)

),

be proved easily
which the growth
clusters generated

nich can be repre-
]. In the above, we
a random walker
w before touching
') can similarly be
which we define as
λ_D from the cathode
ng previously struck

(2.28)

,

.25) in a power series.

w')

$P(w'') + \cdots \Big],$ (2.29)

a "sticking probability"
first strikes the growing
ather its sticks only with
λ' (Fig. 2.7). If it strikes
bounds with probability
h a model is proportional
hat $\lambda' = \lambda$.
numerically for almost as
that at large distances, the
A clusters is similar to that
e structure of these clusters
bserved in Eden growth, as

$$- \exp\left(\frac{-\alpha e v}{k_B T}\right)\Big],$$ (2.22)

t, is a function of the hydration
ie cations, and $\alpha < 1$ is a dimen-
$\approx .025$ volts, the Butler-Volmer

$$\frac{v}{s},$$ (2.23)

charge is accumulated in the
(which serves as the bound-
ined by Laplace's equation)
potential $v(w) = Q(w)/C$,
le layer at w, and C is the
his will lead to corrections
rite

$$)Q(w')\Big\}.$$ (2.24)

he surface is an equi-
ial. $H(w, w')$ gives the
distance λ_D from the
potential. These two

all compared to the
e, then the cationic
n current given by
sition current at w

(2.25)

rtional to the
f the distance
s the relative
double laye
$V \sim 10^7$
red to
en

This is simply our crude electrodeposition model mentioned
opposite limit $W \to \infty$, in which the surface resistance domi
tion resistances, we have

$$\lim_{W \to \infty} i_d(w) = -\frac{\Phi_0 \sigma_e}{W} \int_S dw' \, H^{-1}(w, w') P(w'$$

using the identity $\int dw' H^{-1}(w, w') P(w') = -C$, which may
from Eq. (2.24). This is a version of the "Eden model," in
rate (or probability) is constant over the entire surface. The
by Eden growth are "compact," with $D = d$.

The most interesting case is that of intermediate W, w
sented by a modification of the original DLA model [13, 19
showed that $P(w)$ is proportional to the probability that
starting at the anode will arrive at cathode surface point
the cathode elsewhere or returning to the anode. $H(w, w$
related to a property of random walks. Consider $\Pi(w, w')$
the probability that a random walker starting a distance
surface point w' will strike the cathode at w without havi
either the cathode or the anode. It can be shown that

$$H(w, w') = \frac{1}{\varepsilon} [\Pi(w, w') - \delta(w - w')$$

with ε the dielectric constant of the electrolyte.

The form of $H(w, w')$ suggests that we expand Eq. (2
This yields

$$i_d(w) = \frac{\Phi_0 \sigma_e}{\varepsilon} \left[\lambda P(w) + \lambda(1 - \lambda) \int dw' \, \Pi(w, w') P($$

$$+ \lambda(1 - \lambda)^2 \iint dw' \, dw'' \, \Pi(w, w') \Pi(w', w''$$

with $\lambda = (1 + (W/\varepsilon))^{-1}$.

This can be interpreted as the growth rate for
DLA model [20]. In such a model, when a particle
surface, it does not stick with probability one, but
probability λ', and rebounds with probability $1 -$
again, it again sticks with probability λ' and r
$1 - \lambda'$, and so forth. The growth probability in suc
o the growth rate defined by Eq. (2.29), provided
ticking probability DLA has been explored
s standard DLA, and it is well established
l structure of the sticking probability DL
DLA (Fig. 2.8). At short distances, th
r to the relatively compact shapes
from the above discussion.

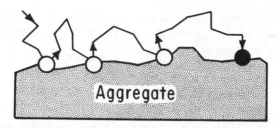

FIGURE 2.7. Random walk for a sticking probability DLA. Open circles represent encounters with the surface in which the walker is reflected. The closed circle indicates the final encounter, in which the particle adheres to the surface at the point of contact.

2.4.2   Diffusive Electrodeposition

In general, the approximation of constant concentrations does not work if there is a large concentration of supporting electrolyte, which does not participate in the surface reactions. This is the case studied by Brady and Ball in a classic experiment; here, diffusion-limited aggregation is also relevant, although in a different way [21]. This experiment is discussed in detail in sec-

FIGURE 2.8. A 150,000 particle aggregate. This aggregate was grown using a sticking probability DLA algorithm, with sticking probability $\lambda' = 0.1$.

tion 9.2.4; here, we only comment that in this case, the diffusive process is precisely diffusion of the cationic species. Electrical migration plays a negligible role in this experiment.

Problems

2.1. How can the Helmholtz equation $(\Delta - s)\phi = 0$ be simulated using random walkers?

2.2. Consider an almost flat surface growing via electrodeposition with secondary current distribution. Calculate the rate of growth of a sinusoidal corrugation of wavelength l on the surface. Can this be extended to the nonlinear regime of the Butler-Volmer equation?

2.3. Using Eq. (2.24), and the fact that the capacitance per unit area of a surface that is smooth on the length scale of λ_D is constant, show that $\int dw' H(w, w') P(w') = -C$.

2.4. Prove the validity of Eq. (2.28).

References

[1] B. Mandelbrot, *The Fractal Geometry of Nature* (Freeman, New York, 1982).

[2] J.P. Eckmann and D. Ruelle, Rev. Mod. Phys. **57**, 617 (1985); and references therein.

[3] D.R. Hofstadter, Phys. Rev. B **14**, 2239 (1976).

[4] G. Grimmett, *Percolation* (Springer-Verlag, New York, 1989); and references therein.

[5] T.A. Witten, Jr. and L.M. Sander, Phys. Rev. Lett. **47**, 1400 (1981).

[6] P. Bak, C. Tang, and K. Wiesenfeld, Phys. Rev. Lett. **59**, 381 (1987); Phys. Rev. A **38**, 364 (1988).

[7] B. Mandelbrot, *Fractals: Form, Chance, and Dimension* (Freeman, San Francisco, 1976).

[8] For reasonably up-to-date reviews of fractal growth, see J. Feder, *Fractals* (Plenum Press, New York, 1989); T. Vicsek, *Fractal Growth Phenomena* (World Scientific, Singapore, 1989).

[9] F. Argoul, A. Arneodo, G. Grasseau, and H.L. Swinney, Phys. Rev. Lett. **61**, 2558 (1988); but see also G. Li, L.M. Sander, and P. Meakin, Phys. Rev. Lett. **63**, 1322 (C) (1989).

[10] R.C. Ball and T.A. Witten, Phys. Rev. A **29**, 2966 (1984).

[11] T.C. Halsey and P. Meakin, Phys. Rev. A **32**, 2546 (1985).

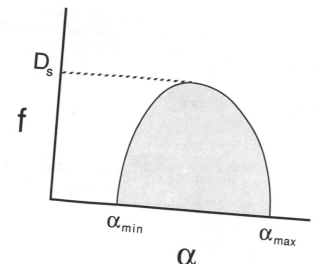

FIGURE 3.1. A typical $f(\alpha)$ curve. The maximum of the curve is the dimension of the surface, and the minimum and maximum values of α determine the maximum and minimum growth probabilities.

value of α, α_{max}. This leads qualitatively to the type of $f(\alpha)$ function seen in Fig. 3.1.

Some special points on this curve possess particularly simple interpretations. At $q = 0$, $\tau(q) = -D_s$, the dimension of the surface of the cluster, as remarked above. But $\tau(q = 0) = -f(\alpha(q = 0))$, with $\alpha(q = 0)$ defined implicitly by $df/d\alpha = 0$. Thus, the maximum value of $f(\alpha)$ corresponds to the dimension of the cluster surface.

If $f(\alpha_{min,max} = 0)$, as is the case in Fig. 3.1, then the minimum and maximum values of α correspond to at most a small number of points on the surface of the cluster; that is, to a number of points that do not scale with the ratio of the cutoff a to the size of the cluster r. (This, again, appears to be the case for almost all systems of physical interest.) Furthermore, the maximum probability region or regions on the surface have a growth probability of $(a/r)^{\alpha_{min}}$, and the minimum probability region or regions have a growth probability of $(a/r)^{\alpha_{max}}$. Note that the smallest value of α corresponds to the largest probability, and vice versa.

3.2 Systematic Definition of $\tau(q)$

The above definition of $\tau(q)$ in terms of sums of probabilities at the ultraviolet length scale, while physically quite useful and illuminating, is not very precise. In addition, we often wish to define $\tau(q)$ for systems where the natural

---

[12] L. Pietronero and H.J. Wiesmann, J. Stat. Phys. **36**, 909 (1984).

[13] T.C. Halsey and M. Leibig, J. Chem. Phys. **92**, 3746 (1990).

[14] L. Niemeyer, L. Pietronero, and H.J. Wiesmann, Phys. Rev. Lett. **52**, 1033 (1984).

[15] M. Eden, in *Proceedings of the Fourth Berkeley Symposium on Mathematical Statistics and Probability*, Vol. 1, edited by J. Neyman (University of California, Berkeley, 1961).

[16] N. Ibl, in *Comprehensive Treatise of Electrochemistry*, Vol. 6, (Plenum, New York, 1983).

[17] J. O'M. Bockris and A.K.N. Reddy, *Modern Electrochemistry* (Plenum, New York, 1970).

[18] T.C. Halsey, Phys. Rev. A **35**, 3512 (1987); T.C. Halsey, Phys. Rev. A **36**, 5877 (1987).

[19] E. Chaissang, B. Sapoval, G. Daccord, and R. Lenormand, J. Electroanal. Chem. **279**, 67 (1990).

[20] P. Meakin, Phys. Rev. A **27**, 1495 (1983).

[21] R. Brady and R.C. Ball, Nature (London) **309**, 225 (1984).

3

Multifractality

Thomas C. Halsey

"Multifractality" is a fancy word for a relatively simple concept. The growth probability distribution $G(w)$ defined in chapter 2 is a probability measure defined upon a fractal support, the surface of the DLA cluster. It has become clear over the past few years that frequently such "fractal measures" can be characterized by the scaling of their moments, with the scaling exponents of different moments being nontrivially related. We shall explain this scaling using the example of $G(w)$, but the method may be applied to any probability measure.

3.1 Definition of $\tau(q)$ and $f(\alpha)$

Suppose that the size of a particle in a DLA cluster is a, and the scale of the cluster itself is r. Because these are the only length scales in the problem, it is natural to define a moment exponent function $\tau(q)$ by [1–6]

$$\int dw\, G^q(w) = \frac{\int dw\, P^q(w)}{[\int dw\, P(w)]^q} \equiv a^{1-q}\left(\frac{a}{r}\right)^{\tau(q)}. \tag{3.1}$$

We are arbitrarily restricting ourselves to two dimensions; the factors of a ensure that this relation is dimensionally correct.

It is clear that for $q = 0$, $\tau(q) = -D_0$, the fractal dimension of the surface of the cluster. However, for $q \to \infty$ the integral will be more and more dominated by the regions of large $G(w)$, and $\tau(q)$ will be determined primarily by the growth measure scaling in these regions. Conversely, for $q \to -\infty$, the integral will be dominated by the regions of especially small $G(w)$.

Let us explore this point in more detail. Because the integrand in Eq. (3.1) will be slowly varying over length scales of a, the integral may be approximated by a sum over regions of size a. If the total growth measure in the ith such region is G_i, then we may write

$$\sum_i G_i^q = \left(\frac{a}{r}\right)^{\tau(q)}. \tag{3.2}$$

Now we wish to break the set $\{G_i\}$ into classes with the same [...]
6]. The G_i vary quite widely over the cluster; the growth probabilit[...]
considerably larger than that deep inside one of the fjords. In fact, a[...]
local growth probability density diverges (a divergence that is cut[...]
scale of a), whereas in an indentation, the probability density goe[...]
(again, this is cut off at the length scale of a particle). Thus it is n[...]
write

$$G_i \equiv \left(\frac{a}{r}\right)^{\alpha_i},$$

defining the local growth measure scaling exponent α_i. The behavior [...]
number of regions with scaling exponent α_i between α and $\alpha + d\alpha$, $N([...]$
defines a second exponent $f(\alpha)$,

$$N(\alpha)\, d\alpha = \left(\frac{a}{r}\right)^{-f(\alpha)}. \tag{}$$

Thus we have that

$$\sum_i G_i^q = \int d\alpha \left(\frac{a}{r}\right)^{q\alpha - f(\alpha)} = \left(\frac{a}{r}\right)^{\tau(q)}. \tag{3.}$$

The integral over α may be approximated using a saddlepoint technique[...]
provided that (a/r) is sufficiently small [4, 6]. The value of α dominating the[...]
integral will be that for which

$$\frac{df(\alpha)}{d\alpha} = q, \tag{3.6}$$

defining implicitly the functions $\alpha(q)$ and $f(\alpha(q))$. Then we obtain immedi-ately that

$$\tau(q) = q\alpha(q) - f(\alpha(q)), \tag{3.7}$$

and, taking the derivative with respect to q of both sides of Eq. (3.7), and using Eq. (3.6),

$$\frac{d\tau(q)}{dq} = \alpha(q). \tag{3.8}$$

Equations (3.6–3.8) are the equations of a Legendre transform. Using them, one can pass from the moment scaling function $\tau(q)$ to the local scaling density function $f(\alpha)$ and back.

Typically, there will be minimum and maximum values of α characterizing the scaling of the most concentrated and the most disperse regions of probability on the measure. If $\tau(q)$ is an analytic function of q, as seems to be the case for most physical systems of interest, then one can draw some further conclusions regarding the form of $f(\alpha)$. In the first place, $f(\alpha)$ must be concave downwards, with $d^2f/d\alpha^2 < 0$. Furthermore, $df/d\alpha$ must go to $+\infty$ at the minimum value of α, $\alpha_{\min}$, and must go to $-\infty$ at the maximum ultraviolet length scale may vary from place to place on the measure (for an example, see the two-scale Cantor set below). For these reasons, it is instructive to consider a more precise definition of $\tau(q)$ [6, 7].

Consider a probability measure $d\mu$. Choose a set of N disjoint regions $\{S_1, S_2, S_3, \ldots\}$, with $\bigcup_i S_i$ covering the probability measure. We require that each region fit within a d-dimensional ball of radius $l_i < l$. The total probability in the ith region is p_i.

Now we define a "partition function" Γ by

$$\Gamma(q, \tau, \{S_i\}, l) = \sum_{i=1}^{N} \frac{p_i^q}{l_i^\tau}. \tag{3.9}$$

Eventually we will want $\lim_{l \to 0} \Gamma \sim 1$. If $q > 1$, then $\sum p_i^q \to 0$ for $l \to 0$, so we will require $\tau > 0$. If $q < 1$, then $\sum p_i^q \to \infty$ for $l \to 0$, so we will require $\tau < 0$. We will treat these two regions separately. For $q = 1$, the choice $\tau = 0$ will give $\Gamma = 1$ by conservation of probability.

For region A, $q > 1$, $\tau > 0$, we define Γ by maximizing over all possible partitions satisfying $l_i < l$.

$$\Gamma(q, \tau, l) = \text{Sup } \Gamma(q, \tau, \{S_i\}, l). \tag{3.10}$$

If there exist constants a and α_0 such that $p_i < a(l_i)^{\alpha_0}$, then it follows that this Sup exists for $\alpha_0(q - 1) - \tau > 0$.

For region B ($q < 1, \tau < 0$), we define Γ by minimizing over all possible partitions satisfying $l_i < l$.

$$\Gamma(q, \tau, l) = \text{Inf } \Gamma(q, \tau, \{S_i\}, l). \tag{3.11}$$

The requirements for this Inf to exist are more subtle than those for the Sup to exist.

Now having defined $\Gamma(q, \tau, l)$ in these two different regions, we take the limit $l \to 0$, and write

$$\Gamma(q, \tau) = \lim_{l \to 0} \Gamma(q, \tau, l). \tag{3.12}$$

Now Γ is monotone nondecreasing as a function of τ, and is monotone nonincreasing as a function of q. We thus define $\tau(q)$ by

$$\Gamma(q, \tau) = \begin{cases} \infty; & \text{for } \tau > \tau(q), \\ 0; & \text{for } \tau < \tau(q). \end{cases} \tag{3.13}$$

3.3   The Two-Scale Cantor Set

An elementary example of the above procedure, which is also illuminating regarding the multiplicative character of multifractality, is provided by the two-scale Cantor set [3, 6, 8]. The two-scale Cantor set is defined entirely analogously to the one-scale Cantor set discussed in chapter 2. Consider a unit interval with total probability measure $\int d\mu = 1$. We replace this interval by two intervals placed respectively on the left- and right-hand sides of the original interval, the lengths of these two intervals being l_1 and l_2. A proportion p_1 if the original probability is assigned to the left-hand interval, and a proportion p_2 to the right-hand interval (Fig. 3.2). We must have

$$p_1 + p_2 = 1. \tag{3.14}$$

Now we proceed in each interval as at the first stage. We divide the interval into subintervals whose lengths are multiplied by l_1 and l_2, the probabilities of these subintervals being p_1 and p_2 times the probability of the "mother" interval. The procedure is carried out an arbitrarily large number of times.

Now we wish to calculate the partition function $\Gamma(q, \tau, l)$ defined above. In general, the optimization of the partitioning is neither feasible nor necessary. We will use instead a partitioning based upon the recursive construction of the measure. Consider $\Gamma_m(q, \tau)$, defined by covering the 2^m subintervals at the mth iteration in the construction of the set.

$$\Gamma_m(q, \tau) = \sum_{i=1}^{2^m} \left(\frac{p_i^q}{l_i^\tau} \right). \tag{3.15}$$

Clearly we have

$$\Gamma_{m+1}(q, \tau) = \Gamma_m(q, \tau) \cdot \left(\frac{p_1^q}{l_1^\tau} + \frac{p_2^q}{l_2^\tau} \right), \tag{3.16}$$

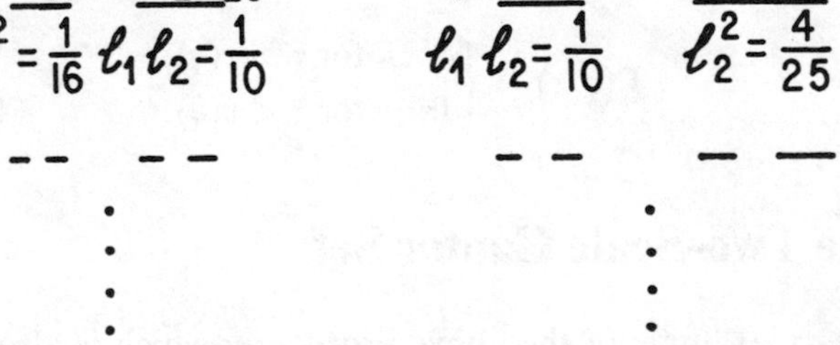

$$\underline{\quad p_1 = \frac{3}{5} \quad} \qquad\qquad \underline{\quad p_2 = \frac{2}{5} \quad}$$
$$\underline{\quad l_1 = \frac{1}{4} \quad} \qquad\qquad \underline{\quad l_2 = \frac{2}{5} \quad}$$

$$\underline{p_1^2 = \frac{9}{25}} \;\; \underline{p_1 p_2 = \frac{6}{25}} \qquad \underline{p_1 p_2 = \frac{6}{25}} \;\; \underline{p_2^2 = \frac{4}{25}}$$
$$\underline{l_1^2 = \frac{1}{16}} \;\; \underline{l_1 l_2 = \frac{1}{10}} \qquad \underline{l_1 l_2 = \frac{1}{10}} \;\; \underline{l_2^2 = \frac{4}{25}}$$

$$-\; -\quad -\; - \qquad\qquad -\; -\quad -\; -$$

$$\vdots \qquad\qquad\qquad\qquad \vdots$$

FIGURE 3.2. The construction of a two-scale Cantor set. For this example the rescalings are $l_1 = 0.25$, $l_2 = 0.4$, and the corresponding density factors are $p_1 = .6$, $p_2 = .4$.

so that

$$\Gamma_m = \left(\frac{p_1^q}{l_1^\tau} + \frac{p_2^q}{l_2^\tau}\right)^m. \tag{3.17}$$

If we require that $\lim_{m\to\infty} \Gamma_m$ go neither to zero nor to infinity, then we must have

$$\left(\frac{p_1^q}{l_1^{\tau(q)}} + \frac{p_2^q}{l_2^{\tau(q)}}\right) = 1, \tag{3.18}$$

which implicitly defines $\tau(q)$.

3.3.1   Limiting Cases

From the above, we have the formula

$$\left(\frac{p_1^q}{l_1^\tau} + \frac{p_2^q}{l_2^\tau}\right) = 1, \tag{3.19}$$

which defines $\tau(q)$. In some simple cases we can solve this explicitly. Consider the case $q = 1$. Because $p_1 + p_2 = 1$, $\tau(q = 1) = 0$ will clearly satisfy this condition.

For $q = 0$, $\tau(q) = -D$ is determined by

$$l_1^{D_0} + l_2^{D_0} = 1, \tag{3.20}$$

which is familiar as the formula for the Hausdorff dimension of a Cantor set [7].

Now consider the case $q \to \infty$. We suppose that $p_1/l_1 > p_2/l_2$. We will then suppose that the first term in Eq. (3.19) dominates in this limit. Then

$$\tau(q) = q \frac{\log p_1}{\log l_1} \equiv q\alpha_{\min}, \tag{3.21}$$

with $\alpha_{\min} = \log p_1/\log l_1$. Now we must confirm our original guess that with this value of τ, the first term in Eq. (3.19) is dominant. The second term will be given by

$$\frac{p_2^q}{l_2^\tau} = \exp\left[q \log l_2 \left(\frac{\log p_2}{\log l_2} - \frac{\log p_1}{\log l_1}\right)\right]. \tag{3.22}$$

Thus if

$$\frac{\log p_2}{\log l_2} < \frac{\log p_1}{\log l_1}, \tag{3.23}$$

our original supposition is correct.

Finally, in the limit $q \to -\infty$, we have

$$\tau(q) = q \frac{\log p_2}{\log l_2} \equiv q\alpha_{\max}. \tag{3.24}$$

Example 3.1 What are $\tau(q), f(\alpha)$ if $p_1 = p_2 = 1/2$, $l_1 = l_2 = l$?

In this case, $\Gamma = (1/2)^{q-1} l^{-\tau}$, so $\tau = (q-1) \log(2)/\log(l^{-1}) = (q-1)D$, with D the Hausdorff dimension. Thus we have $f = D$ and also $\alpha = D$.

3.3.2   Stirling Formula and $f(\alpha)$

Now let us turn to the behavior of $\tau(q), f(\alpha)$ for the 2-scale Cantor set in the general case. We can write

$$\Gamma_m \equiv \sum_{n=1}^{N=2^m} \Gamma_{N,n} = \sum_{n=1}^{N} \binom{N}{n} \left(\frac{p_1^q}{l_1^\tau}\right)^n \left(\frac{p_2^q}{l_2^\tau}\right)^{N-n} = 1. \tag{3.25}$$

The largest term in the sum will dominate in the limit $N \to \infty$. Using Stirling's formula, we find that the value of n for this largest term, which we call n^*, is determined by

$$\left.\frac{\partial \log \Gamma_{N,n}}{\partial n}\right|_{n^*} = q \log(p_1/p_2) - \tau \log\left(\frac{l_1}{l_2}\right) + \log(N/n^* - 1) = 0. \tag{3.26}$$

Solving for $N/n^* - 1$, we have

$$\log(N/n^* - 1) = -q \log(p_1/p_2) + \tau \log\left(\frac{l_1}{l_2}\right). \tag{3.27}$$

The exponent $\tau(q)$ is then determined by solving Eq. (3.27) simultaneously with

$$\log \Gamma_{N,n^*} = 0. \tag{3.28}$$

Equation (3.28) is asymptotically equivalent to

$$0 = (N/n^*) \log(N/n^*) - (N/n^* - 1) \log((N/n^* - 1)$$
$$+ (q \log p_1 - \tau \log l_1) + (N/n^* - 1)(q \log p_2 - \tau \log l_2). \tag{3.29}$$

Now we ask to what values of f, α this value of n^* corresponds. $\alpha(n^*)$ is defined by

$$p_1^{n^*} p_2^{N-n^*} = (l_1^{n^*} l_2^{N-n^*})^\alpha, \tag{3.30}$$

or, equivalently

$$\alpha(n^*) = \frac{\log p_1 + (N/n^* - 1) \log p_2}{\log l_1 + (N/n^* - 1) \log l_2}, \tag{3.31}$$

which interpolates between $\alpha_{\min}$ and $\alpha_{\max}$. The exponent $f(n^*)$ is defined by

$$\binom{N}{n^*} = (l_1^{n^*} l_2^{N-n^*})^{-f}, \tag{3.32}$$

so that, using Stirling's approximation, we have

$$f = -\frac{N \log N - (N - n^*) \log(N - n^*) - n^* \log n^*}{n^* \log l_1 + (N - n^*) \log l_2}. \tag{3.33}$$

If the Legendre transform and steepest descent approximation above are valid, then the value of $\tau(q)$ obtained by using the above values of α and f and the relation $\tau(q) = q\alpha - f$ should be the same as that determined using Eqs. (3.27–3.28). Now

$$q\alpha(q) - f(q) = \frac{1}{n^* \log l_1 + (N - n^*) \log l_2} \{q(n^* \log p_1 + (N - n^*) \log p_2)$$
$$+ [N \log N - (N - n^*) \log(N - n^*) - n^* \log n^*]\}. \tag{3.34}$$

If we use the requirement that $\log \Gamma_{N,n^*} = 0$, which constrains n^*, we have

$$q\alpha(q) - f(q) = \frac{1}{n^* \log l_1 + (N - n^*) \log l_2}$$
$$\times \{q(n^* \log p_1 + (N - n^*) \log p_2)$$
$$- q(n^* \log p_1 + (N - n^*) \log p_2)$$
$$+ \tau(n^* \log l_1 + (N - n^*) \log l_2)\}$$
$$= \tau. \tag{3.35}$$

where we have used Eqs. (3.27–3.28). Thus, within the asymptotic approximation, the relations between $f(\alpha)$ and $\tau(q)$ defined by Eqs. (3.6–3.8) are valid, as we would expect.

3.4 Multifractal Correlations

In the above, we have been considering moments of a probability measure, which we can describe as a measure $p(x) \, dx = d\mu$ on some support. We have a relation between expectation values of moments of p and the exponent function $\tau(q)$,

$$\langle p^q \rangle \alpha \left(\frac{a}{r}\right)^{\tau(q) - D_0}, \tag{3.36}$$

where D_0 is the dimension of the support, and the averaging is over the spatial extent of the measure. We can learn more about the physical interpretation of $f(\alpha)$ by considering two-point correlations of moments of p. Let us define the two-point correlation function $C_{mn}(y)$ by

$$C_{mn}(y) \equiv \langle p^m(x)p^n(x+y) \rangle_x, \tag{3.37}$$

where the brackets $\langle \; \rangle_x$ indicate averaging over x.

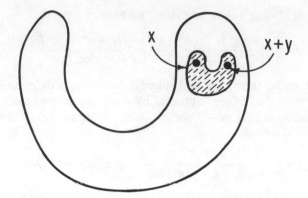

FIGURE 3.3. The "blob" picture of a multifractal measure. The correlation function $C_{mn}(y) = \langle p^m(x)p^n(x+y)\rangle$ factors into a contribution from the length scale y and two contributions from the ultraviolet length scale a.

We can estimate the behavior of $C_{mn}(y)$ using a blob picture (Fig. 3.3) [9]. We expect that the contribution to the local probability arising from scales larger than y will be the same at the two points x, $x+y$, while the contributions arising from scales smaller than y will be independent. If we introduce "smeared" averages denoted by $\langle\ \rangle_{z,r}$, where the effect of fluctuations on scales smaller than z is eliminated from the averaging, while the largest scale is taken to be r (we expect that $\langle p^q\rangle_{z,r} = (z/r)^{\tau(q)-D}$), then in the blob picture we can write

$$C_{mn}(y) \approx \langle p^{m+n}\rangle_{y,r}\langle p^m\rangle_{a,y}\langle p^n\rangle_{a,y} = \left(\frac{y}{r}\right)^{\tau(m+n)-D}\left(\frac{a}{y}\right)^{\tau(m)+\tau(n)-2D}$$

(3.38)

Thus, $C_{mn}(y)$ scales with y as

$$C_{mn}(y) \propto y^{\tau(m+n)-\tau(m)-\tau(n)+D}.$$

(3.39)

3.4.1 Operator Product Expansion and Multifractality

The relation Eq. (3.38) can be interpreted in a manner analogous to field theory [10]. One of the most useful tools in field theory is the *operator product expansion*, which allows products of operators of a theory to be represented as a power series in other operators of a theory [11]. In general, we expect that a field theoretic operator O_i will have an average value

$$\langle O_i\rangle \sim a^{\gamma_i},$$

(3.40)

where γ_i is the scaling index of the operator O_i, and a is the ultraviolet length scale of the theory. Equation (3.40) is analogous to Eq. (3.36) for the behavior of moments of the probability measure. In the operator product expansion, we write

$$O_i(x)O_j(x+y) = \sum_k c_{ijk} \left(\frac{a}{y}\right)^{\gamma_i+\gamma_j-\gamma_k} O_k(x). \qquad (3.41)$$

The structure constants c_{ijk} characterize the expansion, which is ordered in terms of increasing γ_k.

In field theories, the expansion typically starts with the identity operator, so that

$$\langle O_i(x)O_j(x+y)\rangle = c_{ij0} \left(\frac{a}{y}\right)^{\gamma_i+\gamma_j} +\cdots, \qquad (3.42)$$

where the higher order terms are less singular at short distances.

Now we can interpret Eq. (3.38) in a natural way. For multifractal theories this expansion starts with a much higher-order term, so that writing $\gamma_n = \tau(n) - D$,

$$p^m(x)p^n(x+y) = c\left(\frac{a}{y}\right)^{\gamma_n+\gamma_m-\gamma_{n+m}} p^{m+n}(x) +\cdots, \qquad (3.43)$$

is identical to an operator product expansion of the type given by Eq. (3.41), which starts with the $m+n$th term. Again, we expect the higher-order terms (which are corrections to the blob picture of section 3.4) to be less singular at short distances.

3.4.2   Correlations of Iso-α Sets

Now the exponent f cannot be literally understood as the Hausdorff dimension of some set, because in the asymptotic limit of the construction of, for example, the two-scale Cantor set, there will be points with a given α arbitrarily close to any point in the set. However, we can find a geometric intepretation of the exponent f by fixing a particular value of the ultraviolet cutoff a, and asking for the spatial correlations between intervals of size a with given values of α. Knowledge of such correlations is useful in understanding the internal structure of multifractal measures. The $f(\alpha)$ function alone determines the behavior of correlation functions both between such iso-α-sets, and also between sets of differing α, at least at large distances [12].

The main object of study $f(\alpha, \alpha', \omega)$ is defined through the the following relation

$$P(\alpha, \alpha', y) \equiv a^{D_0 - f(\alpha,\alpha',\omega)}, \qquad (3.44)$$

where $P(\alpha, \alpha', y)$ denotes the probability to find pairs of α-sites and α'-sites at a distance $y = a^\omega$. As in the above, a is the lower cutoff scale; the length scale r has been set equal to one.

To relate $f(\alpha, \alpha', \omega)$ to $f(\alpha)$, it is useful to reconsider the 2-point probability moment correlation function $C_{mn}(y)$,

$$C_{mn}(y) = \langle p^m(x)p^n(x+y) \rangle,$$

where the average is taken over all sites with nonvanishing measure. In section 3.4 we showed that

$$C_{mn}(y) = a^{D_0 + \tilde{\tau}(m,n,\omega)}, \tag{3.45}$$

where $\omega = \log(y)/\log(a)$, and,

$$\tilde{\tau}(m, n, \omega) = \tau(m) + \tau(n) + D_0 + \omega[\tau(m+n) - \tau(m) - \tau(n) - D_0]. \tag{3.46}$$

Using the relation

$$C_{mn}(y) = \int_\alpha \int_{\alpha'} a^{m\alpha + n\alpha'} a^{D_0 - f(\alpha, \alpha', \omega)} \, d\alpha \, d\alpha', \tag{3.47}$$

$f(\alpha, \alpha', \omega)$ can be expressed as a double Legendre transformation

$$f(\alpha, \alpha', \omega) = \min_{m,n} [m\alpha + n\alpha' - \tilde{\tau}(m, n, \omega)]. \tag{3.48}$$

In terms of the $f(\alpha)$ spectrum, this can be expressed as,

$$f(\alpha, \alpha', \omega) = f[\alpha(m_0)] + f[\alpha(n_0)] - D_0 + \omega[f[\alpha(m_0 + n_0)] - f[\alpha(m_0)]$$
$$- f[\alpha(n_0)] + D_0], \tag{3.49}$$

where m_0, n_0 denote the values of m, n for which the extremum condition Eq. (3.48) is satisfied.

The above relations imply that for the case $\alpha = \alpha'$, $\lim_{\omega \to 1} f(\alpha, \alpha', \omega) = f(\alpha)$. This accounts for the results of Meakin [13], who found self-similiarity in the correlations of single-α sets for randomly generated Cantor-type probability measures. It is in this sense that $f(\alpha)$ can be regarded heuristically as a dimension corresponding to regions of the measure associated with the exponent α.

3.5 Numerical Measurements of $f(\alpha)$

The $f(\alpha)$ function for the growth probability for DLA, and for various of the other η-models, has been studied numerically by several authors. There are various procedures for determining $f(\alpha)$ for a DLA cluster.

(i) Given a cluster (or ensemble of clusters), random walkers may be launched toward the surface of the cluster, with the proviso that when they strike, they are destroyed, with their last position being recorded. This will provide a snapshot of the growth measure for a particular cluster. This method, although quite efficient for moderately large q, is not effective for $q < 1$, as very few particles will strike in the disfavored

TABLE 3.1. Values of $\tau(q)$ obtained from
the numerical simulations of Halsey et al.
[5]. Only the statistical errors are quoted.

q	$\tau(q)$
2	0.98 ± 0.01
3	1.71 ± 0.01
4	2.43 ± 0.02
5	3.13 ± 0.02
6	3.82 ± 0.03
7	4.45 ± 0.06
8	5.14 ± 0.08

regions. This was the original method of Halsey et al. and of Meakin et al. (see Table 3.1 for results) [5].

(ii) For a cluster grown on a lattice, Green's function methods may be used to calculate the growth probability at any point. This method does access the $q < 1$ behavior, although only very small clusters may be examined in this way. This was the method of Amitrano et al., and also of very interesting recent exact enumeration studies by Lee et al. (see chapter 4) [14, 15].

(iii) Finally, a lattice may be superimposed upon a cluster (which may originally be either on-lattice or off-lattice) and Laplace's equation may be relaxed using standard algorithms. Hayakawa et al. succeeded in obtaining results for 50,000-particle clusters in this way [16]. Their results are clearly in accord with the above phenomenology.

3.6   Ensemble Averaging and $\tau(q)$

DLA clusters are stochastic objects, so some averaging is necessary to define the function $\tau(q)$, or $f(\alpha)$, as an ensemble property [17]. If brackets $\langle \ \rangle$ refer to averaging over a probabilistic ensemble of DLA clusters, then one can choose to define either an "annealed" $\tau(q)$, $\tau_a(q)$, or a quenched $\tau(q)$, $\tau_q(q)$, respectively, by the relations

$$\left\langle \sum_i G_i^q \right\rangle = \left(\frac{a}{r}\right)^{\tau_a(q)}, \tag{3.50}$$

$$\left\langle \log\left(\sum_i G_i^q \right) \right\rangle = \tau_q(q) \log(a/r). \tag{3.51}$$

Most computer studies have relied upon annealed averaging, but quenched averaging probably yields a more physical result. This is because in averaging over the entire ensemble of possible clusters, some clusters will appear, such as a perfectly linear cluster, or a cluster with a "tunnel" of exponentially declining probability, that might alter an annealed average, despite their very low probability of appearance. Quenched averages should prove more robust against these effects. In averaging over a relatively small number of stochastically generated clusters, these questions are probably academic, as only rather typical members of the ensemble will appear.

Example 3.2 Consider a stochastic Cantor set generated in the following way. Pick a number x at random between 0 and 1. Pick y at random between x and 1. Put a probability 1/2 on the interval $[0, x]$, and a probability 1/2 on the interval $[y, 1]$. Now subdivide each interval, following the same procedure with new random numbers. What is D for this measure, using the annealed averaging procedure? What is the formula for $\tau(q)$, using the annealed averaging procedure?

The construction of the set is still hierarchical, so the average partition function $\langle \Gamma \rangle$ may be obtained by taking a product of generators, because the averaging is independent at each step in the hierarchical generation of the set. Thus $\tau(q)$ is determined by

$$\left(\frac{1}{2}\right)^q \langle [x^{-\tau} + (1 - y)^{-\tau}] \rangle = 1.$$

We can write the average explicitly as an integral

$$\langle [x^{-\tau} + (1 - y)^{-\tau}] \rangle = \int_0^1 dx \int_x^1 dy \, \frac{1}{1 - x} [x^{-\tau} + (1 - y)^{-\tau}] = \frac{(2 - \tau)}{(1 - \tau)^2}.$$

Thus, $\tau(q)$ is determined implicitly by

$$\log\left(\frac{(2 - \tau)}{(1 - \tau)^2}\right) = q \log(2).$$

For the case $q = 0$, we have $\tau(0) = -D = (1 - \sqrt{5}/2)$, so that the annealed Hausdorff dimension is simply the golden mean, $D = .618 \ldots$.

Problems

3.1. Calculate and graph the $f(\alpha)$ curve for the two scale Cantor set shown in Fig 3.2.

3.2. Consider a two-scale Cantor set similiar to that defined above, except that only the rightmost interval is subdivided at any stage in the hierarchy, so that uniform measure segments appear on a variety of length scales. Determine $\tau(q)$ for this measure. What is the $f(\alpha)$ spectrum for this measure?

3.3. Consider again the stochastic measure defined in Example 3.2. Sketch the $f(\alpha)$ spectrum. You should see negative values of f appear. Can you suggest a physical interpretation of this result?

3.4. Calculate the quenched spectrum of the stochastic measure defined in Example 3.2. Do negative values of f appear? Comment.

References

[1] B.B. Mandelbrot, Ann. Isr. Phys. Soc. **255** (1977).

[2] A. Renyi, *Probability Theory* (North-Holland, Amsterdam, 1970).

[3] H.G.E. Hentschel and I. Procaccia, Physica D **8**, 435 (1983).

[4] U. Frisch and G. Parisi, in *Turbulence and Predictability in Geophysical Fluid Dynamics and Climate Dynamics*, Proc. of Int. School of Physics "Enrico Fermi" LXXXVIII, edited by M. Ghil, R. Benzi, and G. Parisi (North-Holland, Amsterdam, 1985); R. Benzi, G. Paladin, G. Parisi, and A. Vulpiani, J. Phys. **A17**, 3521 (1984).

[5] T.C. Halsey, P. Meakin, and I. Procaccia, Phys. Rev. Lett. **56**, 854 (1986); P. Meakin, H.E. Stanley, A. Coniglio, and T.A. Witten, Jr., Phys. Rev. A **32**, 2364 (1986).

[6] T.C. Halsey, M.H. Jensen, L.P. Kadanoff, I. Procaccia, B. Shraiman, Phys. Rev. A **33**, 1141 (1986).

[7] J.D. Farmer, Physica D **4**, 366 (1982); and references therein.

[8] P. Grassberger and I. Procaccia, Physica D **13**, 34 (1984).

[9] M. Cates and J. Deutsch, Phys. Rev. A **35**, 4907 (1987).

[10] B. Duplantier and A.W.W. Ludwig, Phys. Rev. Lett. **46**, 247 (1993).

[11] K.G. Wilson, Phys. Rev. **179**, 1499 (1969); L.P. Kadanoff, Phys. Rev. Lett. **23**, 1430 (1969).

[12] C. Meneveau and A. Chhabra, Physica A **164**, 564 (1990); S.J. Lee and T.C. Halsey, Physica A **164**, 575 (1990).

[13] P. Meakin, in *Phase Transitions and Critical Phenomena*, Vol. 12, edited by C. Domb and J. Lebowitz (Academic, London, 1988).

[14] C. Amitrano, A. Coniglio, and F. di Liberto, Phys. Rev. Lett. **57**, 1016 (1986).

[15] J. Lee and H.E. Stanley, Phys. Rev. Lett. **61**, 2945 (1988); J. Lee, Ph.D. thesis, Boston University, (1991).

[16] Y. Hayakawa, S. Sato, and M. Matsushita, Phys. Rev. A **36**, 1963 (1987).

[17] T.C. Halsey, in *Fractals: Physical Origin and Properties*, edited by L. Pietronero (Plenum, London, 1989).

4

Scaling Arguments and Diffusive Growth

Thomas C. Halsey

Of course, the multifractal formalism developed in chapter 3 is, a priori, no better than any of a large number of conceivable phenomenological descriptions at encoding universal scaling information about the behavior of DLA clusters. The advantage of using this formalism for DLA is that a number of scaling laws exist relating the multifractal exponents to one another and to the dimension of the clusters themselves [1–6]. There are essentially three such laws that are in good agreement with numerical results (a fourth proposed law, due to Ball and Blunt, is in poor agreement with numerics [1]). These are (i) the scaling of the information dimension, due to Makarov [2]; (ii) the Turkevich–Scher law [3, 4, 6]; and (iii) the electrostatic scaling law [5, 6].

4.1 The Information Dimension

The information dimension D_I of a probability measure is defined by covering the measure with boxes on a length scale l and forming the sum over the various boxes

$$\sum_i p_i \log p_i = D_I \log(l/r),$$ (4.1)

where p_i is the total probability in the ith box, and r is the scale of the entire set [7]. Clearly,

$$D_I = \left.\frac{\partial \tau(q)}{\partial q}\right|_{q=1} = \alpha(1).$$ (4.2)

This dimension has a natural physical interpretation. If we ask what the total probability is on the part of the measure that has the scaling exponent α, this quantity, which we call $P(\alpha)$, behaves as

$$P(\alpha) = \left(\frac{a}{r}\right)^{\alpha - f(\alpha)},$$ (4.3)

where, as in the above, a is the ultraviolet cutoff. Now, in general, $\alpha \geq f(\alpha)$, with the equality applying only at $\alpha(q = 1)$. Thus the information dimension of a measure corresponds to the dimension of that part of the measure that contains an arbitrarily large proportion of the total measure as the cutoff $a \to 0$.

Makarov has proven a useful theorem for the information dimension of an arbitrary domain in two dimensions, with a probability measure on the surface given by the (normalized) normal electric field given that the surface is an equipotential [2]. This is that

$$D_I = 1, \tag{4.4}$$

or

$$\alpha(q = 1) = f(\alpha(q = 1)) = 1. \tag{4.5}$$

It is not clear that a DLA cluster can be viewed as a domain in the sense of Makarov; nevertheless, it is tempting to apply the Makarov theorem directly to DLA clusters, and presume that their information dimension will also be one. This conclusion is in excellent accord with numerical results [8].

4.2   The Turkevich–Scher Scaling Relation

Turkevich and Scher have recently proposed a simple physical idea, which can be used to yield a scaling relation between the exponent $\alpha_{\min}$ and the dimension of the cluster as defined by Eq. (2.5) [3]. The radius R_g that appears in Eq. (2.5) was defined as the radius of gyration of the cluster. If the clusters are truly self-similar, then they are characterized by no length scale aside from the size of the cluster and the size of the individual particles. Then, different methods of defining the radius should be equivalent, in the sense that for any cluster they will be proportional to each other, with the constants of proportionality independent of the size of the cluster.

It is convenient to take the maximum radius of the cluster r_M as our definition of radius. Now, the maximum radius of a cluster will grow upon the addition of a new particle only if that particle lands at the position that was previously at the maximum radius. If it does so, then the radius will grow by some quantity of order a. Thus, writing the probability of landing upon the maximum radius particle as $G_{\max}$, we have

$$\left\langle \frac{dr_M}{dn} \right\rangle \approx a G_{\max}. \tag{4.6}$$

Now we know from chapter 3 that $G_{\max} \leq (a/r_M)^{\alpha_{\min}}$. Observing the scaling with r_M of both sides of Eq. (4.6), we obtain

$$D \geq 1 + \alpha_{\min}. \tag{4.7}$$

If we further make the plausible assumption that the maximum radius point also corresponds to the maximum probability point, then we obtain

$$D = 1 + \alpha_{\min}. \tag{4.8}$$

Unfortunately, this relation is not quite in agreement with the numerical evidence. Typical measurements of $\alpha_{\min}$ for DLA in two dimensions give $\alpha_{\min} \approx .65$, while the exponent $D \approx 1.71$. Note that these values are still in accord with the inequality Eq. (4.6). This discrepancy may be a consequence of the relation holding only as an inequality, or it may be the result of systematic errors in the calculation of $f(\alpha)$, which remain rather poorly understood.

For other values of η, it is sometimes convenient to write this scaling relation in terms not of the $f(\alpha)$ function for the growth measure, but rather in terms of the $f(\alpha)$ function for the electric field. Recall that $P(w)$ is the normal electric field at the surface point w, given that the cluster surface is at a potential $\phi = 0$, and that a distant conductor (say, at a distance R from the center of the cluster, $R \gg r$) is at a potential $\phi = 1$ (see Fig. 4.1). The total electric field absorbed by the cluster will, in two dimensions, scale logarithmically in (r/R). In fact, this can be used to define an "electrostatic" radius of the cluster, r_E, which will presumably behave in the same way as r_M above,

$$\int dw\, P(w) \equiv 2\pi [\log(R/r_E)]^{-1}, \tag{4.9}$$

so that the flux absorbed by a cluster is the same as that absorbed by a circular conductor of radius r_E inside an outer circular conductor of radius R. Now the growth measure for the η-model is defined the ηth power of the electric field. Defining $\sigma_\eta(q)$ by

$$\int dw [G_\eta(w)]^q \equiv a^{1-q} \left(\frac{a}{r}\right)^{\sigma_\eta(q)}, \tag{4.10}$$

and using for $\tau(q)$ the normalized electric field measure,

$$\frac{\int dw [P(w)]^q}{[\int dw\, P(w)]^q} \equiv \left(\frac{a}{r}\right)^{\tau(q)}, \tag{4.11}$$

we have immediately the relation between $\sigma_\eta(q)$ and $\tau(q)$ [6],

$$\sigma_\eta(q) = \tau(\eta q) - q\tau(\eta). \tag{4.12}$$

The Turkevich and Scher relation should clearly be as valid for arbitrary η as it is for $\eta = 1$, provided that $\alpha_{\min}$ for the growth measure is used. Thus, a simple calculation shows that, in terms of $\alpha_{\min}$ for the electric field measure, the Turkevich–Scher relation (in its strong form) is [6]

$$D = 1 + \alpha_{\min}\eta - \tau(\eta). \tag{4.13}$$

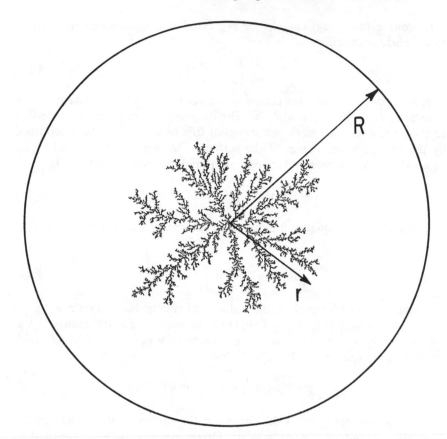

FIGURE 4.1. We imagine that a cluster of radius r is surrounded by a circle of radius R, with $R \gg r$. The circle and the cluster are both equipotentials; $P(w)$ is the normal electric field at the cluster surface in this configuration.

These relations can be easily generalized to arbitrary dimensionality; one finds that the Turkevich–Scher law [Eq. (4.13)] is independent of dimensionality.

4.3 The Electrostatic Scaling Relation

A further scaling relation can be derived using an energetic argument based upon elementary electrostatics [5, 6]. A system with two conductors held at fixed potential is a capacitor, whose energy ε is simply related to the potential across the capacitor and the charge Q stored therein. Because the potential across this capacitor is $\Delta \phi = 1$, we have

$$\varepsilon = +\frac{1}{2} Q = \frac{1}{8\pi} \int dw \, P(w). \tag{4.14}$$

Of course, this energy can also be expressed as the volume integral of the electric field squared,

$$\varepsilon = \frac{1}{8\pi} \int_V dv (\vec{\nabla}\phi)^2. \tag{4.15}$$

Now suppose we move the cluster surface point w normal to the surface by a distance $f(w)$ (Fig. 4.2), while holding its potential, and that of the distant surface, constant. The work performed in this operation will be determined by the change in the charge of the surface. This work is performed by an external charge reservoir at fixed potential upon the system, its value is

$$\delta w = \delta Q = \frac{1}{4\pi} \delta \int dw\, P(w). \tag{4.16}$$

This work must be equal to the change in the energy stored in the system, so that

$$\delta \int dw\, P(w) = \frac{1}{2} \delta \int_V dv (\vec{\nabla}\phi)^2. \tag{4.17}$$

The change in the integral on the right-hand side is due to two effects: the exclusion of a shell of thickness $f(w)$ from the volume, and the change in $\vec{\nabla}\phi$. To *zeroth* order in f, $(\vec{\nabla}\phi)^2$ near the surface may be replaced by $(P(w))^2$; an elementary calculation then shows that

$$\delta \int dw\, P(w) = \frac{1}{2} \delta \int_V dv [P(w)]^2 f(w). \tag{4.18}$$

Now suppose that $f(w)$ is the change in the surface of the cluster arising from the addition of a particle at w'. Of course, $f(w)$ will be localized to within a particle size a about w', and its total area $\int dw\, f(w)$ will be simply the d-dimensional particle volume $k_d a^d$. Because $P(w)$ is slowly varying on scales of a, we may immediately average over the probability $G_\eta(w')$ of a particle accreting at any position upon the surface, to obtain

$$\left\langle \frac{d}{dn} \int dw\, P(w) \right\rangle = \frac{k_d a^d}{2} \left\langle \frac{\int dw [P(w)]^{2+\eta}}{\int dw [P(w)]^\eta} \right\rangle, \tag{4.19}$$

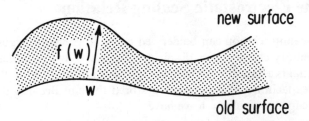

FIGURE 4.2. The shaded area represents the volume excluded from free space in the growth of the surface. The normal change in the surface position at w is $f(w)$.

where the brackets indicate averages over the ensemble of clusters, and k_d is a constant.

To pass from this identity to the electrostatic scaling law, some assumptions are needed. First, the order of differentiation and averaging on the left-hand side of Eq. (4.19) must be interchanged. Second, the average of the ratio on the right-hand side must be replaced by the ratio of the averages. For $\eta = 1$, this should not pose any problems, as the denominator will not fluctuate significantly, because it depends only on the logarithm of r_E. For other values of η, correlated fluctuations between the numerator and denominator could invalidate this assumption. Of course, if one restricts the ensemble on both sides of the equation to "typical" (i.e., well-behaved) clusters, one can still average the numerator and denominator separately. In two dimensions, these assumptions, in concert with the multifractal form for the moments of $P(w)$, yield immediately

$$\frac{dr}{dn} \propto \left(\frac{a}{r}\right)^{\tau(2+\eta)-\tau(\eta)}. \tag{4.20}$$

Thus, the electric field moments in an η model will obey

$$D = \tau(2+\eta) - \tau(\eta). \tag{4.21}$$

This relation can be easily generalized to other dimensionalities.

This relation is in excellent agreement with currently available numerical results [9, 10, 11]. Thus, for two-dimensional off-lattice DLA, Halsey, Meakin, and Procaccia give $\tau(3) = 1.71 \pm .01$, in excellent agreement with $D = 1.71$ [4] ($\tau(1) = 0$ for a normalized measure). For other values of η, there is also good numerical agreement. This electrostatic scaling law, and those of Turkevich and Scher and of Makarov, are listed in Table 4.1. Figure 4.3 displays the comparison between the first two of these scaling laws and the numerical results of Amitrano for a variety of η-models in two dimensions [11].

TABLE 4.1. The definition of $\tau(q)$, and the Turkevich–Scher, electrostatic, and information dimension scaling laws. α_{min} refers to $f(\alpha)$ for the electric field measure.

$$\frac{\int dw[P(w)]^q}{[\int dw \, P(w)]^q} \equiv \left(\frac{a}{r}\right)^{\tau(q)}$$

$$D = 1 + \alpha_{min}\eta - \tau(\eta)$$

$$D = \tau(2+\eta) - \tau(\eta) + (2-d)$$

$$\alpha(q=1) = 1 \quad \text{in } d = 2 \text{ only}$$

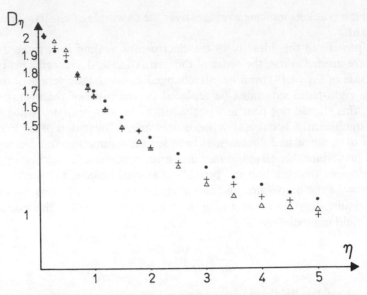

FIGURE 4.3. Comparison between D as a function of η and the Turkevich–Scher and electrostatic scaling laws, Eqs. (4.13 and 4.21). The dots represent D; the crosses represent the electrostatic prediction for D; the triangles the Turkevich–Scher prediction for D. Figure courtesy of C. Amitrano [11].

4.4   Scaling of Negative Moments

All of the scaling laws discussed above refer to properties of positive moments of the growth probability; the behavior of negative moments has been considerably more controversial. In the discussion of chapter 3, we implicitly assumed that the growth probability at the most screened sites $P_{\min}$ scaled as a power law with the size of the cluster,

$$P_{\min} = \left(\frac{a}{r}\right)^{\alpha_{\max}}. \tag{4.22}$$

If this is not the case, then the $f(\alpha)$ formalism will break down for negative values of q. Here again, the issue of averaging raises its head. We can define either a quenched or an annealed average $P_{\min}$,

$$\langle P_{\min}\rangle = P_{\min;a}, \tag{4.23}$$

for an annealed average, and

$$\exp(\langle \log P_{\min}\rangle) = P_{\min;q}, \tag{4.24}$$

for a quenched average [12]. In each case, the brackets $\langle \ \rangle$ refer to averaging over the ensemble of possible clusters, with their appropriate probabilities.

For small clusters, Lee and Stanley have performed the annealed average for DLA directly by enumerating all possible clusters, calculating their probability of occurence, and calculating the minimum growth probability [13]. They obtained the result that

$$P_{\text{min};a} \propto \exp(-r^2) \qquad (4.25)$$

in two dimensions. Such strong decay of the minimum probability will certainly invalidate the formalism above if $q < 0$. Lee and Stanley refer to this as a "phase transition" in the $f(\alpha)$ function.

More recently, Schwarzer et al. have examined the behavior of typical values of the quantity P_{min} [14]. These values will correspond more closely to the quenched average of P_{min}. Their result is that

$$P_{\text{min};q} \propto \exp(\log(-Ar^\beta)), \qquad (4.26)$$

with $\beta \approx 2$, and A a constant. Note that $\beta = 1$ corresponds to traditional multifractal scaling. Although this singularity is much weaker than that observed in annealed averages, it is still sufficient to destroy the multifractal behavior for negative q.

4.5   Conclusions

The $f(\alpha)$ function gives not only an extensive phenomenological characterization of the growth properties of diffusively growing clusters, it also gives an appropriate characterization. The existence of scaling laws for $f(\alpha)$ strongly suggests that the scaling of moments of the growth distribution will be one of the key objects in our eventual quantitative understanding of this problem [15]. Unfortunately, there are many physically relevant properties of the structure of DLA clusters for which $f(\alpha)$ is too crude a measure. One of the most glaring such properties is the hierarchical structure of a DLA cluster, which re-creates branches with roughly the same proportions, and containing roughly the same angles between stems, at every length scale of the cluster. While this structure is encoded in a rough way in $f(\alpha)$, its subtleties must escape such a simple function.

In fact, this hierarchical structure is reminiscent of the structure of strange attractors at the onset of chaos [16]. These phase space objects possess asymptotically an exact hierarchical structure, which may be described by the scaling function for the strange attractors. Recently, it has become clear how to extend the scaling function description to a system that possesses a hierarchical structure "averaged" over fluctuations [17]. This new approach promises to yield a comprehensive description of the structure of DLA clusters, including first principles computations of their quantitative properties. Unfortunately, these developments are beyond the scope of this discussion.

Problems

4.1. Generalize the electrostatic scaling law to arbitrary dimensionality d.

4.2. Consider a cluster growing with an exponential growth law, $G(w) \propto \exp(P(w))$. Is it consistent to assume that the electric field measure $P(w)/\int dw\, P(w)$ exhibits multifractal scaling?

4.3. In the limit $\eta \to 0$, what do the scaling laws derived above imply for the behavior of $\tau(q)$ near $q = 0$ for $d = 2$?

References

[1] R. Ball and M. Blunt, Phys. Rev. A **41**, 582 (1989).

[2] N.G. Makarov, Proc. London Math. Soc. **51**, 369 (1985).

[3] L. Turkevich and H. Scher, Phys. Rev. Lett. **55**, 1026 (1985); Phys. Rev. A **33**, 786 (1986).

[4] T.C. Halsey, P. Meakin, and I. Procaccia, Phys. Rev. Lett. **56**, 854 (1986); R. Ball, R. Brady, G. Rossi, and B.R. Thompson, Phys. Rev. Lett. **55**, 1406 (1985).

[5] T.C. Halsey, Phys. Rev. Lett. **59**, 2067 (1987).

[6] T.C. Halsey, Phys. Rev. A **38**, 4789 (1988).

[7] T.C. Halsey, M.H. Jensen, L.P. Kadanoff, I. Procaccia, and B. Shraiman, Phys. Rev. A **33**, 1141 (1986); and references therein.

[8] M. Matsushita, Y. Hayakawa, S. Sato, and K. Honda, Phys. Rev. Lett. **59**, 86 (1987).

[9] C. Amitrano, A. Coniglio, and F. di Liberto, Phys. Rev. Lett. **57**, 1016 (1986).

[10] Y. Hayakawa, S. Sato, and M. Matsushita, Phys. Rev. A **36**, 1963 (1987).

[11] C. Amitrano, Phys. Rev. A **39**, 6618 (1989).

[12] T.C. Halsey, in *Fractals: Physical Origin and Properties*, edited by L. Pietronero, (Plenum, London, 1989).

[13] J. Lee and H.E. Stanley, Phys. Rev. Lett. **61**, 2945 (1988).

[14] S. Schwarzer, J. Lee, A. Bunde, S. Havlin, H.E. Roman, and H.E. Stanley, Phys. Rev. A **65**, 603 (1990).

[15] But for an opposing view, see L. Pietronero, A. Erzan, and C. Evertsz, Phys. Rev. Lett. **61**, 861 (1988).

[16] See, for example, M.H. Jensen, L.P. Kadanoff, and I. Procaccia, Phys. Rev. A **36**, 1409 (1987).

[17] T.C. Halsey and M. Leibig, Phys. Rev. A **46**, 7793 (1992); T.C. Halsey, Phys. Rev. Lett. **72**, 1228 (1994).

II

Chaos and Randomness

5

Introduction to Dynamical Systems

Stephen G. Eubank and J. Doyne Farmer

5.1 Introduction

Chaos provides a link between determinism and randomness. It demonstrates that even very simple systems are capable of random behavior, and that randomness does not necessarily depend on the complexity of initial data. Instead, nonlinear geometrical relationships in the laws of motion cause mixing of nearby initial conditions, so that the states of the system are "shuffled," much like a deck of cards. Even though the geometric relationships dictated by the laws of motion may be quite simple, the resulting trajectories can be highly complex. Small changes in initial conditions are amplified into very large changes in long-term behavior, making the relationship between cause and effect so complicated as to be effectively random. This complexity is generated internally, rather than externally. From any practical point of view the result is random.

5.2 Determinism Versus Random Processes

There are traditionally two mathematical approaches to the problem of prediction. On one hand we can assume perfect knowledge of both the laws of motion and the initial data. The states of a system can then be regarded as points whose future evolution occurs along deterministic trajectories. Models of this type are called *dynamical systems*. On the other hand we can assume that both the laws of motion and the initial data are uncertain, treating this uncertainty as "noise." Models of this type are called *random processes*.

Traditionally, the theory of dynamical systems and the theory of random processes have been presented as separate subjects, with little or no overlap. For regular motion, in which the behavior of nearby trajectories does not get mixed up, this makes sense; if we can model the system deterministically for short times, we can also do so for long times. For chaos, however, this is *not*

Part II is an updated version of the article by the same authors in *1989 Lectures in Complex Systems*, ed. by I. Jen, Addison-Wesley, Menlo Park, CA, 1990.

sensible because determinism and randomness are inextricably linked. For accurate measurements, with sharply defined states, a deterministic dynamical model may be an appropriate approximation for short times. However, because even the smallest uncertainties are amplified exponentially, after a finite amount of time the ensemble of possible states consistent with the initial measurement error becomes so spread out that this ceases to be a good approximation in any sense. Even if we begin with a deterministic model, we are rapidly forced to consider a random process. Long-term prediction becomes impossible, and determinism gives way to statistical description.

However, chaos is not entirely bad news. It presents the possibility that what previously seemed hopelessly random may be predictable, at least for short times. Traditionally, it was implicitly assumed that random behavior was due to extreme complexity. If a system is sufficiently complicated, with a large number of irreducible independent degrees of freedom, then from a practical point of view it becomes impossible to model deterministically; it would simply not be feasible to make enough measurements, much less simulate the model. A random process model allows us to do the best job we can, lumping many degrees of freedom into probability distributions involving only a few variables. Because some degrees of freedom are neglected, predictability is quite limited, but the model is at least tractable. Under favorable circumstances (such as large autocorrelations in the motion), models based on the theory of random processes allow us to make better predictions than we could make with no model at all.

Chaos tells us that randomness does not necessarily involve an enormous number of independent degrees of freedom. In the presence of nonlinear dynamics, only a few independent variables are sufficient to generate chaotic motion. When only a few degrees of freedom are involved we can model the short-term behavior deterministically. In such cases we can make predictions that are far better than those we would expect from a traditional model based on the theory of random processes. Chaos is thus a double edged sword; on one hand, it implies that long-term predictions are impossible, but on the other hand, it implies that in some cases short-term predictions *are* possible.

Chaos also tells us that we must consider *nonlinear* models if we are to take advantage of short-term predictability. Chaos does not occur in linear dynamical systems, and it is not possible to capture it with linear models in even an approximate sense. Chaos is an inherently nonlinear phenomenon.

Because chaos is inherently nonlinear, conventional linear statistical measures such as correlation functions are inadequate to describe it. A chaotic process can be uncorrelated over short times, even though it is quite deterministic; the property of linear correlation simply does not provide us with the proper information to adequately characterize chaotic behavior. Other statistical descriptions, based on nonlinear statistical averages such as the entropy or mutual information, are more appropriate. Through the introduction of new statistical measures, such as dimension, dynamical entropy, and Lyapunov exponents, dynamical systems theory has provided a new language for the description of random behavior.

5.3   Scope of Part II

Our central aim is to present the topics of chaotic dynamical systems and random processes in an integrated way, making it clear how each relates to the other. We assume a minimum of background knowledge, and in particular we assume no knowledge of either dynamical systems theory or the theory of random processes.

Chapter 5 begins by reviewing basic concepts and terminology of dynamical systems theory. It ends with a discussion of chaos, demonstrating how it leads from a deterministic to a probabilistic description. Chapter 6 discusses statistical descriptions, including basic concepts from probability and measure theory. We introduce linear statistical measures, such as the correlation function and power spectrum, and contrast them to nonlinear measures such as entropy and information.

We often model nature as though it were continuous. However, in any real experiment we are forced to make measurements that reduce the observations to a stream of discrete symbols. In statistical mechanics this is called *coarse graining*. Changing the level of coarse graining is equivalent to changing the resolution of our instruments. The way statistical measures, such as information, scale as the resolution changes gives us important information about inherent geometrical properties. Chaotic dynamics naturally generates sets with complicated properties, called fractals. The notion of dimension, which arises from considering scaling properties under changes in resolution, provides a natural language to discuss fractals; it provides at least one notion of "complexity," which is related to the number of independent degrees of freedom.

It is important to distinguish the notions of chaos and complexity. Chaos is independent of dimension; it can occur with only a few degrees of freedom, or with a large number. The fact that a system is chaotic implies that nearby trajectories separate exponentially, at least on average. The Lyapunov exponents provide a precise quantification of this. The overall rate of separation is measured by the metric entropy. The metric entropy and the dimension are in general quite independent; from a certain point of view the metric entropy measures "how chaotic" a trajectory is, whereas the dimension measures "how complex" it is. This is discussed in chapter 6.

In analyzing data from experiments, it is often necessary to reconstruct dynamics based on only partial information. For example, to model a fluid flow using the Navier–Stokes equations requires several functions in three dimensions to specify uniquely the future evolution of the fluid. The fluid thus has an infinite dimensional state space, because each function contains an infinite list of potentially independent numbers. In an experiment, however, we may have to make do with only partial data, such as the output of a single probe at a fixed position in the flow. In this case, analyzing the dynamics requires the reconstruction of a state space. We discuss several aspects of this problem in chapter 7.

Finally, in chapter 7 we discuss methods for exploiting the presence of low dimensionality to make predictions that are better than those one would get with the conventional approaches indicated by the theory of random processes.

5.4   Deterministic Dynamical Systems and State Space

This section is a brief review of the terminology and basic ideas of dynamical systems theory.

Dynamical systems theory is the study of things that change, of phenomena that vary in time. It attempts to answer the question: Based on the past and the present, how can we predict the future? It does so by abstracting some of the properties of the phenomena and making a mathematical model. Such models, which provide a mathematical rule for how the relevant properties change in time, are called *dynamical systems*.

The first, and often most difficult, aspect of the modeling process is deciding what information is important—what are the causes that produce the effects? This information is encapsulated in the *state*, which contains the information upon which the dynamical system acts. The state can be a number, a vector, a function, a set of functions, or any of a variety of different mathematical objects; it can be discrete or continuous. In a good model the state contains enough information to make the phenomena as predictable as possible. Ideally it contains the minimal information for which this is so.

If the evolution of the state through time is unique, then the dynamical system is *deterministic*. If there are random influences, uncertainties that prevent it from being unique, then it is *stochastic*, or equivalently, a *random process*.

The evolution of a deterministic system from time t_1 to t_2 can be written in terms of the rule F:

$$\vec{x}(t_2) = \vec{F}(\vec{x}(t_1)), \tag{5.1}$$

where F is a single-valued function and $\vec{x}(t)$ is the state at time t. This is called a *mapping*. The time increment for mappings is often chosen to be unity. Thus the state $\vec{x}(t+1)$ is the successor to $\vec{x}(t)$. Furthermore, we will denote the times either by parentheses, as above, or subscripts (e.g., $\vec{x}_{t+1}$). The meaning should be clear from context. Alternatively, the rule may be given in terms of infinitesimal time evolution; that is

$$\frac{d\vec{x}}{dt} = \vec{F}(\vec{x}(t)). \tag{5.2}$$

Dynamical systems that are continuous in time are called *flows*. We shall denote differentiation with respect to time by a dot (i.e., $\dot{x} = dx/dt$).

The state can be visualized as a point. The set of all possible states of a system is called the *state* (or *phase*) *space* of the system. The function F can be represented by a vector field in state space and the evolution of a system by a curve in state space, known as a *trajectory* or *orbit* (see Fig. 5.1). Figure 5.2 shows trajectories arising from various initial conditions.

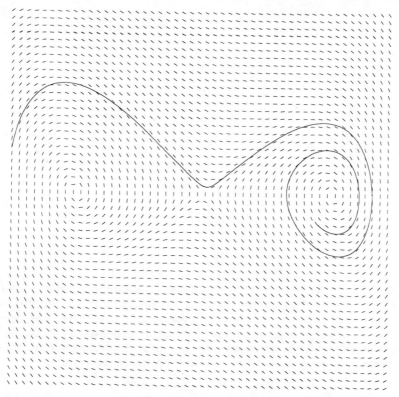

FIGURE 5.1. A deterministic dynamical system represented by a vector field in a two-dimensional state space. The solid line is an integral curve of the vector field and represents a trajectory of the dynamical system.

For instance, for a point particle Newton's laws define a dynamical system. If the forces depend only on position and velocity, then the position and velocity define the state of the particle. To be even more specific, consider a simple pendulum constrained to move in a plane. Its state is completely specified by two numbers: the position θ and velocity $\dot{\theta}$. In appropriate units, the rule governing its evolution is:

$$\frac{d^2\theta}{dt^2} + \sin(2\pi\theta) = 0. \tag{5.3}$$

This second-order equation can be reduced to a system of two first-order equations:

$$\frac{d\theta}{dt} = \dot{\theta}$$

$$\frac{d\dot{\theta}}{dt} = \sin(2\pi\theta). \tag{5.4}$$

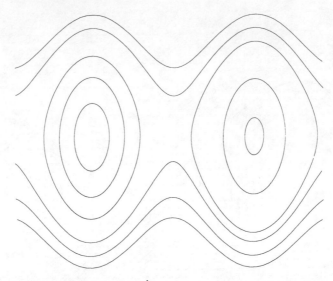

FIGURE 5.2. Trajectories in the $(\theta, \dot{\theta})$ phase space of a simple pendulum, Eq. (5.4). The roughly elliptical trajectories come from libration, in which the pendulum swings back and forth about the straight down position. The other trajectories represent oscillations, in which the pendulum passes through the upside-down position. Because it is impossible to distinguish an angle θ from $\theta + 2\pi$, $\theta + 4\pi$, and so forth, the state space can be wrapped onto a cylinder by identifying the lines $\theta = \pm\pi$.

Note that this trick works in general—an nth-order differential equation can always be reduced to system of n first-order equations. The converse is not always true.

In a system with many particles, the state is a list of the positions and velocities of each of the particles. The space in which each individual particle moves is the *configuration space*. For N particles moving in a D dimensional configuration space, the dimension of the state space[1] is $2DN$. We can also let N approach infinity, so that the state is a function over configuration space. A fluid, for example, can be modeled in terms of a set of partial differential equations. The state of the fluid is described by the velocity, temperature, pressure, and so forth, as functions of position. Because specifying a function requires an infinite set of independent numbers, functions are infinite dimensional objects. The evolution of the fluid through time can be visualized as the trajectory of a point in an infinite dimensional space, where each point in the space is a possible function describing the entire fluid.

In an ideal setting, the rule F that specifies the dynamics is fixed, given to us by God at the beginning of the experiment and then never changed again. In practice this is seldom the case. The dynamical system F is usually an

---

[1] In statistical mechanics, this is called the Gibbs description of the phase space.

approximation to the true dynamics, valid for only a limited period of time. It can be altered by interactions with other systems, or by changes that occur on time scales that are too slow to be incorporated easily into the model.

The slowly changing aspects of a model are often incorporated as *parameters*. The parameters are the aspects of F that are fixed, but can easily be changed. For example, imagine a sequence of experiments. In each experiment the parameters are constant, but between experiments they are varied. The notion of parameters is often quite natural. For a pendulum, for example, the natural parameters are the mass, length, and strength of the gravitational field. Assigning these constant values is an approximation; as the pendulum swings, the local gravitational field varies from place to place, and the changes in stress cause the length of the pendulum to change. The "parameters" are the aspects of the dynamics that change slowly enough that they can be approximated as constants, at least for a period of time.

The set of all possible values for the parameters is called the *parameter space*, or the *control space*. There is usually a good deal of arbitrariness in the assignment of a parameter space; we could suppress some of the parameters by setting them to constants, or add parameters by including more complicated expressions. Often, however, there is a natural choice of free parameters, as in the case of the pendulum above.

5.5 Classification

For what follows it is important that the reader understand certain elementary descriptive terms from dynamical systems theory. We give an overview of the basic concepts, and give an intuitive, nonrigorous discussion. Our motive is to get the basic ideas across. The interested reader is urged to consult other references for a more careful development [1, 2].

5.5.1 Properties of Dynamical Systems

Linearity

A *linear* system is one for which the principle of superposition holds. That is, for any two states $\vec{x}_a$ and $\vec{x}_b$:

$$\vec{F}(\vec{x}_a + \vec{x}_b) = \vec{F}(\vec{x}_a) + \vec{F}(\vec{x}_b) \tag{5.5}$$

Roughly speaking, if F contains no nonlinear functions such as $x_i x_j$ or $\sin(x)$, it is linear. Note that nonlinear functions of time alone do not imply nonlinearity. For example, $\dot{x} = \sin(x)$ is a nonlinear equation, but $\dot{x} = \sin(t)$ is not. Also, solutions of linear equations may be nonlinear functions of time; for example, the equations of motion for a simple linear harmonic oscillator are $\ddot{x} + x = 0$; the solution is $x(t) = \sin(2\pi t)$.

Autonomy

An *autonomous* dynamical system does not depend explicitly on time. That is, the dynamical system does not contain time as an explicit variable.

It is generally possible to convert a nonautonomous system to an autonomous one by expanding the state space and making appropriate substitutions. For example, the linear, nonautonomous system

$$\dot{x} = \sin \omega t \qquad (5.6)$$

can be converted to the nonlinear autonomous system

$$\dot{x} = \sin \theta;$$
$$\dot{\theta} = \omega. \qquad (5.7)$$

The opposite is not true in general, however—typical nonlinear autonomous systems cannot be reduced to linear nonautonomous systems (this is not surprising, because linear nonautonomous systems are much simpler).

Determinism

The trajectories of a deterministic dynamical system may not split apart going forward in time. Because there would be two possible futures for a single state at an intersection point, the future could not be uniquely determined by the state. In some cases, however, such as noninvertible maps, trajectories can merge; that is, they can split apart going backward in time.

Invertibility

A dynamical system is invertible if each state $\vec{x}(t)$ has a unique predecessor $\vec{x}(t-1)$. Thus, a dynamical system is invertible if trajectories never merge. Continuous deterministic flows are always invertible,[2] and so are maps derived from flows.

Reversibility

If the dynamical system obtained by letting $t \rightarrow -t$ is equivalent to the original one, it is reversible. This is a very special property.

Orientation Preserving

An *orientable manifold* is one that has a well-defined difference between its inside and its outside. A circle in the plane, for example, is orientable, but a line segment is not. An orientation-preserving dynamical system is one that cannot turn an orientable manifold inside out. Continuous flows must be

---

[2] For a flow to be noninvertible, there would have to be a point x where a trajectory followed backward in time split apart. For this to happen, however, requires that the flow field takes on two different values at x. This violates determinism.

orientation-preserving, for in order to deform the outside into the inside continuously, they must pass through each other, which violates invertibility. Thus continuous deterministic flows are always orientation preserving, and so are maps constructed from such flows. A map is orientation preserving if its Jacobian has a positive determinant (see section 5.7.1).

Invariance Under Coordinate Transformations

The most fundamental properties of dynamical systems are those that are independent of the particular coordinates used to describe them; that is, those that are invariant under changes of coordinates. A coordinate transformation is a one-to-one invertible mapping that takes each state x into a state y. A coordinate transformation is continuous if both it and its inverse are everywhere continuous, and smooth if they are everywhere differentiable. Continuous transformations preserve the topological properties of the trajectories, whereas smooth ones preserve the geometry. A property that is not invariant under coordinate transformations is less meaningful than a property that is.

5.5.2   A Brief Taxonomy of Dynamical Systems Models

Dynamical systems can be trivially classified according to the continuity and locality of the underlying variables. A variable can be either discrete (i.e., describable by a finite integer) or continuous. There are three essential properties:

- **Time.** All dynamical systems contain time as either a discrete or continuous variable.
- **State.** The state can either be continuous, as in ordinary differential equations, or take on discrete values, as for automata.
- **Space.** Space plays a special role in dynamical systems. Some dynamical models, such as automata or ordinary differential equations, do not contain the notion of space. Other models, such as lattice maps or cellular automata, contain a notion of locality and therefore space without doing so in a continuous fashion. Partial differential equations or functional maps have continuous spatial variables

This is summarized in Table 5.1.

5.5.3   The Relationship Between Maps and Flows

Physical laws are usually stated in terms of continuous time, and models in the natural sciences are much more commonly formulated as flows than as maps. However, for many purposes maps are easier to analyze and understand. Fortunately, flows can often be reduced to maps, at least conceptually. There are an infinite number of ways to do this—each flow corresponds to an infinite number of equivalent maps.

TABLE 5.1. Classification of dynamical systems according to whether the basic elements incorporate a notion of continuity or whether they are discrete.

Model	Space	Time	State
Ordinary differential equations	*	cont.	cont.
Maps (difference equations)	*	discrete	cont.
Delay differential equations	*	cont.	cont.
Automata	*	discrete	discrete
Partial differential equations	cont.	cont.	cont.
Computer representation of a p.d.e.	local	local	local
Functional maps	cont.	discrete	cont.
Lattice models	local	discrete or cont.	cont.
Cellular automata	local	discrete	discrete

As an intermediate category we include "locality" (1), for quantities that are discrete but where there are neighborhood relationships and the relevant quantities are approximately continuous. In a computer representation of a partial differential equation, for example, the state is a collection of variables $f(\vec{x}, t)$. $f, \vec{x}$, and t are all discrete, but there is nonetheless a rough notion of continuity; typically points that are nearby in space or time have values that are close to each other. We designate this as "local" in the state f, the spatial variable $\vec{x}$, and the time t. An asterisk indicates that the dynamical system contains no explicit notion of space. Some of these relationships are inclusive; for example, a cellular automaton is a particular kind of automaton with spatial structure, and a computer representation of a p.d.e. is a particular kind of map with locality relationships for all its elements.

A common method for reducing a flow to a map is to use a *surface of section*. This is done by specifying a surface of dimension one less than that of the state space; that is, a surface of codimension one.[3] When a trajectory crosses the surface, its position is recorded, as shown in Fig. 5.3.[4] The resulting sequence of states $\ldots, \vec{x}(t-1), \vec{x}(t), \vec{x}(t+1), \ldots$ defines a map, which is often called a Poincaré map. Because the dimension of the map is one less than that of the flow, for many purposes the map is easier to analyze. Because there are an infinite number of ways to specify surfaces of section, there are an infinite number of possible Poincaré maps.

Each Poincaré map is equivalent to the original flow in that its trajectories are in a one-to-one correspondence with those of the flow. The topological structure of the map and the flow are the same, and the geometrical properties, such as the dimension and the Lyapunov exponents, are the same once the reduction of dimension is taken into account.

However, in the process of reducing the flow to a map, the timing relationships between the orbits are lost. In general, trajectories consume very different lengths of time between crossings of the section, but in a Poincaré map

---

[3] A surface of dimension d in a space of dimension N has codimension $N - d$.
[4] Sometimes it is clearer to make the Poincaré map only when the trajectory crosses the section in a given direction.

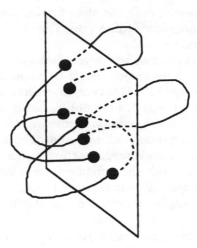

FIGURE 5.3. A trajectory of a continuous flow can be reduced to a map by recording the position of the trajectory on successive crossings of a surface of codimension one. The resulting map is topologically and in many respects geometrically equivalent to the original flow.

they all correspond to one iteration. Furthermore, a trajectory that corresponds to an orbit with a given period in the constructed map may give rise to an orbit with a different period when a different surface of section is used. For example, consider the periodic trajectory shown in Fig. 5.4. For section (a) this periodic orbit is reduced to a fixed point, whereas for (b) it is reduced to a

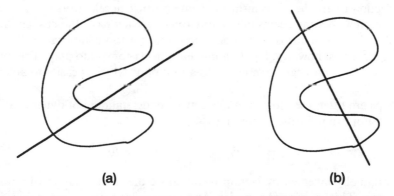

FIGURE 5.4. Two different Poincaré sections of a single periodic orbit. Points are included in the Poincaré section as the plane is traversed from left to right. There is one point in section (a), which is thus a fixed point of the corresponding Poincaré map; there are two points in section (b), which form a period two cycle in the corresponding Poincaré map.

period two cycle. However, once this identification is made, the trajectories are in a one-to-one correspondence. A good choice of the surface of section may make analysis simpler.

In some cases, such as periodically driven oscillators, a surface of section is equivalent to a stroboscopic section, made by recording the position of an orbit at periodic time intervals. However, in general, unless there are strict periodicities, stroboscopic sections are not as useful as surfaces of section, because they can complicate the motion needlessly. For example, in a surface of section a continuous periodic orbit is reduced to a sequence of discrete points. In contrast, for a stroboscopic section, unless the period P_1 of the strobe is *commensurate* with the period of the cycle P_2; that is, unless P_1/P_2 is rational, the cycle generates a closed curve in the stroboscopic section. In this case the cycle is no simpler in the map than it was in the original flow.

Except for special cases, the map derived from a flow cannot be written down as a simple closed form expression. The construction of Poincaré maps is usually done with a computer. In many cases maps can also be constructed qualitatively as an aid to mathematical analysis.

Just as every flow corresponds to an infinite number of maps, as long as a map *could be* a valid Poincaré map of a flow, then it corresponds to an infinite number of flows. Maps constructed from flows are always invertible and orientation preserving. As long as a map satisfies these two conditions, it is possible, at least in principle, to construct a flow by inventing trajectories that continuously connect the iterates of the map. Such a flow is called a *suspension*. In practice, of course, constructing a suspension may be quite difficult, and it may be impossible to do so with a closed form expression.

In some cases, it is possible to go from a continuous time model to a Poincaré mapping by analytically integrating the equations of motion from one section to another. For instance, some particle accelerators are circular, with counter-rotating beams interacting once on each pass. In between interactions, their behavior is well understood and not too complicated. During the interactions, however, highly nonlinear effects come into play. The interaction of the beams can be modeled as a mapping, which is easier to analyze [3].

Maps are often used to gain a qualitative understanding of dynamical behavior. For example, the Hénon map [4],

$$x_{t+1} = 1 + y_t - ax_t^2; \qquad y_{t+1} = bx_t, \qquad (5.8)$$

was originally invented by Hénon to illustrate the fractal nature of strange attractors. This map is not a model of any particular system, and in fact, it is not orientation preserving and so cannot correspond to a flow. However, many of its properties are representative of those of a wide class of dynamical systems. It has received a great deal of attention because it is so easy to study.

5.6 Dissipative Versus Conservative Dynamical Systems

Whether a dynamical system is dissipative or conservative makes an important difference in its properties. This terminology comes from physics, where the notion of energy is well defined. A physical system that conserves energy corresponds to a dynamical system that preserves volumes in phase space, whereas a physical system that loses energy with time corresponds to a dynamical system that contracts volumes in phase space. This terminology is carried over to the more general context in dynamical systems where energy is no longer defined. The word is usually used in slightly different sense, however: Any dynamical system in which state space volumes are asymptotically (in time) invariant under the dynamics is *conservative*; any dynamical system that is not conservative is *dissipative*.

There are a few subtle points about this that deserve more discussion. A general dynamical system can have some regions of state space where volumes are contracting, some regions where volumes are expanding, and others where volumes are conserved. The term "conservative" is usually reserved for dynamical systems where the state space volume is preserved everywhere, although in principle it could also be applied to a subset of a dynamical system.

Whether or not a dynamical system preserves state space volumes locally can be altered easily by choosing a different metric or by making a nonlinear coordinate transformation. However, the asymptotic, or "long-time" properties do not depend on the metric and are invariant under coordinate transformations. This is the reason for use of the word "asymptotic" in the definition given above.

This distinction is particularly confusing for noninvertible maps. Although they may expand volumes locally, they can be used as approximate models valid in the limit of infinite dissipation. We will return to explain this in the discussion of reduction to one dimensional maps (section 5.9.3).

Conservative dynamical systems arise often in physics because of Newton's laws, which can be rewritten in terms of Hamilton's equations as

$$\frac{\partial p_i}{\partial t} = \frac{\partial H}{\partial q_i}; \qquad \frac{\partial q_i}{\partial t} = -\frac{\partial H}{\partial p_i}, \tag{5.9}$$

where p_i and q_i are canonically conjugate state space variables, and $H(\{p, q\})$ is a function called the Hamiltonian, which is related to the energy. This relation between partial derivatives of conjugate variable is known as *symplectic*. Liouville's theorem guarantees that phase space volumes are invariant under time evolution of a Hamiltonian system. Furthermore, energy conservation constrains the trajectories of a Hamiltonian system to lie on a hypersurface of constant energy. Though these may seem like strong constraints, they do not prevent complicated trajectories from appearing even in low-dimensional Hamiltonian dynamics.

5.7   Stability

Stability is a generic word characterizing response to perturbations. If a perturbation is amplified, resulting in a large change, then the system is *unstable*. If the perturbation is damped out, resulting in little or no change, then the system is *stable*. The intermediate case, where the perturbation persists at roughly its initial level, is called *marginal stability*.[5]

The word stability is used in many different senses. First, a system may be stable to one class of perturbations, and unstable to another. Stability under all possible perturbations, of any size, is called *absolute stability*. Stability is often analyzed by expanding in a Taylor series. Linear stability analysis is done by studying the linear dynamical system obtained by neglecting all the nonlinear terms. For sufficiently small perturbations the linear terms dominate, so that a linear stability analysis correctly describes the stability properties. If the linear dynamical system is marginally stable, however, or the perturbations are too large, then in general the stability can be determined only by considering the full dynamical system, or possibly by studying higher order nonlinear terms.

Second, stability applies to different aspects of a dynamical system. It can refer to individual points (local stability), to trajectories (asymptotic local stability), to families of trajectories (attractors), or to an entire dynamical system (whether there is a unique attractor). The perturbations can be perturbations of states (dynamical stability), or perturbations of control parameters (structural stability). We will begin by discussing the dynamical stability of points and trajectories, distinguishing other notions of stability as we go along.

5.7.1   Linearization

The stability of states and their trajectories to small perturbations is studied by linearization. The linearization of a dynamical system about a given point describes the local evolution of two nearby states. For a map $\vec{F}$ let $\vec{y}_t = \vec{x}_t + \vec{\varepsilon}_t$, where $\vec{\varepsilon}_t$ is infinitesimally small. This implies

$$\vec{\varepsilon}_{t+1} = \vec{y}_{t+1} - \vec{x}_{t+1} = \vec{F}(\vec{y}_t) - \vec{F}(\vec{x}_t) = J(\vec{x}_t)(\vec{\varepsilon}_t) + O(\vec{\varepsilon}_t^2), \qquad (5.10)$$

where $J(\vec{x}_t)$ is the derivative or Jacobian $\vec{\nabla}\vec{F}$, or:

$$J = \begin{pmatrix} \dfrac{\partial F_1}{\partial x_1} & \dfrac{\partial F_1}{\partial x_2} & \cdots \\[2mm] \dfrac{\partial F_2}{\partial x_1} & \dfrac{\partial F_2}{\partial x_2} & \cdots \\[2mm] \cdots & \cdots & \cdots \end{pmatrix}, \qquad (5.11)$$

in Euclidean coordinates.

---

[5] In some contexts, such as conservative dynamical systems, marginal stability is simply called "stability"; this is natural because for conservative systems marginal stability is the strongest possible form of stability. In other cases, such as older treatments of limit cycles, marginal stability is called "instability," because it is less stable than fixed point behavior.

The *local stability* of a map over a given interval of time can be determined by iterating the linearized equations. For example, the growth of a perturbation $\vec{\varepsilon}_0$ after time t is

$$\vec{\varepsilon}_t = J(\vec{x}_{t-1})J(\vec{x}_{t-2})\ldots J(\vec{x}_0)\vec{\varepsilon}_0 = \prod_{i=t-1}^{0} J(\vec{x}_i)\vec{\varepsilon}_0. \tag{5.12}$$

The perturbation $\vec{\varepsilon}_t$ grows (and hence the perturbation is unstable over this interval of time) if $\|\vec{\varepsilon}_t\|/\|\vec{\varepsilon}_o\| > 0$, where $\|\vec{e}\|$ is the norm of $\vec{e}$. If the largest eigenvalue, λ_1, of the product of the Jacobian matrices,

$$J(\vec{x}_0 \ldots \vec{x}_{t-1}) \equiv \prod_{t-1}^{0} J(\vec{x}_i), \tag{5.13}$$

has modulus greater than one; that is, if $|\lambda_1| > 1$, then there exists *some* perturbation about $\vec{x}_0$ that will grow during the interval from 0 to t. When $t = 1$, this can be determined by simply examining the eigenvalues of $J(\vec{x}_0)$.

For a flow $\dot{\vec{x}} = \vec{F}(\vec{x})$ the situation is somewhat different. Linearizing about $\vec{x}$ yields $\dot{\vec{\varepsilon}} = J(\vec{x})\vec{\varepsilon}$. The solution is[6]

$$\vec{\varepsilon}(t) = \exp \int_0^t J(\vec{x}(t'))dt' \; \vec{\varepsilon}(0). \tag{5.17}$$

Again, a particular perturbation $\vec{\varepsilon}$ grows if $\|\vec{\varepsilon}_t\|/\|\vec{\varepsilon}_0\| > 0$. There exists *some* perturbation that will grow if the largest eigenvalue λ_1 of the matrix $\exp \int_0^t J(\vec{x}(t'))dt'$. satisfies $\|\lambda_1\| > 1$. Equivalently, there is a perturbation that grows if the largest eigenvalue of the matrix $\int_0^t J(\vec{x}(t'))dt'$ has a positive real part.

The rate at which volumes shrink can also be quantified in terms of the Jacobian matrix (See Problem 5.4). For a map, for example, the rate at which the volume changes is given by the determinant of the Jacobian; if $\det |J| = 1$, then volume is conserved. For a flow, on the other hand, for reasons similar to those in the discussion of linear stability, whether volumes expand or contract is given by the *trace* of the Jacobian matrix.

---

[6] The exponential of a matrix M is defined by analogy with the Taylor expansion of the usual exponential:

$$\exp(M) \equiv \sum_{i=0}^{\infty} M^i/i!, \tag{5.14}$$

as long as it converges. The exponential of a diagonal matrix is the matrix of exponentials of its elements:

$$\exp(\mathrm{diag}(\lambda_1, \lambda_2, \ldots, \lambda_n)) = \mathrm{diag}(e^{\lambda_1}, e^{\lambda_2}, \ldots, e^{\lambda_n}). \tag{5.15}$$

The logarithm of a matrix can be defined by a similar analogy with the usual logarithm. Non-integral powers of a matrix can be defined through use of the exponential and logarithm:

$$M^p \equiv e^{p \ln M}. \tag{5.16}$$

Example 5.1 The Hénon mapping. The Jacobian matrix for the Hénon mapping (Eq. 5.8) is:

$$J = \begin{pmatrix} -2ax & 1 \\ b & 0 \end{pmatrix}. \tag{5.18}$$

The linearized map at the point (X, Y) is:

$$x_{t+1} = -2aXx_t + y_t; \qquad y_{t+1} = bx_t. \tag{5.19}$$

The determinant is $-b$, a constant. When $|b| < 1$ the mapping is dissipative. When $b > 0$, the mapping is orientation reversing.

Note that local stability over any fixed interval of time can be altered by a coordinate transformation; for example, by making a local nonlinear distortion of the flow. However, as the length of time becomes longer, from an intuitive point of view this becomes more difficult, because coordinate changes in different places tend to cancel each other out. For invariant sets the asymptotic stability properties, as the time interval goes to infinity, are invariant under coordinate transformations.

5.7.2   The Spectrum of Lyapunov Exponents

The Lyapunov exponents provide a coordinate-independent measure of the asymptotic local stability properties of a trajectory. The concept is very geometrical. Imagine a small infinitesimal ball of radius $\varepsilon(0)$, centered on a point $\vec{x}(0)$. Under the action of the dynamics the center of the ball may move, and the ball will be distorted.[7] Because the ball is infinitesimal, this distortion is governed by the linear part of the flow. The ball thus remains an ellipsoid. Suppose the ith principal axis of the ellipsoid at time t is of length $\varepsilon_i(t)$. The spectrum of Lyapunov exponents for the trajectory $\vec{x}(t)$ can be defined as

$$\lambda_i = \lim_{t \to \infty} \lim_{\varepsilon(0) \to 0} \frac{1}{t} \frac{\varepsilon_i(t)}{\varepsilon(0)}. \tag{5.20}$$

For fixed $\varepsilon(0)$, as t increases the ball is folded over and can no longer be adequately represented by an ellipsoid. On the other hand, the long time limit is important for gathering enough information to represent the entire trajectory. Thus the $t \to \infty$ limit must be taken while the inital radius of the ball is simultaneously being shrunk.

Note that the Lyapunov exponents depend on the trajectory $\vec{x}(t)$. Their values are the same for any state on the same trajectory, but may be different

---

[7] Unless the flow happens to be constant!

for states on different trajectories. The trajectories of an N dimensional state space have N Lyapunov exponents. This is often called the *Lyapunov spectrum*. It is conventional to order them according to size. The qualitative features of the asymptotic local stability properties can be summarized by the sign of each Lyapunov exponent. A positive Lyapunov exponent is denoted by $+$, indicating an unstable direction, and $-$ denotes a negative exponent, indicating a stable direction. Thus $(+, -, -)$ implies a trajectory in a three-dimensional state space with one positive Lyapunov exponent.

The Lyapunov exponents provide a generalization of the eigenvalues of a linearized map at a point. For a map, for example, a small perturbation ε_0 grows after time t to:

$$\vec{\varepsilon}_t = \prod_{i=t-1}^{0} J(\vec{x}_i)\vec{\varepsilon}_0. \tag{5.21}$$

Let $j_i(t)$ be the eigenvalues of $\prod_{i=t-1}^{0} J(\vec{x}_i)$. The *Lyapunov numbers* are:

$$\Lambda_i = \lim_{t \to \infty} |j_i(t)|^{1/(t-1)}. \tag{5.22}$$

The Lyapunov exponents are just the logarithm of the Lyapunov numbers:

$$\lambda_i = \log \Lambda_i. \tag{5.23}$$

A numerical algorithm for computing the Lyapunov exponents was proposed independently by Bennetin et al. [5] and by Shimada and Nagashima [6]. This was developed into an algorithm for computing the Lyapunov exponents from data by Wolf et al. [7]. An alternative to this algorithm was independently proposed by Eckmann and Ruelle [8] and by Sano and Sawada [9].

5.7.3 Invariant Sets

Sets that stay invariant under the action of the dynamical system play an important role in dynamical systems. A set Ω is invariant under the dynamical system F if $F^t(\Omega) = \Omega$ for all times t.

A fixed point, as the name suggests, is an invariant point, one that "stays fixed" under the action of the dynamical system. The fixed points of a mapping are the solutions of the equation $\vec{F}(\vec{x}) = \vec{x}$, which can be solved graphically by looking for the intersection of $\vec{F}$ with the identity mapping. (See example 2.) For a flow the fixed points are the solutions of

$$\frac{d\vec{x}}{dt} = \vec{F}(\vec{x}) = 0. \tag{5.24}$$

The stability of a fixed point $\vec{x}_f$ is easy to determine. Because $\vec{x}_f(t) = \vec{x}_f(t+1)$, the Jacobian is constant in time, $J(\vec{x}_f(t)) = J(\vec{x}_f(0)) \equiv J(\vec{x}_f)$. The growth of small perturbations is determined by:

$$\vec{\varepsilon}(t) = J(\vec{x}_f)^t \vec{\varepsilon}(0). \tag{5.25}$$

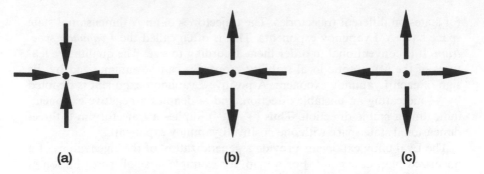

(a)   **(b)**   **(c)**

FIGURE 5.5. Possible stability properties for a fixed point in two dimensions. (a) stable or elliptic point; (b) saddle or hyperbolic point; and (c) repellor.

If the largest eigenvalue of $J(\vec{x}_f)$ satisfies $|\lambda_1| > 1$ then the fixed point is linearly unstable. Similarly, for a flow the growth of small perturbations is determined by

$$\vec{\varepsilon}(t) = \exp\{J(\vec{x}_f)t\}\vec{\varepsilon}(0); \qquad (5.26)$$

if the largest eigenvalue of $J(\vec{x}_f)$ has a positive real part then the fixed point is linearly unstable (see Eq. 5.14).[8]

A fixed point whose eigenvalues are all unstable is called a *repellor*; one with some stable and some unstable eigenvalues is *hyperbolic* or a *saddle point*. For Hamiltonian systems, a fixed point whose largest eigenvalue is marginally stable is called an *elliptic point*, because small perturbations result in elliptic orbits around the fixed point (see Fig. 5.5).

The next simplest class of invariant sets are *cycles*, or *periodic orbits*. The cycles of a map can be found from the equation $\vec{F}^p(\vec{x}) = \vec{x}$, where $\vec{F}^p$ means the pth iterate of the map,

$$\vec{F}^p(\vec{x}) = \underbrace{\vec{F} \circ \vec{F} \circ \cdots \vec{F}}_{p \text{ times}}(\vec{x}). \qquad (5.27)$$

The cycles are thus the fixed points of the map $\vec{F}^p$. Everything said above for fixed points applies to cycles, if $\vec{F}^p$ is substituted for $\vec{F}$.

The eigenvalues of fixed points are invariant under coordinate transformations, so the stability of a fixed point cannot be altered by simply changing coordinates. Similarly, for a period p cycle the eigenvalues of the derivative of F^p (i.e., of $J(\vec{x}_0 \ldots \vec{x}_{p-1})$ defined in Eq. 5.13), are invariant. The "average eigenvalue per iteration," or the *Floquet numbers*, can be defined by taking

---

[8] For simplicity in what follows, we refer to the eigenvalues themselves as stable, unstable, or marginal. What we mean is that the eigenvalue's modulus is greater than, less than, or equal to unity for a map or that the eigenvalue has positive, negative, or zero real part for a flow.

the pth root of this matrix.[9] The eigenvalues of fixed points and the Floquet numbers are useful quantities for characterizing the coordinate invariant properties of dynamical systems.

The next simplest type of invariant set is a torus, which may be visualized as the surface of a doughnut. Just as a square is the Cartesian product of a line segment with a line segment, a torus is the Cartesian product of two circles. In dynamical systems a torus arises from the Cartesian product of two cycles with incommensurate frequencies. Because the frequencies have no rational relationship, the state wanders over the entire torus. An event in which a state returns to one of its previous values is called a *recurrence*. Although the orbits on a torus come arbitrarily close to recurring, the never do so exactly. If the two fundamental frequencies of the cycles are rationally related, however, then they will combine to form a single cycle, which is "wrapped around" the underlying torus.

The Cartesian product of n different limit cycles with incommensurate frequencies is an n-torus. With this terminology, a point is a 0-torus, a cycle is a 1-torus, the object described above is a 2-torus, and so forth. Motion on an n-torus is often called *quasiperiodic* or *multiply periodic* motion.

More complicated invariant sets with the geometry of fractals are also possible. (An invariant set of the Hénon map, shown in Fig. 5.6, is one example.) Such complicated invariant sets are one of the main subjects in the next two chapters.

5.7.4   Attractors

We are often interested in what happens after a long time has elapsed, when short-term transient behaviors have died out, so that a dynamical system exhibits its own "typical behaviors." By choosing a special initial condition, we may force the system to do something that it would never do on its own; with the passage of time it may "settle down" and display its natural behavior. For dissipative systems there are essentially two possibilities: Either the motion is unstable, and flies off to infinity, or the motion is bounded, and approaches an attractor.[10] Thus the existence of an attractor is a form of stability associated with families of trajectories.

Roughly speaking, an attractor is an invariant set that "attracts" nearby states. More formally, Ω is an attractor if there is an open neighborhood N about it such that $F^t(N) \to \Omega$ as $t \to \infty$, and Ω cannot be broken into pieces $\Omega_1, \Omega_2, \ldots$ such that $F^t(\Omega_1) \cap F^t(\Omega_2) = \emptyset$. The *basin* of an attractor is the set

---

[9] The pth root of a diagonalizable matrix M is defined as the matrix $M^{1/p}$ whose eigenvalues are the pth roots of the eigenvalues of M.

[10] There are of course exceptional possibilities. For instance, if an initial state is on the boundary between two basins, it may either remain fixed or wander over the boundary, without reaching either attractor. Such trajectories are unstable, however, and are exceptional for typical dynamical systems.

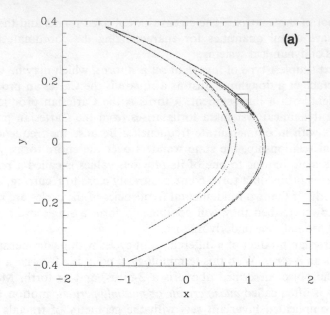

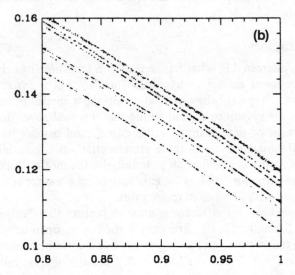

FIGURE 5.6. An attractor in the Hénon map (Eq. 5.8) for the parameters $a =$ 1.4, $b = 0.3$. The region outlined in (a) is blown up in (b) to show details of the complicated structure. Further blowups show similar structure at all scales.

of points that are attracted to it; that is, $\{x : \lim_{t \to \infty} F^t(x) \subset \Omega\}$. A given dynamical system may have many attractors, each with its own distinct basin of attraction. The boundary between two basins of attraction is sometimes called a *separatrix*. For invertible dissipative systems the fact that volumes shrink (at an exponential rate) ensures that in the limit as time goes to infinity any volume shrinks to zero. Thus, the attractors of invertible systems have zero volume. Conservative dynamical systems do not have attractors.

A state that is initially not on an attractor will never reach it in finite time—it only approaches its attractor, getting exponentially closer with the passage of time. The state eventually comes close enough that from a practical point of view it is indistinguishable from a state on the attractor. Once this happens it is common to say that the trajectory is "on the attractor." In physical systems, with their everpresent background noise, it is impossible to distinguish orbits that are on the attractor from those that are within the characteristic scale of the noise.

Relaxation onto an attractor can be viewed as a simplification of the motion, as a reduction of the number of irreducible degrees of freedom. This reduction can be extreme. Consider, for example, the bulk motion of a fluid that comes to rest—although the state space has an infinite number of initial degrees of freedom, its final state has none. In other words, although the state space is infinite dimensional, its attractor is zero dimensional.

There is often no obvious correspondence between the physical degrees of freedom in a problem and the degrees of freedom on the attractor. Identification of the physical modes of oscillation on an attractor is an important, nontrivial problem.

5.7.5   Regular Attractors

A regular motion is one whose trajectories are asymptotically locally stable.[11]

A *regular attractor* is an attractor that is a smooth Euclidean manifold whose trajectories are regular (or at least almost all of them are regular). Because the trajectories are locally stable, nearby states remain nearby as they evolve. Regular motion is thus predictable far into the future.

Fixed Points

Any of the invariant sets we have discussed so far can be attractors. The simplest case is a fixed point. Except for possible exceptional cases,[12] a fixed

---

[11] More precisely, a regular motion has no positive Lyapunov exponents, as discussed in section 5.7.2.

[12] In some cases the Taylor expansion may be singular, so that the linear term does not correctly describe the leading behavior.

point is an attractor whenever the largest eigenvalue of the derivative is stable, because this guarantees that sufficiently small perturbations will damp out. A fixed point attractor has a Lyapunov spectrum of the form $(-, -, - \ldots)$.

Example 5.2 The logistic map illustrates many of the features of simple dynamical systems. It is a noninvertible mapping from the real numbers into themselves:

$$x_{t+1} = F(x_t) = \mu x_t (1 - x_t), \tag{5.28}$$

and for $\mu \leq 4$ maps the interval $0 \leq x \leq 1$ into itself. The fixed points x_f satisfy $x_f = \mu x_f (1 - x_f)$. This implies

$$x_f = 0 \quad \text{or} \tag{5.29}$$

$$x_f = 1 - 1/\mu. \tag{5.30}$$

The Jacobian of Eq. (5.28) is just the derivative of $F: J(x) = \mu(1 - 2x)$. Evaluated at the fixed points, the Jacobian is:

$$J(x_f) = \mu \quad \text{or} \quad 2 - \mu, \tag{5.31}$$

respectively. The trivial fixed point $x_f = 0$ is thus stable for $0 < \mu < 1$, but becomes unstable for $\mu > 1$. The fixed point at $1 - 1/\mu$ is stable for $1 < \mu < 3$. Figure (5.7b) shows the approach to this fixed point when $\mu = 2$.

Example 5.3 Linear dynamical systems. The attractors of a linear system $\vec{x}_{t+1} = J\vec{x}_t$ are easily understood by examining the eigenvalues of J. If the largest eigenvalue λ_1 satisfies $|\lambda_1| > 1$, then the entire system is unstable. If $|\lambda_1| < 1$, then it has a fixed point attractor. If $|\lambda_1| = 1$, then it is marginally stable, with no attractor. Thus, the only possible attractor of a linear dynamical system is a fixed point. Furthermore, this fixed point is unique—a linear dynamical system cannot have more than one basin of attraction. There are only two generic asymptotic behaviors of a linear system: Either there is a unique fixed point attractor, whose basin of attraction is the entire state space, or the system is unstable. When it is marginally stable, it can have invariant sets that are cycles or tori. These are not attractors, however. Furthermore, the parameters where this occurs are very special and such behavior is unstable to small changes in the parameters and/or initial conditions.

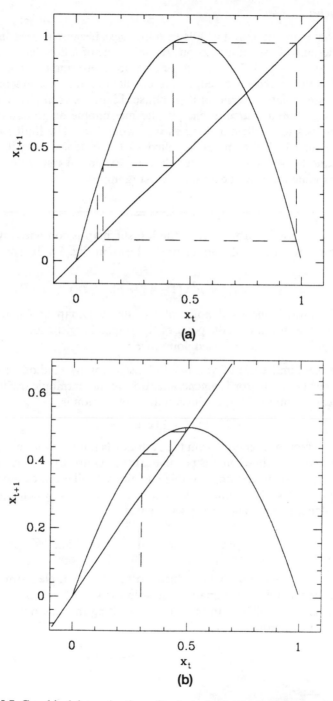

FIGURE 5.7. Graphical determination of trajectories for the logistic map. (a) Evaluation of a trajectory for $\mu = 4$. From the starting value of x_t, travel vertically to the intersection with the curve $x_{t+1} = \mu x_t(1 - x_t)$ (to find the value of the map), then horizontally to the intersection with the line $x_{t+1} = x_t$ (to move on to the next iterate), and repeat. (b) Approach to a stable fixed point attractor for $\mu = 2$.

Limit Cycles

A cycle that is an attractor it is called a *limit cycle*, because it is the limit of the trajectories in its basin of attraction. A limit cycle of a flow has a Lyapunov spectrum $(0, -, -, \ldots)$. Familiar examples are metronomes or grandfather clocks,[13] which make sustained, stable oscillations that are independent of how they are initiated except for their phase. Limit cycles in physical systems ultimately require a source of energy; the metronome or grandfather clock only continues to oscillate as long as it is wound up. The limit cycle is the trajectory for which the energy supplied to the system during its traversal exactly cancels the energy dissipated. In fact, for simple systems one can find an approximate form for the limit cycle using this idea.

Example 5.4 Periodic orbits. When the logistic map is composed with itself, the result is a function with two maxima, shown in Fig. 5.8. Its analytic form is given by:

$$x_{t+2} = \mu x_{t+1}(1 - x_{t+1}) = \mu^2 x\{1 - (1 + \mu)x + 2\mu x^2 - \mu x^3\}. \tag{5.32}$$

Figure 5.8 shows one fixed point of the iterated map and demonstrates graphically that it is a periodic point of the original logistic map. Notice that the fixed point of F is also a fixed point of F^2.

Example 5.5 Limit cycle of van der Pol oscillator by method of averages. The van der Pol oscillator is a generalization of the harmonic oscillator with a nonlinear "damping" term. The equation of motion is:

$$\ddot{x} + \mu(x^2 - 1)\dot{x} + x = 0. \tag{5.33}$$

The $(x^2 - 1)$ removes energy from the system when it is negative, and pumps energy into the system when it is positive. It thus indirectly acts as a source of energy to the system, creating a stable sustained oscillation, or limit cycle as shown in Fig. 5.9. The limit cycle is the trajectory for which the average energy balance is zero. The energy gain is just:

$$\Delta E(t) = \int_{x(t_0)}^{x(t_0+t)} \mu(x^2 - 1)\dot{x}\, dx = \int_{t_0}^{t_0+t} \mu(x^2 - 1)\dot{x}^2\, dt. \tag{5.34}$$

In the limit as $\mu \to 0$, Eq. (5.33) reduces to that for a simple harmonic oscillator. Hence for sufficiently small μ, the solution should be close to that of a simple harmonic oscillator. In particular, rewriting the equation as a system of first-order equations:

$$\dot{x} = y; \qquad \dot{y} = -\mu(x^2 - 1)y - x \tag{5.35}$$

---

[13] Another familiar example is the screaming sound produced by the feedback loop induced by holding a microphone next to a speaker, a common cause of eardrum damage among rock musicians.

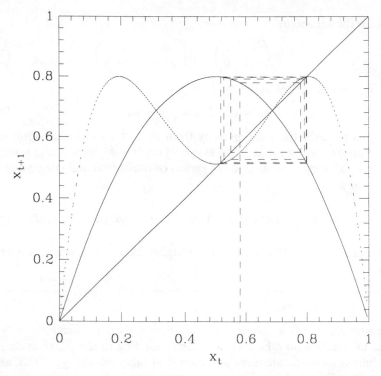

FIGURE 5.8. A limit cycle of the logistic map. The logistic map $F(x)$ is shown as a solid curve, and its second iterate, $F^2(x)$, is shown as a dotted curve. The fixed points are the intersection with the identity line. The fixed point of $F(x)$, near $x = 0.68$ is unstable. The two fixed points of $F^2(x)$ at roughly $x = 0.45$ and $x = 0.85$ are a period two limit cycle of $F(x)$.

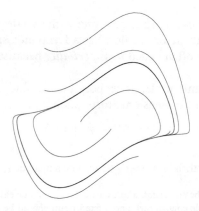

FIGURE 5.9. Trajectories from six initial conditions approach a roughly rectangular limit cycle of the van der Pol oscillator.

yields for the Jacobian:

$$J = \begin{pmatrix} 0 & 1 \\ -1 & 0 \end{pmatrix} - \mu \begin{pmatrix} 0 & 0 \\ 2xy & x^2 - 1 \end{pmatrix}. \qquad (5.36)$$

Thus the linearized flow at the origin is

$$\dot{x} = y; \qquad \dot{y} = -x + \mu y. \qquad (5.37)$$

Clearly, μ perturbs the solution away from that of a simple harmonic oscillator. Approximating the solution as $x = A\cos(t + \phi)$, with A and ϕ varying on time scales long compared to the period of oscillation and integrating over one period gives:

$$\Delta E(2\pi) \approx \mu A^2 \int_0^{2\pi} (A^2 \cos^2 t - 1) \sin^2 t \, dt = \mu A^2 (A^2/2 - 1)/2. \qquad (5.38)$$

So for small μ the limit cycle is given roughly by $A^2 = 2$, which is independent of μ.[14]

Because a curve is a one-dimensional object, there is effectively only one degree of freedom on a limit cycle—corresponding to the phase angle ϕ. In fact, there is a coordinate transformation that takes advantage of this, allowing one to write the phase $\phi(t)$ as a linear function of t:

$$\phi(t + \Delta t) = \phi(t) + \omega \Delta t, \qquad (5.39)$$

where ω is the oscillator's frequency. The transformation essentially flattens a complicated closed curve into a loop parameterized in such a way that it is traversed at a constant speed. By analogy with Hamiltonian systems, these coordinates can be called action-angle variables.

Tori

The next most complicated attractor is one with two degrees of freedom, or a *two-torus*.[15] A torus attractor of a flow has a Lyapunov spectrum $(0, 0, -, \ldots)$. Motion on a torus is often called *quasiperiodic*, because it contains two frequencies.

Just as for the limit cycle, it is possible to find angle variables $\vec{\phi} = \{\phi_1, \phi_2, \ldots, \phi_n\}$ on an n dimensional torus such that:

$$\phi_i(t + \Delta t) = \phi_i + \omega_i \Delta t. \qquad (5.40)$$

---

[14] Of course, the entire analysis assumes μ is small. The amplitude is independent of μ only for small μ.

[15] For the terminology to be consistent, a torus attractor should be called a "limit torus," to distinguish it from its unstable counterpart, and a fixed point should be called a "limit point." As usual, the commonly used terminology is not consistent.

Two nearby points on a torus must remain nearby, because given two such points at time t, $\phi_i(t)$ and $\psi_i(t) = \phi_i(t) + \varepsilon_i$,

$$\psi_i(t + \Delta t) = \phi_i(t) + \varepsilon_i + \omega_i \Delta_t = \phi_i(t + \Delta t) + \varepsilon_i. \qquad (5.41)$$

Thus the separation between the points does not grow on average as Δt increases.

For a flow, the Poincaré sections of a toroidal attractor T_n are tori of one lower dimension, T_{n-1}. Because the n-torus is an invariant set of the flow, its section is an $n - 1$ dimensional attracting invariant set of the Poincaré map. The Poincaré section of a two-torus, for example, is topologically equivalent to a circle, and is usually called an *invariant circle*.

Models of coupled oscillators are very common in the physical sciences. The oscillators might be masses on springs, or more abstract objects such as vibrational modes in a fluid. Imagine a dynamical system consisting of n independent oscillators, each with a limit cycle. If the fundamental periods of the oscillators are irrationally related, then the attractor of the combined dynamical system is an n-torus. When the oscillators are coupled together, however, there is a tendency for the fundamental frequencies of their limit cycles to shift, so that more of them become rationally related. When all of them are rationally related they are said to be *phase-locked*, or *entrained*. This phenomenon was originally observed by Huygens, who noticed that clocks placed next to each other would synchronize. The regions of parameter space where phase-locking occurs are sometimes called the *Arnold tongues*.

Until the early 1970s higher-dimensional tori were usually assumed to be the cause of fluid turbulence, through a theory due to Landau [10]. Stressing a fluid can induce oscillations; Landau hyphothesized that as stress on a fluid increases, more and more modes of oscillation are excited, and that this increase in the number of independent degrees of freedom is the cause of turbulence.

This theory was challenged by Ruelle and Takens, who proved that torus attractors in more than three dimensions are structurally unstable, or in other words, that they can be destroyed by an arbitrarily small change of parameters. In contrast, they demonstrated that a strange attractor *can* be structurally stable in more than three dimensions. This result is somewhat difficult to interpret—it does not imply directly that higher-dimensional tori cannot occur, but it does suggest that they may not be as likely as other kinds of attractors. Ruelle and Takens's work persuaded Gollub and Swinney and others to look at the transition to turbulence in simple fluid systems such as Taylor–Couette flow, and Rayleigh–Bénard flow [11, 12]. Indeed, in Taylor–Couette flow they saw a transition from a trivial fixed point, to a nontrivial fixed point, to a limit cycle, to a two-torus, and then to an attractor that clearly had a broadband power spectrum, suggesting a strange attractor. This is discussed in more detail later.

5.7.6   Review of Stability

As mentioned already, stability occurs in many different guises, and applies to many different aspects of a dynamical system. Now that we have introduced all the basic components of dynamical systems, it is worth reviewing the different kinds of stability. We will do this by moving from stability on the smallest scale to stability on the largest scale.

First, we can ask about the properties of nearby trajectories. If a small perturbation tends to grow, at least initially, then the system is locally unstable. As noted in section 5.7.1, this notion of stability is not coordinate invariant. Nonetheless, it can have a significant effect in a real experiment, and can even lead to phenomena that are easily mistaken for chaos [13].

We can then ask about *asymptotic local stability*: After an infinite amount of time, does the perturbed trajectory approach the original trajectory, or does it move away? At what rate? This is measured by the Lyapunov exponents, discussed in section 5.7.2.

On a bigger scale, we can ask about the stability of families of trajectories, and whether they approach attractors. If a state is perturbed away from an attractor, does it return? How large can the perturbation be before it does not return? This clearly depends on the number of attractors and their basins of attraction. If there is only one attractor, and its basin of attraction is the entire state space, then it is stable to any perturbation. This is "absolute attractor stability." If there is more than one attractor, or if there are regions of the state space where the trajectories tend to infinity, then the attractor is no longer stable to all perturbations.

Yet another concept of stability has to do with changes in parameters. To discuss this in more detail, however, we must first discuss bifurcations.

5.8   Bifurcations

Changing parameters can cause qualitative changes in behavior. Attractors may appear or disappear, or change their qualitative properties. Such an event is called a *bifurcation*.[16]

A common type of bifurcation occurs when an attractor becomes unstable. A fixed point of a flow, for example, loses its stability when the real part of the leading eigenvalue of the Jacobian changes from negative to positive; that is, when it crosses the imaginary axis in the complex plane. Bifurcations are classified according to how stability is lost; that is, according to what sort of invariant sets exist in state space before and after the bifurcation. The particular way in which stability is lost depends on symmetry properties of the Jacobian. These in turn can be determined by where the eigenvalue crosses

---

[16] Bifurcation is Latin for "forking into two," and probably derives from the fact that in some cases, such as a pitchfork bifurcation, the orbit seems to "fork" from Period 1 to Period 2.

the imaginary axis. For example, consider the algebraic properties of the Jacobian matrix: an antisymmetric Jacobian has pure imaginary eigenvalues, whereas a symmetric Jacobian has pure real eigenvalues. An antisymmetric matrix rotates vectors, but a symmetric one does not. Thus complex eigenvalues for the Jacobian of a fixed point mean the flow nearby is rotational and hence that perturbations of the fixed point spiral around it. Typically, perturbations in this case spiral into the equilibrium before the bifurcation and out to a limit cycle afterwards. In contrast, real eigenvalues for the Jacobian imply an absence of rotation. Typically, perturbations about such an equilibrium flow directly in to the equilibrium before the bifurcation and directly out to another equilibrium afterward. Thus, the bifurcations of equilibria in flows are often categorized into two cases according to where the leading eigenvalue crosses the imaginary axis:

- Hopf bifurcation—a pair of complex conjugate eigenvalues crosses the imaginary axis away from the real line. As a result, the fixed point becomes a limit cycle.
- Saddle-node—a real eigenvalue passes through the origin. A stable fixed point and a saddle emerge.

The Hopf and saddle-node bifurcations are illustrated in Figs. 5.10 and 5.11.

Example 5.6 The saddle-node bifurcation in a one-dimensional flow. Consider the following dynamical system: $\dot{x} = F(x) = \mu - x^2$, with μ real. It has fixed points at $x_\pm = \pm\sqrt{\mu}$, for $\mu > 0$. The Jacobian in this one-dimensional case is a scalar: $\frac{\partial F}{\partial x} = -2x$. Evaluated at the fixed points, we find the "eigenvalue" at $x_\pm$ to be $\lambda_\pm = \mp 2\sqrt{\mu}$. Thus x_+ is stable and x_- unstable. As μ increases through 0, the eigenvalues approach the origin along the imaginary axis, then move away from the origin along the real axis. In particular λ_+ crosses the imaginary axis into the right half-plane along the real line. Summarizing, for μ just below zero, there are no fixed points. At zero, two fixed points appear, one stable (a node) and the other unstable (a one-dimensional analogue of a saddle).

For a map, there is another possibility. A fixed point in a map loses its stability when the norm of the leading eigenvalue equals unity, $|\lambda_1| = 1$; that is, when the leading eigenvalue passes through the unit circle.[17] The Hopf bifurcation in a mapping changes a fixed point into an invariant circle. Because of the discrete nature of a mapping, an eigenvalue that is a rational

---

[17] As explained in Eq. (5.14), the exponential relates properties of linearized flows and mappings. The exponential maps the imaginary axis to the unit circle.

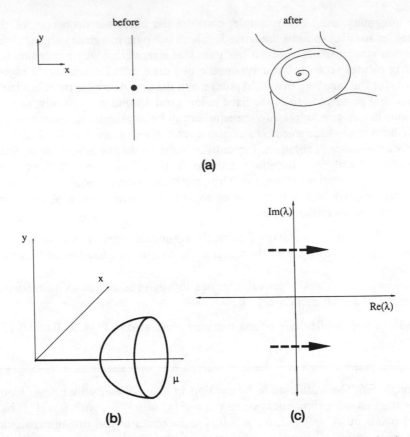

FIGURE 5.10. (a) A Hopf bifurcation in a flow creates a limit cycle from a fixed point in state space: before, a stable fixed point; after, trajectories approaching a limit cycle. (b) Bifurcation diagram. A control parameter axis is added to the state space axes and limit sets like those in (a) are plotted for each value of the control parameter. (c) Eigenvalues of the Jacobian cross the imaginary axis away from the origin as the bifurcation occurs.

root of unity; that is, one for which $\lambda^{p/q} = 1$ for some integers p and q, could generate a periodic orbit instead of an invariant circle. With one important exception, such bifurcations do not occur in typical systems. The exception is the *subharmonic* or *period doubling* bifurcation, when λ passes through $(-1, 0)$ in the complex plane and a period two orbit results.

Example 5.7 Subharmonic bifurcation in the logistic map. Consider a linear stability analysis of the fixed points determined in Example 2 : $x_f = 0$ for all μ or $x_f = 1 - 1/\mu$, for $1 \le \mu \le 4$. The derivative of the mapping evaluated

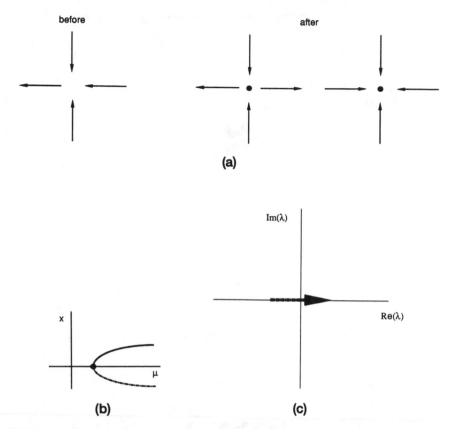

before

after

(a)

Im(λ)

Re(λ)

x

μ

(b)

(c)

FIGURE 5.11. Sketches for a saddle-node bifurcation in a flow analogous to those in the previous figure. (a) Effect of a saddle-node bifurcation on the state space: before, no fixed point exists; after, a saddle and a stable fixed point. (b) Bifurcation diagram. (c) An eignvalue of the Jacobian crosses the imaginary axis at the origin as the bifurcation occurs.

at 0 is μ; at the other fixed point it is $2 - \mu$. Thus the trivial fixed point is stable for $\mu < 1$ and the nontrivial one is stable for $1 < \mu < 3$. At $\mu = 3$ the system undergoes a subharmonic bifurcation. Figure 5.12 shows the loss of stability at $\mu = 3$. What happens for $\mu > 3$? The periodic point of Fig. 5.8 becomes stable in the sense that it is a stable fixed point of the iterated map. Thus at $\mu = 3$ the stable orbit has doubled its period from 1 to 2. In other words, the frequency has halved. Hence the names "period doubling" or "subharmonic" for this bifurcation.

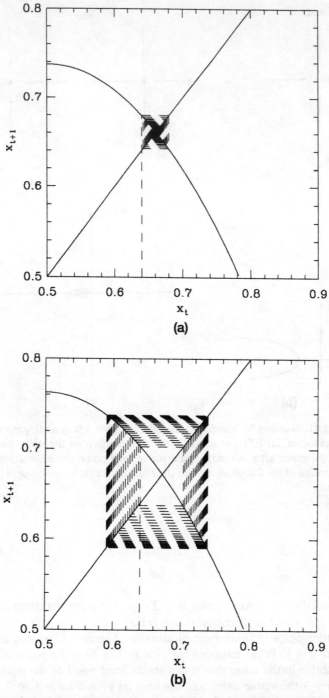

FIGURE 5.12. A period doubling bifurcation in the logistic map at $\mu = 3$. (a) close up view of a trajectory near the fixed point for $\mu = 2.95$. (b) a trajectory with the same initial condition as (a) for $\mu = 3.05$. Note that the trajectory is now drawn to a period two orbit.

Example 5.8 Saddle-node bifurcation in a one-dimensional map. Consider the following dynamical system: $x_{t+1} = \mu - x_t^2$, with μ real. The fixed points are $x_{\pm} = -[1 \pm \sqrt{1+4\mu}]/2$. And the Jacobian at any point x is $J = -2x$. Thus, the fixed point at $x = -1/2$ undergoes a saddlenode bifurcation at the point $\mu = -1/4$. This bifurcation is shown in Fig. 5.13.

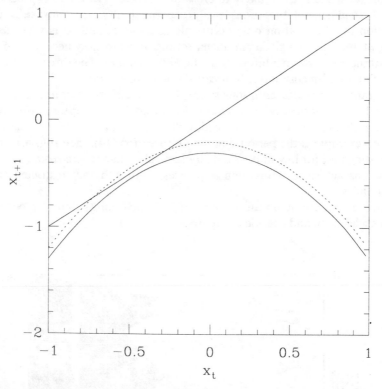

FIGURE 5.13. A saddle node bifurcation in the one dimensional map $x_{t+1} = \mu - x_t^2$. There are no fixed points for $\mu < -1/4$ (solid line). For $\mu > -1/4$ a pair of fixed points appear at $x_{\pm} = -1/2(1 \pm [1 + 4\mu]^{1/2})$ (dotted lines). x_+ is unstable; x_- stable.

Symmetries in the dynamics (e.g., spatial reflection symmetry), introduce special bifurcations different from the generic ones listed above. Perturbations to the dynamics that break the symmetries will change such bifurcations into combinations of the generic ones. This process is known as *unfolding* a special bifurcation.

In a system with more than one control parameter, a given bifurcation typically occurs at more than one value of the control parameter. The set of parameters where the bifurcation occurs form a surface, sometimes called a *bifurcation surface*. The bifurcation can be classified according to the codimension of the bifurcation surface. The codimension is an important aspect of a bifurcation, because in some sense it measures "how likely" the bifurcation is to occur. In a three-dimensional parameter space, for example, a codimension two bifurcation forms a sheet, so that an arbitrary curve in the parameter space is quite likely to cross the surface. Thus, one expects to be able to find such bifurcations easily experimentally. In contrast, a codimension two bifurcation only occurs along a curve, and so is very easy to miss. Starting from a given parameter setting, it is typically necessary to vary m parameters to reach a bifurcation of codimension m. The three cases above classify the codimension one bifurcations of a fixed point.

A graph of the surfaces in state space along which bifurcations occur versus control parameters is called a *bifurcation diagram*. Very often, bifurcations come in characteristic sets, or cascades, that define a bifurcation sequence. A famous example is the period-doubling bifurcation [14]. (See Fig. 5.14.) It is interesting that for low-dimensional dynamical systems there seem to be only a few characteristic bifurcation sequences, with a highly restricted set of possibilities.

Bifurcations are by no means restricted to fixed points, but also occur for limit cycles, tori, and chaotic attractors.

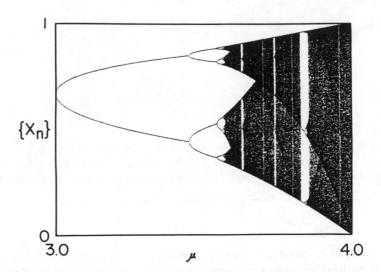

FIGURE 5.14. Bifurcation diagram of the logistic map. The parameter μ is set to a constant, plotted on the horizontal axis. The map is iterated until transients have died out, then the iterates are plotted on the y axis. The map bifurcates from a fixed point to a two-cycle at the extreme left, and then period doubles until it becomes chaotic. The chaotic region is interlaced with stable periodic orbits.

A bifurcation can be *subtle*, meaning that for a small change in the parameters there is only a small change in the behavior, or *catastrophic*, meaning that a small change in parameters can lead to a very large change in behavior. A typical example of a subtle bifurcation is the saddle-node bifurcation, in which a node (a sink or source) and a saddle (a hyperbolic fixed point) coalesce, leaving no fixed point.

When a new attractor appears that is unrelated to existing attractors, the result can be a catastrophic change in trajectories. For instance, if there are two distinct stable fixed points in a system and one loses its stability, the system will jump immediately from that one to the other.

In particular, consider the sequence of events pictured in Fig. 5.15. As a control parameter p increases in a system with a stable fixed point x_0, a second stable fixed point x_1 appears at p_0, and x_0 loses its stability at $p_1 > p_0$. This sequence generates *hysteresis*. As p is increased through p_0, the system stays at x_0, because it remains stable. At p_1 the system jumps from x_0 to x_1. If the parameter is now decreased, the system stays at x_1 until $p = p_0$ instead of jumping back at p_1. Hysteresis can occur whenever there is more than one attractor.

5.9 Chaos

Chaos is defined by the presence of positive Lyapunov exponents. For regular motion, nearby trajectories remain close to each other with the passage of time. For chaotic dynamics, in contrast, nearby trajectories separate at an

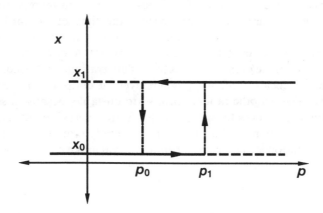

FIGURE 5.15. Bifurcation diagram for a hysteretic system. The solid lines represent stable fixed points. As the control parameter p is increased, the system hops from one fixed point to another. As the control parameter is decreased, the hop takes place at a different value of p.

exponential rate, so that the motion is asymptotically locally unstable. Naively, one might think that this is only possible for trajectories that go to infinity. Indeed, this is the case for linear systems; orbits that are locally unstable are also globally unstable, and tend to infinity. For nonlinear systems this is no longer true—local instability can coexist with bounded motion. Nearby trajectories separate exponentially, but are eventually folded back together so that they are contained in a bounded region. It is not surprising that chaotic dynamics often leads to invariant sets with rather complicated geometry.

Chaos connects determinism and randomness. Because nearby trajectories separate exponentially, the smallest error is amplified at an exponential rate, and quickly reaches macroscopic proportions. Imperceptible differences in the present rapidly turn into glaring discrepancies in the future.[18] Although chaotic dynamical systems are completely deterministic *in principle*, in practice chaos forces us to confront the inevitable underlying uncertainties of measurements. Over any long period of time determinism is defeated, and only statistical predictions are possible.

5.9.1   Binary Shift Map

The binary shift map, illustrated in Fig. 5.16, is perhaps the simplest example of a chaotic dynamical system. It corresponds to multiplication by two modulo one.

$$x_{t+1} = 2x_t \quad (\text{mod } 1). \tag{5.42}$$

Although this mapping is piecewise linear on $(0, 1/2)$ and on $(1/2, 1)$, because of the modulo operator there is a discontinuity at $1/2$ so that it is *not* a linear map. Because the slope is two, the map doubles the size of any small uncertainty interval at every iteration. This is easily visualized by representing the state x in base two. Iteration of the map shifts the digits of x to the left by one place, and then truncates the integer part. If x is a rational number, its expansion is asymptotically periodic. Thus, any iterate of a rational number generates a cycle of the map. The trajectory of an irrational initial condition, in contrast, never repeats itself, and wanders over the entire unit interval, filling it uniformly. The periodic rational orbits are unstable, because a slight perturbation typically makes them irrational. "Almost every" point is irrational, and so chaotic irrational orbits are overwhelmingly more common. Because any initial condition is mapped to the unit interval and never escapes, the unit interval is an attractor.

---

[18] Humans seem to have poor intuition for exponential rates of change. As noted by Malthus, during its early states the geometric progression is deceptively slow, and during later stages is alarmingly rapid. For a chaotic system any uncertainty, no matter how small, is amplified to macroscopic proportions in finite time. This time varies as the logarithm of the initial uncertainty, and is surprisingly insensitive to the initial uncertainty level.

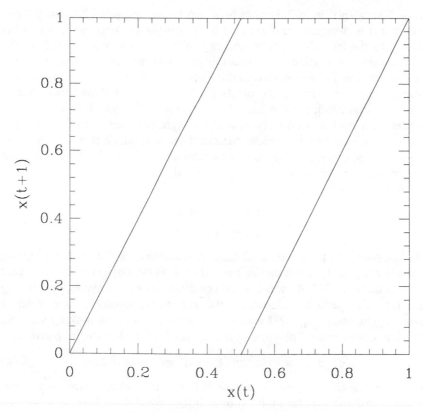

FIGURE 5.16. The binary shift map.

The binary shift map destroys microscopic information. Suppose, for example, that we know the initial condition to 10 bits of precision. Every iteration of the map shifts the uncertainly level one position to the left. After 10 iterations we know nothing at all, as shown in Table 5.2.

TABLE 5.2. Loss of information
in the binary shift map.

0.1100101001
0.100101001?
0.00101001??
0.0101001???
0.101001????
0.01001?????
0.1001??????
0.001???????
0.01????????
0.1?????????
0.??????????

The binary shift map illustrates how errors are amplified by chaotic systems, but it is deceptive and atypical in its simplicity. Because it just shifts patterns to the left, all of the complexity of its orbits seems to reside in the initial conditions. Rational initial conditions generate cycles, and only irrational initial conditions lead to chaotic behavior. In more typical dynamical systems the complexity of the orbits is intrinsic to the dynamical system. Rational initial conditions typically lead to irrational orbits, orbits are mixed together in a complicated fashion, and the dynamics seems to have a complexity of its own. Making precise the exact sense in which this complexity is intrinsic to the dynamics, and not just a limitation of the observer, brings up subtle questions that are still not fully resolved.

The tent map, defined as

$$x_{t+1} = \begin{cases} 2x, & \text{if } x \le 1/2; \\ 1 - 2x, & \text{else,} \end{cases} \tag{5.43}$$

is a close cousin of the binary shift map. It is identical to the binary shift map on the interval $(0, 1/2)$, but on the interval $(1/2, 1)$ the slope is reversed so that it is continuous at $1/2$. Rational initial conditions generate cycles, just as they do in the binary shift map. However, the tent map is topologically equivalent to the logistic map (Eq. 5.28) with $\mu = 4$, as can be seen by making a simple change of coordinates. Substituting $\sin^2 \pi\theta$ for x in the logistic map gives:

$$\sin^2 \pi\theta_{t+1} = 4 \sin^2 \pi\theta_t (1 - \sin^2 \pi\theta_t) = \sin^2 2\pi\theta_t. \tag{5.44}$$

There is a one-to-one correspondence between the orbits of the two maps. However, for the logistic map rational initial conditions lead to irrational values under a single iteration. Also, for the logistic map the absolute value of the derivative is greater than one in some places and less than one in others, so that errors alternately grow or shrink depending on the state. Nonetheless, on average the errors double with every iteration, just as they do for the tent map.

Note that these maps are not invertible. To go backward in time requires a choice between one of two possibilities. Thus, at every iteration two trajectories are folded together. This compensates for the separation of nearby trajectories.

5.9.2   Chaos in Flows

Chaos is caused by stretching and folding. The stretching is local, causing separation of nearby trajectories; the folding is global, and keeps the motion bounded. For noninvertible mappings, stretching happens when the magnitude of the derivative is greater than one. Folding occurs when trajectories merge.

To understand how the analogous behavior occurs for a flow, imagine a dynamical system in three dimensions, and a two-dimensional ribbon of parallel trajectories within this system, as shown in Fig. 5.17. Suppose that this ribbon is stretched at one end, so that the trajectories separate going forward in time. If they separated indefinitely, the motion would tend to infinity. To keep the motion bounded, the ribbon is folded over onto itself. The result is a motion that is locally unstable, because nearby trajectories separate exponentially, but globally stable because they are trapped on an invariant set.

There is a catch in the above description. Flows are always invertible—trajectories cannot merge. Thus, the ribbon cannot fold down onto itself exactly, but can only come close to doing so. The invariant set cannot be a

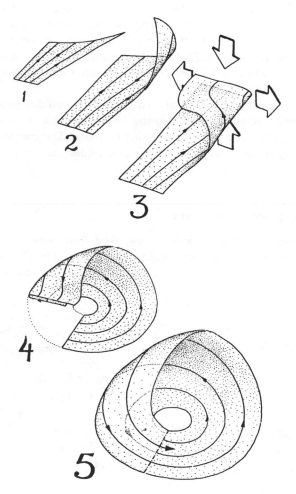

FIGURE 5.17. Folding and stretching in a flow, as time progresses, from 1 to 5 [15].

simple two-dimensional manifold. Instead, after each fold, the number of ribbons increases. A single sheet is stretched and folded. Then it is folded again so that locally it looks like four sheets, then eight, and so forth, until eventually it has been folded an infinite number of times. Because the dynamics is dissipative, after each iteration the folds are pressed closer and closer to each other. The result is that the sheet is "thickened" into a complicated geometric object, reminiscent of Greek pastry, with a fine structure consisting of an infinite number of thin sheets (see Fig. 5.18). Such objects are called *fractals* [16].

Although this fractal set is quite complicated, it is nonetheless invariant: if we start with this set, it maps into itself. Furthermore, such sets can be attractors. It is easy to imagine how this is possible, by simply arranging the flow field off the attractor so that all of the trajectories tend toward it. Attractors for which the motion is asymptotically locally unstable are called *chaotic attractors* or *strange attractors*. The chaotic attractors of flows or invertible maps are typically fractals; the chaotic attractors of noninvertible maps may or may not be fractals. The chaotic attractors of the logistic equation, for example, are not fractals.

Note that for flows the only possible attractors in two dimensions are fixed points and limit cycles. This is called the *Poincaré-Bendixson theorem*. Thus chaos is only possible in three dimensions or more. The construction above makes it fairly obvious why: stretching and folding is not possible in two dimensions.

5.9.3   The Rössler Attractor

Chaotic attractors with the basic form described above arise in many different dynamical systems. One such was originally discovered by Otto Rössler, who set out to look for the "simplest example" of a dynamical system with a

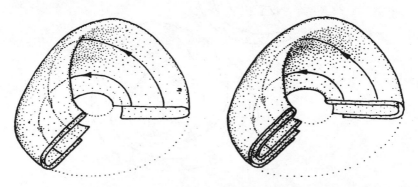

FIGURE 5.18. Folding and stretching makes a fractal [15].

chaotic attractor [17]. Rössler's equations are

$$\dot{x} = -(z+y); \qquad \dot{y} = x + ay; \qquad \dot{z} = b + (x-c)z, \qquad (5.45)$$

where the x and y equations are equivalent to a those of a linear damped harmonic oscillator.[19] All the nonlinearity comes from the xz term in the third equation. The attractor of the Rössler system is shown in Fig. 5.19. It has a Lyapunov spectrum of the form $(+, 0, -)$. The attractor lies in a thin disk near the $x - y$ plane everywhere except at the fold, in the upper right corner. The trajectories are stretched, and then the outside is folded down on top of the inside.

By using a surface of section, the Rössler equations can be reduced to a two-dimensional mapping of the plane into itself. Fig. 5.20 is a plot of the x versus z coordinates for downward crossings of the plane $y = 0$. This figure is deceptively simple, for several reasons. First, while this seems like a simple curve, it is a collection of discrete points. Watching successive crossings, the

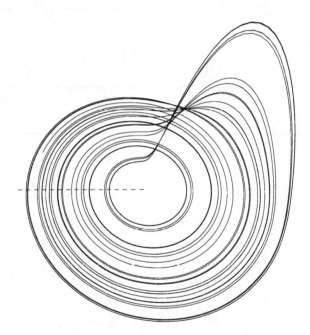

FIGURE 5.19. A single trajectory of the Rössler dynamical system, projected onto a two-dimensional plane, using the parameters $a = 0.15$, $b = 0.20$, and $c = 10.0$. This is perhaps the simplest example of a chaotic attractor in a flow. The dashed line shows the surface used to construct the Poincaré section for Fig. 5.20.

---

[19] Consider, for example, the second derivative of y.

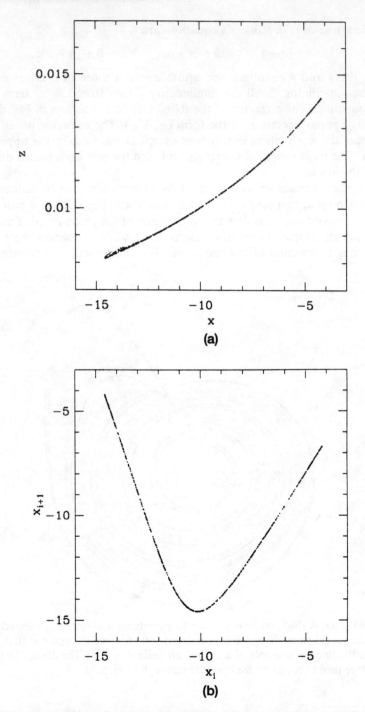

FIGURE 5.20. (a) Poincaré section of the Rössler attractor, made by recording the x and z coordinates for downward crossings of the plane $y = 0$. (b) Reduction to a one-dimensional map, made by plotting x_t versus x_{t+1}.

points fill the curve in what seems like random order. Second, although the attractor looks like a simple one-dimensional curve, it is not. With finer resolution we would see that it has the complicated fractal structure of the previous figure. For the Rössler equations the damping is so strong that the folds of the attractor are enormously compressed, and the fractal structure is not easily visible to the eye.

Because the attractor in the surface of section is approximately one dimensional, knowledge of one coordinate, say x, permits approximate determination of the other. For motion on the attractor this makes it possible to approximate the two-dimensional Poincaré map by a one-dimensional map. This is apparent by plotting x_{t+1} versus x_t, shown in Fig. 5.20b. Note the similarity of this to the logistic map. Keep in mind, however, that while the reduction to a two-dimensional Poincaré map is exact, reduction to a one-dimensional map is just an approximation.

5.9.4   The Lorenz Attractor

Probably the most famous chaotic attractor was originally discovered by Lorenz [18]. As a meteorologist, Lorenz had long been perplexed that, even though the equations governing the atmosphere are deterministic, many aspects of the atmosphere nonetheless are highly unpredictable. Studying a complicated model of fluid convection due to Saltzman, he observed that three of the modes made sustained irregular motions, while all the rest settled down to a fixed point. He stripped Saltzman's model down to the critical three modes, getting the following simple set of nonlinear equations:

$$\dot{x} = \sigma(y - x)$$
$$\dot{y} = -xz + rx - y \tag{5.46}$$
$$\dot{z} = xy - bz$$

Integrating these equations numerically, he observed the attractor shown in Fig. 5.21. In his 1963 paper, "Deterministic Nonperiodic Flow," he clearly stated how the stretching and folding in these equations leads to unpredictable behavior and fractal structure, and conjectured that this was the underlying cause of the unpredictability of the atmosphere.

Fig. 5.22 shows sensitive dependence on initial conditions on the Lorenz attractor. A small cloud of points, representing the possible values of an initial measurement, start in the upper left hand side in the first frame. As they evolve through time this cloud is stretched into a long thin filament by the local instabilities on the attractor. As this filament passes the saddle point at the bottom of the figure, it is apparently broken into two pieces; the number of pieces doubles on every successive pass. Eventually the points are so spread apart that they cover the entire attractor. Thus, what began as a fairly precise measurement of an initial condition becomes highly uncertain. What began as deterministic behavior becomes purely statistical behavior.

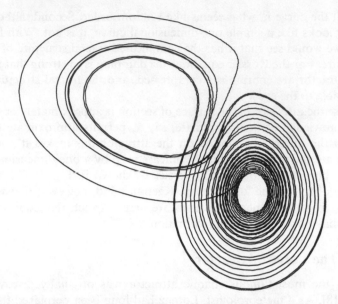

FIGURE 5.21. Part of a single trajectory of the Lorenz equations, Eq. (5.47), projected onto a two-dimensional plane. The parameters are $\sigma = 10$, $b = 8/3$, and $r = 28$.

5.9.5   Stable and Unstable Manifolds

The *stable manifold* of a given point is the set of points that asymptotically approach this point as it evolves through time, in the limit as $t \to \infty$. In other words, for a dynamical system F the stable manifold of x is:

$$\left\{ y : \lim_{t \to \infty} \|F^t(x) - F^t(y)\| = 0 \right\}. \tag{5.47}$$

Similarly, the *unstable manifold* of a point consists of the set of points that exponentially diverge away from it. This can be stated more precisely going backwards in time. The unstable manifold is:

$$\left\{ y : \lim_{t \to -\infty} \|F^t(x) - F^t(y)\| = 0 \right\}. \tag{5.48}$$

For a linear dynamical system the stable manifold is the same as the hyperplane associated with the stable eigenvalues, and the unstable manifold is the same as the hyperplane associated with the unstable eigenvalues. The stable and unstable manifolds in the vicinity of a fixed point can be found by linearizing the dynamical system around the fixed point. Further away from the fixed point, however, when the linear dynamical system ceases to be a good approximation, the stable and unstable manifolds deviate from these hyperplanes.

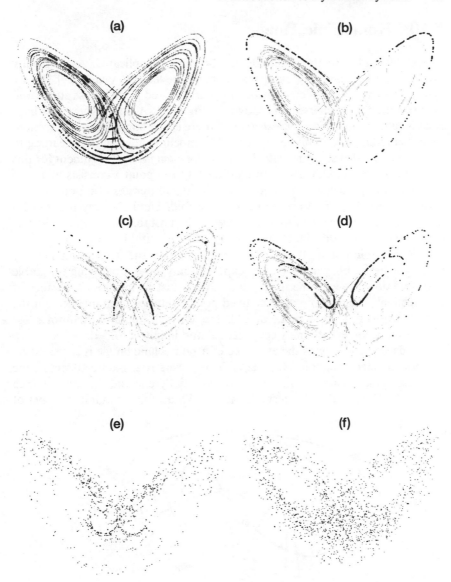

FIGURE 5.22. Sensitive dependence for the Lorenz attractor. This figure illustrates the evolution of an ensemble of points, all of which were initially so close together that they were visually indistinguishable from a single point. Each frame shows an xz projection of the ensemble after varying intervals of time. A single trajectory is shown lightly in the background to outline the attractor. Time is given in units where the average time between maxima of z equals one, so that one unit may be thought of as "number of passes around the attractor." (a) Enough time has passed so that the spreading becomes visible to the eye. Ten different strobes are shown, at times varying from $t = 6.9$ to $t = 7.8$. (b) A single strobe, at $t = 8.2$. (c) $t = 8.5$. (d) $t = 12.1$. (e) $t = 18.2$. (f) $t = 27.7$. By the last two frames, the points are scattered all over the attractor, and it is no longer necessary to display a background trajectory to see its outline.

5.10   Homoclinic Tangle

The stable and unstable manifolds can be quite complicated. As originally observed by Poincaré, this complicated structure leads to chaos [19]. To see how this happens, consider a *homoclinic point*, a fixed point whose stable and unstable manifolds intersect, as shown in Fig. 5.23. A point where this happens is called a *homoclinic intersection*. If there is one homoclinic intersection, then there must be an infinite number of homoclinic intersections, leading to a very complicated structure called a *homoclinic tangle*. The argument for this is as follows: because the stable manifold of a fixed point x consists of all the points that approach x as time goes to infinity, all iterates of a point on the stable manifold are also on the stable manifold. Similarly, any iterate of a point on the unstable manifold remains on the unstable manifold. A homoclinic point is on both the stable and unstable manifolds, so its iterates are also homoclinic points. But a point that is not on the fixed point initially never reaches it, but can only come exponentially close. Following the stable manifold down toward the fixed point, there must be an infinite number of intersections, accumulating on the fixed point. Between each intersection the unstable manifold must make a fold; the unstable manifold cannot cross itself, so it becomes extremely complicated and infinitely folded.

For dissipative systems the existence of a homoclinic tangle is a necessary, but not sufficient, condition for chaos. When there *is* a chaotic attractor, the unstable manifold and the attractor are essentially one and the same.[20] (See Fig. 5.24.) The fact that the unstable manifold must fold an infinite number of

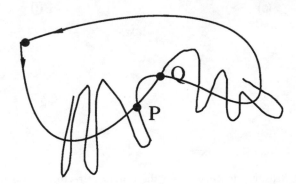

FIGURE 5.23. A homoclinic tangle. The solid curves represent stable and unstable manifolds of the fixed point shown in the upper left hand corner. (This is a map, not a flow, and the solid curves are not trajectories!) The points P and Q are successive iterates of a homoclinic intersection. The iterates of P accumulate on the fixed point; the unstable manifold must cross and recross an infinite number of times, creating a very complicated, "tangle."

---

[20] More precisely, the attractor is the closure of the unstable manifold.

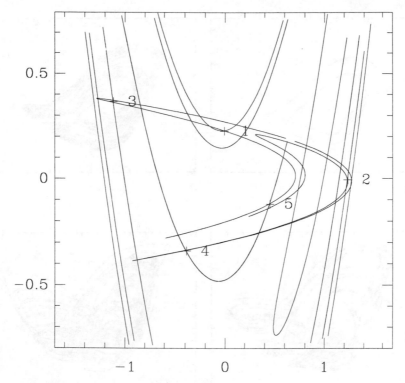

FIGURE 5.24. Stable and unstable manifolds of the Hénon map (Eq. 5.8) at $a = 1.4$ and $b = 0.3$. The curves represent portions of the stable and unstable manifolds of a fixed point. The unstable manifold is the one that resembles the attractor in Fig. 5.6. Iterates of the homoclinic intersection at $(-1.122374\ldots, 0.3722727\ldots)$ are labeled in order. Note that each iterate falls on an intersection of the two manifolds.

times is precisely the reason that the attractor must have a fractal structure. But in a dissipative system the existence of a homoclinic tangle is only a necessary and not a sufficient condition for chaos, because the unstable manifold of a homoclinic tangle is not necessarily an attractor. In this case, a point precisely on the unstable manifold will behave chaotically, but the slightest perturbation will send it toward an attractor somewhere else. Thus, although the possibility for chaotic motion exists, stability prevents it from manifesting itself.

5.10.1   Chaos in Higher Dimensions

Chaos can occur in any number of dimensions. Chaotic attractors can have any number of positive Lyapunov exponents.

Finite dimensional chaotic attractors occur even when the state space dimension is infinite. For example, Fig. 5.25 shows a limit cycle and three

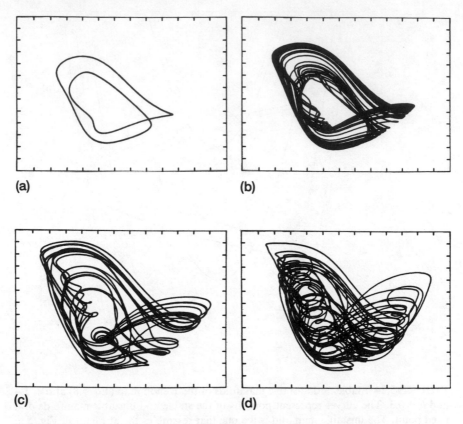

FIGURE 5.25. Phase plots for some attractors of the Mackey–Glass equation.

chaotic attractors of the Mackey–Glass differential delay equation

$$\frac{dx(t)}{dt} = a\,\frac{x(t-\tau)}{1+x(t-\tau)^c} - bx(t). \qquad (5.49)$$

Because the derivative depends both on x at time t and on x at time $t-\tau$, to make the evolution well posed it is necessary to specify $x(t)$ over an interval of length τ. The state space of initial conditions consists of functions; thus, this is an infinite dimensional dynamical system. Nonetheless, its attractors are finite dimensional. In some cases this is apparent from geometric analysis, but in general it requires computing the Lyapunov exponents [20].

5.10.2   Bifurcations Between Chaotic Attractors

It is also possible for chaotic attractors to make bifurcations between different kinds of behavior. For example, Fig. 5.26 shows a subtle bifurcation between two different states of the Rössler attractor.

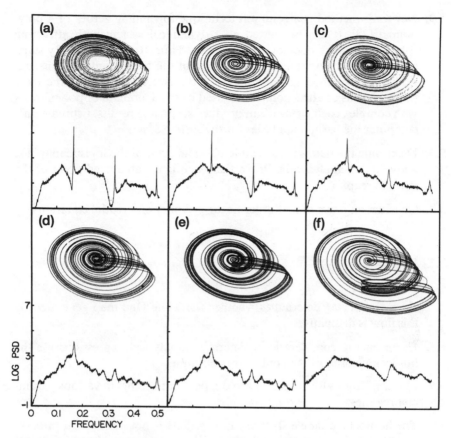

FIGURE 5.26. A bifurcation from the simple Rössler attractor to "the funnel." The power spectrum of the motion is shown below each frame.

There is a repellor near the origin, in the center of "the hole" in the simple Rössler attractor. As a parameter is varied the attractor approaches the repellor until it encompasses it. The attractor rolls up around the repellor, developing a structure that devotees call "the funnel." As seen from the power spectrum, the simple Rössler attractor contains a sharp periodic motion superimposed on otherwise chaotic behavior; this is called *phase coherence*. As the funnel develops more turns the phase coherence is lost [21, 22].

Problems

5.1. Using a standard integration technique, for example Runge–Kutta, plot trajectories for systems such as the simple harmonic oscillator, the van der Pol oscillator, and the simple pendulum.

5.2. Newton's method is a common technique for finding zeroes of a polynomial [23]. It can be viewed as a dynamical system with attracting fixed points corresponding to the zeroes. Thus, from an arbitrary starting point, trajectories are drawn to one of the fixed points. The usefulness of the method is determined by the size and shape of the fixed points' basins of attraction. When used over the domain of polynomials with complex coefficients it can produce surprising results. Estimate on a computer the basin boundaries for the system $z^3 + i = 0$.

5.3. Determine the points in the stable periodic orbit of the logistic map just above $\mu = 3$ analytically. Prove that each is a stable fixed point of the iterated map.

5.4. Consider a system of first order linear ordinary differential equations:

$$\dot{\vec{x}} = M\vec{x}.$$

Using the identity

$$\det M = \exp(\text{Tr} \ln M)$$

show that $\text{Tr} \, M$ determines whether the Time One map generated by this flow is dissipative.

5.5. Using energy considerations, determine analytical expressions for the homoclinic orbit of the nonlinear pendulum.

5.6. Plot trajectories for the Lorenz and Rössler attractors in various parameter regimes.

5.7. The homoclinic tangle sketched in Fig. 5.23 is not orientation preserving. To see this, notice that the loop formed by the stable and unstable manifold joining P and Q must map into the loop formed by Q and the next homoclinic intersection. When it does so, the orientation of this loop is reversed. Sketch a homoclinic tangle that *is* orientation preserving.

5.8. Reconstruct Fig. 5.24. First find the fixed points of the Hénon map. Linearize the map about one of those points. Find the stable and unstable eigenspaces at the fixed points. Iterate a set of points in the unstable eigenspace to find the unstable manifold. Calculate the time-reversed map. Iterate a set of points in the stable eigenspace under the time-reversed map to find the manifold. Try to find a point on the intersection of the two. If you iterate a fixed set of points, you will find that you quickly lose any ability to resolve regions of high curvature on the manifold. The smooth curves in the figure were obtained using the following trick: given a set of points on a smooth curve, interpolate enough new points on the curve so that the iterate of the curve is also smooth.

References

[1] J.-P. Eckmann and D. Ruelle, Rev. Mod. Phys. **57**, 617 (1985).

[2] J. Guckenheimer and P. Holmes, *Nonlinear Oscillations, Dynamical Systems, and Bifurcations of Vector Fields* (Springer-Verlag, Berlin, 1983).

[3] B.V. Chirikov, Phys. Rep. **52**, 263 (1979).

[4] M. Hénon, Comm. Math. Phys. **50**, 69 (1976).

[5] G. Bennetin, L. Galgani, A. Giorgilli, and J.M. Strelcyn, Meccanica, **15**, 9 (1980).

[6] I. Shimada and T. Nagashima, Prog. Theor. Phys. **61**, 1605 (1979).

[7] A. Wolf and J. Swift, in *Statistical Physics and Chaos in Fusion Plasmas*, edited by W. Horton and L. Reichl (Wiley, New York, 1983).

[8] J.-P. Eckmann, S. Oliffson Kamphorst, D. Ruelle, and S. Ciliberto, Phys. Rev. A **34**, 4971 (1986).

[9] M. Sano and Y. Sawada, Phys. Rev. Lett. **55**, 1082 (1985).

[10] L.D. Landau and E.M. Lifschitz, *Fluid Mechanics* (Addison-Wesley, Reading, Mass., 1959).

[11] J.P. Gollub and H.L. Swinney, Phy. Rev. Lett. **35**, 927 (1975).

[12] H.L. Swinney. Physica D **7**, 3 (1983).

[13] R.J. Deissler and J.D. Farmer. Physica D **55**, 155 (1992).

[14] M.J. Feigenbaum. J. Stat. Phys. **19**, 25 (1978).

[15] R. Abraham and C.D. Shaw, *Dynamics: The Geometry of Behavior* (Aerial, Santa Cruz, CA, 1983).

[16] B. Mandelbrot, *Fractal Geometry of Nature* (Freeman, New York, 1982).

[17] O. Rössler, Ann. NY Acad. Sci. **316**, 376 (1978).

[18] E.N. Lorenz, J. Atmos. Sci. **20**, 130 (1963).

[19] H. Poincaré, Science et Méthode, Bibliotèque Scientifique, 1908. English translation by F. Maitland (Dover, New York, 1952).

[20] J.D. Farmer, Z. Naturforsch. **37a**, 1304 (1982).

[21] D. Farmer, J. Crutchfield, H. Froehling, N. Packard, and R. Shaw, Ann. NY Acad. Sci. **357**, 453 (1980).

[22] J.D. Farmer, Phys. Rev. Lett. **47**, 179 (1981).

[23] W.H. Press, B.P. Flannery, S.A. Teukolsky, and W.T. Vettering, *Numerical Recipes* (Cambridge University, Cambridge, 1986).

6

Probability, Random Processes, and the Statistical Description of Dynamics

Stephen G. Eubank and J. Doyne Farmer

This chapter links the theory of random processes with the statistical description of dynamics. The reader may also wish to consult the references [1].

6.1 Nondeterminism in Dynamics

The real world is fraught with uncertainty. Any real measurement has only finite precision, and does not determine initial conditions precisely; reality inevitably contains unknown factors that we cannot hope to take into account, so that any dynamical law is only an approximation. Deterministic models are good approximations when there are only a few degrees of freedom.

When there are many degrees of freedom deterministic models become intractable. This can happen because it is impossible to measure all the degrees of freedom, because the necessary model is too big to run on a computer, or because the interactions are so complicated or so poorly understood that it is impossible to model them.

Even if there are only a few degrees of freedom, if the system is chaotic a deterministic description is useful only over short periods of time. Over longer periods of time, uncertainties are amplified exponentially, and deterministic models break down. Uncertainties may be due to the inevitable imprecision of initial measurements, or to the background of unknown fluctuations. In either case, the time over which a deterministic model remains valid to a given degree of approximation, or *prediction time*, depends logarithmically on the uncertainty level. If the uncertainty level can be reduced, the time over which a deterministic model is valid can be extended. Because of the logarithmic dependence, however, the prediction time increases only very slowly as the precision is increased. For example, for the binary shift map the uncertainty doubles every iteration. If the initial uncertainty is one part in a thousand, then meaningful predictions are possible for roughly 10 iterations into the future. If we improve the precision by a factor of a thousand, so that the initial states are accurate to one part in a million, then predictions are possible for 20 iterations; prediction time only doubles. It is only a matter of time before determinism is lost, and the time required is fairly insensitive to precision.

When deterministic dynamical models break down, for any of the reasons above, we must turn to nondeterministic dynamical systems, also called *stochastic processes* or *random processes*. Instead of modeling the deterministic evolution of states through time, we must now model the evolution of an ensemble of *possible states* through time. This is more cumbersome, because a single point in a deterministic model becomes a function in a nondeterministic model, describing an entire ensemble of possible states. All the factors that are impossible to model in detail are lumped into this function. The random process model attempts to minimize the effects of our ignorance by making as much use as possible of limited information.

6.2   Measure and Probability

This section introduces basic terminology from probability theory.

A variable whose value is uncertain is called a *random variable*.[1] The relative frequency of its values can be described by a probability function or a measure. These two descriptions are equivalent—the choice is purely a matter of convenience. If several random variables each have the same probability distribution, then they are *identically distributed*.

A *probability density function* $p(x)$ gives the relative likelihood of the possible values of the continuous variable x. The probability of finding x in any particular domain D is the *measure* of D. The measure of D is the integral of $p(x)$ over D:

$$\mu(D) = \int_D p(x)dx. \tag{6.1}$$

The measure is a function that assigns a nonnegative number to every possible subset—it tells how much "weight" each subset should have. The word "measure" is thus used in two slightly different senses: as the value of the function on a particular subset and as the function itself. The measure induced by a probability density is also called a "probability measure." Eq. (6.1) can also be written

$$\mu(D) = \int_D d\mu(x). \tag{6.2}$$

This is just a matter of notation. A measure makes it possible to take the weighted average of a function $f(x)$. This can be written in several ways.

$$\langle f(x) \rangle \equiv \int f(x)p(x)dx \equiv \int f(x)d\mu(x). \tag{6.3}$$

It is common practice to normalize $p(x)$ so that the probability of finding x *somewhere* is one; that is, so that $\int p(x)dx = 1$.

---

[1] In any *realization*, the value is known.

The probability distribution function $P(x)$ is the integral of the density function. For example, if x is a continuous variable on the real line, then

$$P(x) = \int_{-\infty}^{x} p(x')dx'. \tag{6.4}$$

This is easily extended to any number of dimensions.

The common notion of "volume" is called *Lebesgue measure*. For example, on the unit interval the Lebesgue measure of a line segment S of length s is just $\mu_L(S) = s$. Lebesgue measure is usually denoted "dx." Individual points have zero Lebesgue measure. A measure μ is *singular* if there is a set S such that its Lebesgue measure $\mu_L(S) = 0$ but $\mu(S) > 0$. Often this means that individual points have positive measure.[2] When a measure is singular, the probability distribution function is discontinuous. The probability density function, which is the derivative of the probability distribution function, contains nonintegrable infinities and so is not well defined. A notational trick that sometimes gets around this is writing the probability density function in terms of the *Dirac delta function*, which is defined so that

$$\int f(x)\delta(x)dx = f(0) \tag{6.5}$$

($\delta(x) = 0$ except at $x = 0$, where it is undefined, except in terms of this integral.) However, to avoid the mathematical problems associated with singular behavior many people prefer to work with either measures or probability distribution functions.

The *support* of a measure is the union of all the subsets in its domain with positive measure. Loosely speaking, a point x is in the support of μ if any tiny ball around it will have at least a tiny bit of measure. Measures are additive over a countable number of unions of disjoint subsets; that is, $\mu(B \cup C) = \mu(B) + \mu(C)$.

The process of making measurements reduces a continuous random variable x to a discrete random variable. This is often called *coarse graining*. For example, imagine measuring x with a ruler. The ruler implicitly divides the domain into a collection of nonintersecting sets that cover the domain. This is called a *partition*. A ruler makes a special partition that is a regular grid. Each grid box is an "element" of the partition. Labeling the elements with an index i induces a mapping from x to the discrete variable i. The probability that x is in the ith element of the partition is $\mu(S_i)$. The *resolution* of the coarse graining is the size of its biggest element. For example, if the boxes of the grid are all ε on a side, then ε is the resolution of the coarse-graining.

When x is not derived from a continuous variable, the notion of a probability density no longer makes sense. Unfortunately, for discrete variables the term "probability distribution" is used in a different sense than it is for continuous variables. If the probability of any given discrete value of x is $P(x)$, the set of probabilities $\{P(x)\}$ is often called a *probability distribution*. This

---

[2] Singular measures are also commonly encountered when the support of a measure is a Cantor set or, more generally, a fractal.

does not involve an integral, and is not at all the same as the probability distribution of a continuous variable. The term "measure" is used consistently, however; the measure $\mu(x)$ of any given value of x is $P(x)$. "Measure" can also be used to refer to the entire collection $\{P(x)\}$, in which case "measure" and "probability distribution" are synonyms. When x is a discrete variable it is sometimes called an *event*.

We will sometimes use the generic term *probability function* to blur the distinction between continuous and discrete variables, and among probability densities, probability distributions, and measures. The precise interpretation should be clear from context.

Joint and Conditional Probability

Probabilities can be functions of more than one variable. It is often useful conceptually to separate the different variables, to emphasize their interactions with each other.

- The probability of events x and y occurring together is given by their *joint* probability, denoted $P(x, y)$. The joint probability is symmetric in x and y.
- The probability for just one of the variables, independent of the value of the others, is called the *marginal* probability. It can be determined by summing the joint probability over all the other variables; for example, $P(x) = \sum_y P(x, y)$.
- The probability that y occurs given that x has occurred is called the *conditional* probability and is often denoted $P(y|x)$. The conditional probability is defined by

$$P(y|x) \equiv \frac{P(x, y)}{P(x)}. \qquad (6.6)$$

The conditional probability $P(y|x)$ is undefined when $P(x)$ vanishes. This is sensible, because the question, "What is the probability that y will occur given that x has occurred?" is ill-posed if x cannot occur.
- We can equivalently define all of the concepts above in terms of density functions, replacing sums by integrals.

Statistical Independence

The joint probability characterizes the degree of dependence between random variables. Two random variables x and y are *statistically independent* only if their joint distribution factorizes, that is, if $P(x, y) = P(x)P(y)$. This definition says that the value of one variable has no influence on the other, because

$$P(y|x) = P(y, x)/P(x) = P(y)P(x)/P(x) = P(y). \qquad (6.7)$$

The abbreviation IID is often used to refer to random variables that are independent and identically distributed.

It is common practice to use the same symbol to refer to many different probability functions, and distinguish them only by their arguments. For example, it is common to abbreviate the probability functions $P_x(x)$ and

$P_y(y)$, as $P(x)$ and $P(y)$, even when P_x and P_y are different functions. Another possibly confusing practice is that of denoting a conditional probability function as $P(y|x)$. This is best thought of as a family of functions of one variable, y; the variable x is an index that labels *which* element of the family is being considered.

6.2.1   Estimating a Density Function from Data

To build models of the real world from observations, it is necessary to decide how the observations are related to actual events in the real world. This problem often reduces to estimating a probability function for the true state given the observations. Although "meta-modeling" principles, such as the Central Limit Theorem and Jaynes's maximum entropy principle, can be invoked to provide the rationale for particular schemes, their relevance to the problem must be accepted axiomatically. Rather than enter a metaphysical argument over the "best" criteria, we present a technique that we have found useful in applications.

The simplest way to estimate a density function from data is with a histogram. Simply partition the domain using, say, a grid, and count the number of points in each grid box. An estimate of the measure of each grid box is $\frac{n_i}{N}$, where n_i is the number of points in the box and N is the total number of points. Histogramming is fast, but it does not usually give a good estimate when the number of data points is limited. To get a histogram requires making a good choice for the size of the bins, which involves a compromise between statistical stability and spatial resolution.

Kernel density estimation is less time-efficient but usually more data-efficient for smooth probability densities. The basic idea is to let each point "influence" surrounding points through a *kernel function*. Kernel density estimators are of the form:

$$p(x) = \frac{1}{N} \sum_{i=1}^{N} k\left(\frac{\|x - x_i\|}{w}\right), \tag{6.8}$$

where the sum is over the entire data set, k is the kernel (a normalized probability density), $\| \; \|$ represents an appropriate norm, and w is a window width. To make a good estimate the kernel function and width must be properly matched to the density function and the number of data points.

Example 6.1 Comparison of density estimation techniques (see Fig. 6.1). A histogram with 128 bins is used to estimate the density function for both the logistic map time series and the quasiperiodic time series of Fig. 6.2. This is compared to a kernel density estimator of the form

$$p(x) = \sum_i \exp(-(x - x_i)^2/\sigma^2), \tag{6.9}$$

with $\sigma = 0.0625$.

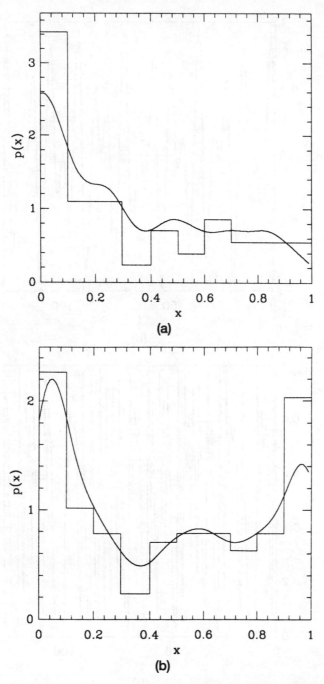

FIGURE 6.1. A histogram and a kernel density estimate for the probability of states x in a quasiperiodic time series (a) and a logistic map time series (b). The time series are shown in Fig. 6.2.

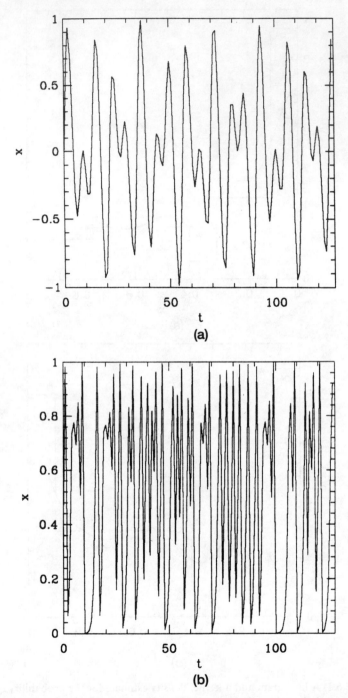

FIGURE 6.2. A pair of time series: (a) quasiperiodic, $x_t = (\sin(2\pi t/7) + \sin(2\pi t/11.36))/2$. (b) logistic, $x_{t+1} = 4.0x(1.0 - x)$. Curves are drawn through the discrete points to guide the eye.

6.3 Nondeterministic Dynamics

The theory of random processes, originally due to Slutsky and Kolmogorov [2–4], is essentially the theory of nondeterministic dynamical systems. Imagine a series of measurements x_t at discrete times t, and for simplicity let $t = 1$, $t = 2$, and so forth. The set of measurements $\{x_t\}$ is called a *time series*. If x is a scalar variable it is a *univariate* time series; if x is of higher dimension it is a *multivariate* time series. One way to define a random process is in terms of the joint probability of the values of the time series at successive times:

$$p(x_1, \ldots, x_d). \tag{6.10}$$

If the joint probabilities of order higher than d factorize in the following way:

$$p(x_1, \ldots, x_{d+1}) = p(x_1)p(x_2, \ldots, x_{d+1}), \tag{6.11}$$

then this is a dth order *Markov process*. The process is *deterministic* if there is some value of d such that the probability density function approaches a delta function in the limit of perfect measurements of $\{x_t\}$.

When the dynamics are formulated this way the dynamical laws, if any, are not apparent. An alternative way to define a random process is as a *stochastic dynamical system*, in either discrete time,

$$\vec{x}_{t+1} = \vec{F}(\vec{x}_t, n_t), \tag{6.12}$$

or continuous time,

$$\frac{d\vec{x}}{dt} = \vec{F}(\vec{x}(t), n(t)), \tag{6.13}$$

where $n(t)$ is a random variable representing the the uncertain aspects of the time evolution, often called the *noise* and $\vec{x}_t$ is the state. The state no longer gives us precise information about the behavior of the system, but it summarizes as much as we can know, assuming we cannot know the noise. The degree to which the process is deterministic or random depends on the relative dependence of $\vec{F}$ on the state $\vec{x}$ relative to the noise n. It also depends on whether the noise is completely random, or whether it is correlated (either to $\vec{x}$ or to the noise at other times).

One source of noise is fluctuations in the environment. Nothing is completely isolated from its surroundings. There can be no complete isolation from long-range forces of nature such as gravitation and electromagnetism, thermal coupling to the environment is always present, and so forth. Rather than try to incorporate the entire universe into a dynamical system, interactions with the environment are lumped into a term that is labeled "noise." When the noise influences future dynamical behavior, as it does in Eq. (6.13), it is often called *dynamical noise*. When it is independent of the state x_t it can be taken as *additive* dynamical noise,

$$\vec{x}_{t+1} = \vec{F}(\vec{x}_t) + n_t. \tag{6.14}$$

Another source of noise is measurement. It is impossible to measure any quantity with infinite precision. No measurement ever pins down the state of a system to a point, but only restricts it to a region of state space as in Fig. 6.3. A ruler, for example, rounds a real number to an integer. The unknown remainder can be considered noise. Alternatively, the output of an electronic measuring instrument is uncertain because of complicated microscopic processes such as thermal motion. In the classical limit, a good measuring instrument does not effect the dynamics. In this case the "true" state $\vec{x}$ remains deterministic,

$$\vec{x}_{t+1} = \vec{F}(\vec{x}_t), \tag{6.15}$$

but the observed state $\vec{y}$ does not:

$$\vec{y}_t = \vec{x}_t + \varepsilon_t. \tag{6.16}$$

We will assume that the measuring instruments are decoupled from the dynamics, and call any random process in this form *observational noise*. Obser-

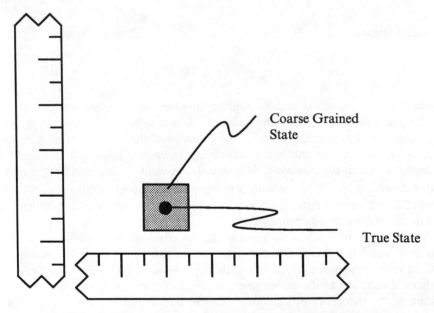

FIGURE 6.3. Observational noise arising from imprecise measurements.

vational noise is obviously much simpler than dynamical noise, but it is still by no means trivial. Iterated once, an observational noise process looks like

$$\vec{y}_{t+1} = \vec{F}(\vec{y}_t - \varepsilon_t) + \varepsilon_{t+1}. \tag{6.17}$$

The dynamics of $\vec{y}$ is much more complicated than that of $\vec{x}$. Though the dynamics of additive dynamical noise may seem similar to those of observational noise, observational noise does not change the underlying deterministic state of a system, whereas dynamical noise does. For example, dynamical noise can knock a system from one basin of attraction into another. Observational noise cannot.

Example 6.2 Pure white noise. A time series whose values x_t are IID is sometimes called *pure white noise*. White noise implies that the random variables x_t are linearly independent (or uncorrelated; see section 6.5). The two should not be confused, as they can be quite different. The iterates of the logistic map at $r = 4$, for example, are white noise, but certainly not pure white noise.

6.4   Averaging

6.4.1   Stationarity

A random process is stationary if its statistical properties are independent of time. This is true if the joint probability measure that characterizes it is independent of time:

$$p(x_t, \ldots, x_{t+d}) = p(x_{t'}, \ldots, x_{t'+d}). \tag{6.18}$$

A limit cycle, for example, is not stationary, because its properties depend on the phase of the cycle. However, by taking $t - t'$ to be the period of the cycle (which is equivalent to reducing the cycle to a map through a strobo-scopic section), it becomes stationary. A more fundamental type of nonstationary behavior occurs for systems with systematic trends. For example, so far the population of the earth has generally increased, and is not stationary.[3]

Stationarity is often a question of time scales. In most climates, a sequence of temperature measurements made over a 12-hour period is not stationary—

---

[3] Some nonstationary processes can be reduced to stationary processes by an appropriate trans-formation. For a positive random variable with stochastic exponential growth, for example, this can be done by taking the difference of the logarithms of successive values.

there is a very clear difference between the beginning and the end of the time series. Temperature measurements made over millions of years, on the other hand, might look quite stationary—providing the earth is not systematically heating up or cooling down over that period of time. In practice it is often difficult, if not impossible, to distinguish between a temporary "fluctuation" and a real trend. Motion on a chaotic attractor is stationary, but this may not be apparent for observations made over only a short period of time.

For a deterministic dynamical system stationarity is related to the existence of invariant sets. Loosely speaking, except for the possibility of periodicity, motion on invariant sets is stationary, whereas transient motion is nonstationary.

Example 6.3 A random walk. The linear random process $x_{t+1} = x_t + z_t$, where $\{z_t\}$ is pure white noise, is called a *random walk*. If the mean and variance of $\{z_t\}$ are μ and σ^2, respectively, the mean and variance of x_t are $t\mu$ and $t\sigma^2$. The process is nonstationary, because the mean and variance are time dependent.

6.4.2   Time Averages and Ensemble Averages

When determinism breaks down we must resort to statistical averages. This can be done either in terms of time averages or ensemble averages. A time average is taken over a single experiment, whereas an ensemble average is taken over many identical experiments.

The time average of a function f along a trajectory $\vec{x}(t)$ is

$$\langle f(\vec{x}(t)) \rangle = \lim_{T \to \infty} \frac{1}{T} \int_0^T f(\vec{x}(t)) dt. \qquad (6.19)$$

The time average often depends on initial conditions—different trajectories produce different time averages. Nonetheless, in many cases there are families of trajectories with the same statistical properties; for example, when the dynamical system has attractors.

It then becomes useful to define a probability measure, and use it to compute ensemble averages.

Ergodicity and Dynamical Measures

Probability measures play an important role in deterministic as well as stochastic dynamical systems. From one point of view it might seem that probability is irrelevant to deterministic dynamics, because the properties of a deterministic dynamical system are "determined" by the initial conditions. This neglects measurement.

An *invariant measure* is one that does not change under the action of the dynamics. More precisely, a measure μ is invariant under the map F if, for any set S in the support of μ,

$$\mu(S) = \mu(F^{-t}(S)). \tag{6.20}$$

This is really nothing more than a statement that the combination μ and F preserve probability. In other words, a measure is invariant if the total probability associated with a given set is equal to the probability of the sets that are mapped into it.

A dynamical system may have many invariant measures. An *ergodic* measure μ is one that corresponds to time averages, so that

$$\langle f(\vec{x}(t)) \rangle = \lim_{T \to \infty} \frac{1}{T} \int_0^T f(\vec{x}(t)) dt = \int f(\vec{x}) d\mu(\vec{x}). \tag{6.21}$$

For example, any cycle has an invariant ergodic measure whose probability density is a delta function centered on the cycle, multiplied by a weighting function that is inversely proportional to the speed with which the trajectory traverses the cycle. Using this measure to take an average over the points on the cycle will give the same result as an infinite time average along the cycle.

The word "ergodic" is used in many different senses, and is often confusing. The term originates in statistical mechanics. For a physical system, the "ergodic hypothesis" states that in the limit as the number of particles goes to infinity, ensemble averages taken with respect to Lebesgue measure on the energy surface are equal to time averages. For the ergodic hypothesis to be true, the orbit must cover the energy surface uniformly. The ergodic hypothesis is not true in general, even for infinite dimensional systems. There are many examples, both from theoretical analysis and numerical experiments, of Hamiltonian systems that are not ergodic in this sense. The chaotic orbits of finite dimensional Hamiltonian systems often cover much of the energy surface, but they do not usually cover all of it.[4] There are, nonetheless, invariant sets on the energy surface. Relative to Lebesgue measure on these invariant sets, the trajectories are ergodic.

The word ergodic is applied to trajectories, measures, and invariant sets. Ergodic is sometimes defined to mean "covering" a given set. An orbit is often called "ergodic" if it is recurrent; that is, if it comes back arbitrarily close to itself after a sufficient period of time. Quasiperiodic flow on the torus, for example, is in this sense ergodic on the torus. An orbit that originates outside the torus and is attracted onto it is not ergodic in this sense, because the early parts of the orbit never return near to their starting value. Orbits on

---

[4] For the limit of small nonlinearity this can be proven by the KAM theorem [5].

the torus may not be ergodic relative to a higher dimensional energy surface in which the torus is embedded.

Ergodicity is strongly associated with stationarity, and under some definitions they are essentially equivalent [4]. Behavior that is ergodic under one definition may not be ergodic under another. For example, according to the definition of ergodicity we have presented, the invariant measure on a cycle is an ergodic measure. There are other definitions of "ergodic" such that cycles are *not* ergodic, because of the dependence on phase.[5] There is no universally acceptable definition of the word "ergodic." When we say that a measure is ergodic, we always mean this in the sense defined by Eq. (6.21). We will say that a trajectory is ergodic if it has an ergodic measure.

Just because a measure is ergodic does not mean that it is relevant to the time averages *that describe typical orbits*. For example, a chaotic attractor typically contains an infinite number of invariant measures, associated with each of the unstable cycles embedded in the attractor. Because these unstable cycles are repellors rather than attractors, they do not characterize typical orbits. Only initial conditions that begin on a cycle remain on a cycle—the slightest perturbation sends the state off the cycle and onto the attractor. The time average of typical trajectories are described by the natural measure.

A *natural measure* is an ergodic measure that describes time averages for a set of initial conditions of positive Lebesgue measure in the state space. One dynamical system may have many natural measures, associated with different families of trajectories. In a dissipative dynamical system, for example, each attractor has its own natural measure. Invariant sets that occupy positive Lebesgue measure on the energy surface of a Hamiltonian dynamical system also have a natural measure. For an attractor the natural measure is distinguished from other measures because it is stable to low-level dynamical noise [6]. That is, if the deterministic dynamical system is replaced by one with some additive dynamical noise, as long as the noise is small enough, the natural measure of the noisy system will remain similar to that of the deterministic system.

For example, suppose we suddenly observe a dynamical system for the first time. What is the probability of finding it in a given state? If the motion is on an invariant set, such as a chaotic attractor, then the frequency with which the orbit visits a given region of the attractor tells us the probability of measuring it to be in that region.

This frequency defines a measure on the attractor. The natural measure can be estimated numerically in terms of the frequency with which trajectories visit different parts of the state space; for example, by sampling at uniform time intervals and using a histogram or kernel density estimator.

---

[5] As already mentioned, the nonstationarity of a cycle is really a technical complication that is removed by making observations at multiples of the fundamental period, and of less fundamental importance than other types of nonstationarity.

Example 6.4 The invariant measure for the tent map (Eq. 5.43). Consider a set A in the interval $(0, 1/2)$ with Lebesgue measure $\mu(A)$. It is the image of two disjoint sets, each with Lebesgue measure $\mu(A)/2$: B_1 in the interval $(0, 1/2)$ and B_2 in the interval $(1/2, 1)$. That is, both B_1 and B_2 are stretched out to twice their original size by the map, because the magnitude of its derivative is two everywhere. Because there are exactly two preimages of A each of half the size, the measure of A is the same as that of its preimage for Lebesgue measure. Thus, Lebesgue measure is invariant under the map. Notice that the measure of the set B_1 is not invariant. That is, $\mu(F(B_1)) \neq \mu(B_1)$. This is the reason for defining an invariant measure in terms of the preimage $F^{-1}(S)$ instead of the image $F(S)$.

Example 6.5 The invariant measure for the logistic map at $r = 4.0$. For this parameter value, the invariant measure can be obtained analytically using the map's equivalence to the tent map (Eq. 5.44). Under the transformation $x = \sin^2 \pi\theta$, Lebesgue measure on θ transforms as:

$$d\theta = \frac{dx}{\sqrt{x(1 - x)}}. \tag{6.22}$$

Normalizing, the invariant probability density function is

$$p(x) = \frac{1}{\pi\sqrt{x(1 - x)}} \tag{6.23}$$

as plotted in Fig. 6.4. This density function is infinite at $x = 0$ and $x = 1$. The infinity is caused by the critical point, where the slope is zero, at $x = 1/2$. Linearizing the map at the critical point gives:

$$x_{t+1} \approx 1 + \left(x_t - \frac{1}{2}\right)\frac{dx_{t+1}}{dx_t}\bigg|_{1/2} = 1, \tag{6.24}$$

independent of x_t. Thus, in the linear approximation, an *interval* around the critical point is mapped to unity, so the density becomes large there. The density at 1 is carried by the mapping to $F(1) = 0$. At $r = 4$ the point $x = 0$ is an unstable fixed point, so "the infinity stops here." At other values of r, however, the orbit of the critical point is itself chaotic, and the structure of the measure is quite complicated,[6] as is evident from the histogram approximation to the invariant measure for $r = 3.7$, shown in the top left corner of Fig. 6.6 (see Example 6.6).

---

[6] Note that the natural invariant measure of the logistic map for chaotic parameter value is not singular, because the infinities of the density function are integrable.

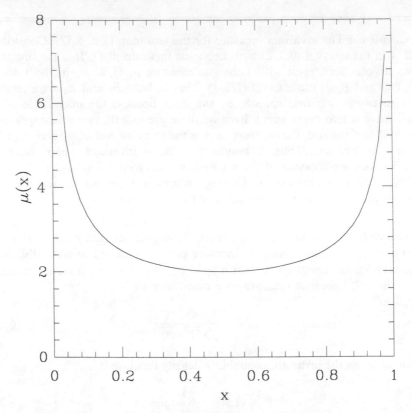

FIGURE 6.4. The invariant density for the logistic map at $r = 4$.

6.4.3 Mixing

To make ensemble averages equal time averages, it may be necessary to prepare the ensemble in a very special way. Consider, for example, regular
motion on a torus. In action-angle variables the natural ergodic measure is
equivalent to Lebesgue measure. Suppose we wish to compute a statistical
average on a parallel computer, and we want to do this as accurately and
efficiently as possible. To speed up the computation, we integrate the equations separately on each processor, choosing a different initial condition for
each. *Providing we pick the initial conditions according to the natural measure*,
we can then average the result for each processor together, and our final
answer will be more accurate than that for one processor by itself. If, however, we choose the initial conditions according to some other measure, our
answer may not be as accurate. For example, if we happen to choose our
ensemble of initial values so they are bunched together on the torus, because
the motion is regular they will remain bunched together as in Fig. 6.5. The

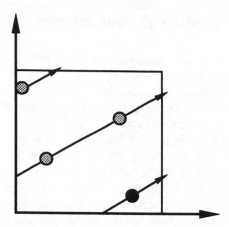

FIGURE 6.5. An ergodic, but not mixing, system. Here a torus has been unfolded onto the plane. Trajectories are regular, so the ball of initial conditions in black does not spread out as it evolves (gray). All initial conditions in the ball follow similar trajectories.

averages computed by the individual points in the ensemble will therefore be highly correlated with each other, because each point in the ensemble samples more or less the same part of the torus. In contrast, when the initial conditions are chosen properly, the samples will be statistically independent, and the ensemble average will be more accurate than that of a single point.

If, in contrast, we performed the experiment for an invariant set that is mixing, the answer would be insensitive to the choice of initial conditions. For a chaotic attractor, for example, even if we choose the points of the initial ensemble very close together, they are soon "mixed over the attractor" by the action of the dynamics, as shown in Fig. 5.22. Once the sensitive dependence on initial conditions has done its work, the trajectories are dispersed "randomly" throughout the attractor. Because they all sample different parts of the attractor, the averages they compute are statistically independent.

A map F is *mixing* over an invariant set S if for any two subsets A and B,

$$\lim_{t \to \infty} \mu(F^t(B) \cap A) = \mu(A)\mu(B). \tag{6.25}$$

If we pick two points in S at random, *and the choices are independent of each other*, then the probability that one point is in A and the other is in B is given by the right-hand side of this equation. The term on the left describes the extent to which the map "mixes up" B with A, so that its intersection is what one would expect by choosing a point at random. Mixing occurs when these two are equal; that is, when the flow causes the statistical properties of the ensemble to be independent of its initial properties. Mixing over finite times only occurs in a coarse-grained description.

6.5   Characterization of Distributions

Probability distributions are functions. Functions have an infinite number of degrees of freedom, and they are cumbersome to work with. A function contains too much information to summarize in a few words. A function in one or two dimensions can perhaps be qualitatively understood through a picture, but in more than two dimensions even the eye is unable to communicate the information needed to understand a function. Whenever possible it is much more convenient, both conceptually and computationally, to summarize the properties of a function by a few numbers. In some cases this can be quite effective. For example, for a sharply peaked function the mean and the width give a very good idea of its properties. The domain where an accurate measurement results in a density function that remains sharply peaked is precisely that in which deterministic models work well. For a complicated density function, with many peaks and valleys spread over a large domain, the mean and variance give little useful information.

A function that reduces a probability measure to a number is called a *statistic* or *statistical property*. The most widely studied statistical properties are the first and second moments and their Fourier transforms. These are both linear statistics. For studying nonlinear behavior, they are inadequate.

6.5.1   Moments

The most common way to characterize a distribution is through its moments. The nth moment of a random variable x is $\langle x^n \rangle$, the average value of x raised to the nth power. The first moment is known as the *mean*: $\bar{x} \equiv \langle \bar{x} \rangle$. For convenience it is often implicitly assumed that the mean is zero, or that all moments are calculated about the mean. In this case the *variance* $\sigma^2 \equiv \langle (x - \bar{x})^2 \rangle$ reduces to the second moment. The *standard deviation* σ is the square root of the variance.

Each moment gives some information about a distribution. For smooth distributions, the infinite set of moments gives complete information about the distribution via the characteristic function,

$$\phi(k) \equiv \sum_{n=0}^{\infty} \frac{(ik)^n}{n!} \langle x^n \rangle \tag{6.26}$$

$$= \langle e^{ikx} \rangle = \int e^{ikx} p(x) dx. \tag{6.27}$$

When the series converges, $\phi(k)$ is just the Fourier transform of the probability density $p(x)$, and the density can be recovered from it via the inverse transform.

Example 6.6 The Gaussian and the Central Limit Theorem. The Gaussian density function is:

$$G(x) = \frac{1}{\sqrt{2\pi}\sigma} \exp(-(x-\bar{x})^2/2\sigma^2) \tag{6.28}$$

with mean $\bar{x}$ and standard deviation σ. Because these are the only two parameters in the definition, they determine all higher moments.[7] The Gaussian is important because of the Central Limit Theorem [7], which states that the distribution of a sum of IID random variables with variance σ^2 and mean $\bar{x}$ approaches a Gaussian in the limit as the number of terms goes to infinity:

$$\lim_{n \to \infty} P\left(\frac{1}{n}(x_1 + x_2 + \ldots + x_n)\right) = G(x). \tag{6.29}$$

The usual interpretation of this is that for large n, the sum above will be a random variable drawn from a Gaussian distribution with mean μ and variance σ^2/n. In practice, the central limit theorem remains valid under much more general conditions. Practically speaking, the sum of almost any random variables approaches a Gaussian as long as each is drawn from a distribution with a finite mean.

Example 6.7 The Central Limit Theorem for the logistic map. The iterates of the logistic map are identically distributed, but they are far from independent. The density function for a single iterate in some parameter regimes contains an infinite number of integrable singularities. Nonetheless, at any fixed level of coarse graining, the density function of a sum of iterates of the logistic map, defined as:

$$P_N(x) = \int_0^1 \sum_{t=1}^N f^t(x_1)\mu(x_1)dx_1, \tag{6.30}$$

appears to approach a Gaussian in the limit as $N \to \infty$. N is the number of iterates, f is the map, and μ is the natural measure. This is illustrated in Fig. 6.6. For any given value of N, providing the map has been iterated sufficiently accurately, it would be possible to see fine structure by taking a finer coarse graining. However, by increasing N this structure disappears if the resolution of the coarse graining is held fixed.

---

[7] The Gaussian is by no means the only distribution with this property.

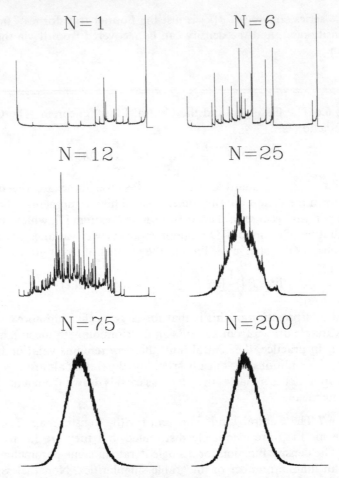

FIGURE 6.6. The central limit theorem. A histogram of the sum of iterates of the logistic map for $r = 3.7$. The density function for any given N is defined by Eq. (6.30). The x axis is renormalized for each value of N to keep the width of the distribution constant. For large N the histogram approaches a Gaussian, suggesting that the central limit theorem holds for the coarse grained distribution even though the random variables are not statistically independent.

Perhaps the two most important linear statistics used to characterize joint probabilities are the two-point correlation (or normalized covariance), and the power spectrum.

Correlation Functions

The *covariance* between two zero-mean random variables is the mean value of their product, where the mean is calculated with respect to their joint distribution:

$$Cov(x, y) = \langle xy \rangle. \tag{6.31}$$

For random variables with nonzero mean, the relevant mean is subtracted from each random variable first:

$$Cov(x, y) \equiv \langle (x - \bar{x})(y - \bar{y}) \rangle. \tag{6.32}$$

As its name implies, the covariance describes how the variance about the mean of one variable is related to that of another. It is useful to scale the covariance so that it is between 1 and -1. This normalized variable is called the *correlation*.[8]

$$C(x, y) \equiv \frac{\langle xy \rangle}{\sqrt{\langle x^2 \rangle \langle y^2 \rangle}} = \frac{\langle xy \rangle}{\sigma_x \sigma_y}. \tag{6.33}$$

The correlation of two statistically independent variables is zero. The converse is *not* true: the fact that the correlation is zero does not imply statistical independence. Two random variables whose correlation is zero are *linearly independent*. Note that the "degree of correlation" depends only on the absolute value of the correlation; variables with a correlation of -1 are just as correlated as those with a correlation of 1.

The *autocorrelation* is the correlation between a variable at one time and the same variable at another time.[9] The autocorrelation for a stationary time series $x(t)$ is:

$$C(\tau) = C(x(t), x(t - \tau)) = \frac{\langle x(t)x(t - \tau) \rangle}{\langle x^2 \rangle}. \tag{6.34}$$

Example 6.8 Temporal autocorrelations. The estimated autocorrelation functions for the data in Fig. 6.2 are displayed in Fig. 6.7. Note that the autocorrelation for the quasiperiodic system does not decay to 0 whereas that for the chaotic system does. For some chaotic systems it can be proved that the autocorrelation function decays exponentially as $\tau \to \infty$, but in general the time behavior of the autocorrelation function is still an open question.

Example 6.9 The correlation between linearly related variables. Suppose $y = \beta x + \eta$, where x and η are independent zero mean random variables with standard deviations σ_x and σ_η, respectively. Note that "taking the expectation value" is a linear operation, so

$$\langle xy \rangle = \beta \langle x^2 \rangle + \langle x\eta \rangle = \beta \langle x^2 \rangle. \tag{6.35}$$

The last term vanishes because x and η are independent. Similarly, σ_y is related to σ_x and σ_η:

$$\sigma_y^2 = \langle y^2 \rangle = \langle (\beta x + \eta)^2 \rangle = \beta^2 \sigma_x^2 + \sigma_\eta^2. \tag{6.36}$$

---

[8] For convenience we assume zero mean unless otherwise stated.
[9] The variable t need not be time; it can be any variable that places the values of x in sequential order.

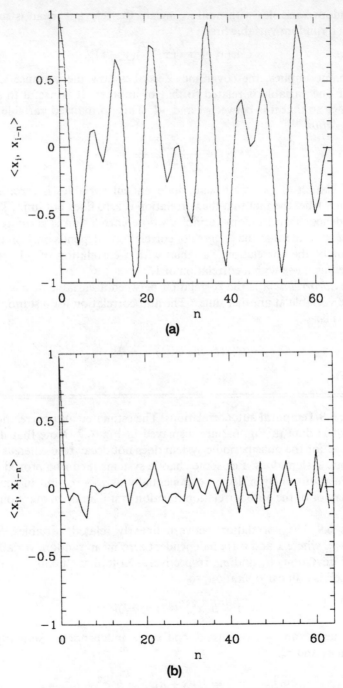

FIGURE 6.7. Autocorrelation functions for data in Fig. 6.2. (a) quasiperiodic; (b) logistic.

Finally, substituting into the definition of the correlation yields:

$$C(x, y) = \frac{\langle xy \rangle}{\sigma_x \sigma_y} = \left\{ 1 + \frac{\sigma_\eta^2}{\beta^2 \sigma_x^2} \right\}^{-1/2}. \tag{6.37}$$

Example 6.10 Linear random processes. A *linear random process* is one whose deterministic part is a linear map. For a *moving average process of order q* (abbreviated $MA(q)$), x_t is written as a finite average of a pure white noise process Z_t

$$x_t = \sum_{i=0}^{q} \beta_i z_{t-i}. \tag{6.38}$$

The auto-correlation between x_t and x_{t-m} is:

$$C(\tau) = \sum_{i=0}^{\infty} \beta_i \beta_{i-\tau} \sigma_z^2. \tag{6.39}$$

For an *autoregressive process of order q*, abbreviated $AR(q)$, x_t is written as a linear combination of its previous q values, together with some additive noise:

$$x_t = \sum_{i=1}^{q} \alpha_i x_{t-i} + z_t. \tag{6.40}$$

An ARMA process of order (p, q) is a linear combination of an $AR(p)$ process and an $MA(q)$ process. In principle these are all equivalent. AR and MA processes can be transformed into each other. The transformation usually causes the order to become infinite, however.

Example 6.11 Gaussian random process. A Gaussian random process is one for which the joint distribution function is Gaussian. The Gaussian random process is a linear process, as shown in detail in the example below. Gaussian processes have the important property that a linear combination of Gaussian variables is itself a Gaussian random variable. A linear transformation of a Gaussian random process remains a Gaussian process, as shown in the next example.

Example 6.12 Linearity of Gaussian processes. The bivariate Gaussian density is

$$G(x, y) = \frac{\sqrt{\alpha\gamma + \beta^2}}{\pi} \exp\{-[\alpha x^2 + 2\beta xy + \gamma y^2]\}. \tag{6.41}$$

It is easy to show by direct integration that

$$\sigma_x^2 = \gamma/2(\alpha\gamma - \beta^2);$$

$$\sigma_y^2 = \alpha/2(\alpha\gamma - \beta^2); \tag{6.42}$$

$$\text{and} \quad \sigma_{xy} \equiv \langle xy \rangle = -\beta/2(\alpha\gamma - \beta^2).$$

Notice that the density can be written as:

$$\frac{(\text{Det} \sum)^{-1/2}}{2\pi} \exp\left\{ -\frac{1}{2}(x \quad y) \sum^{-1} (x \quad y)^T \right\}, \tag{6.43}$$

where

$$\sum = \frac{1}{2(\alpha\gamma - \beta^2)} \begin{pmatrix} \gamma & -\beta \\ -\beta & \alpha \end{pmatrix} \tag{6.44}$$

$$\Rightarrow \sum^{-1} = 2 \begin{pmatrix} \alpha & \beta \\ \beta & \gamma \end{pmatrix}. \tag{6.45}$$

Then $\sigma_x^2 = \sum_{xx}$, $\sigma_{xy} = \sum_{xy}$, and so forth. The matrix $\sum$ is the *covariance matrix*. As in the one-dimensional case, linear correlations define the entire distribution. In particular, note that *for a Gaussian*, linear independence implies statistical independence; for example:

$$\langle xy \rangle = 0 \Rightarrow \beta = 0 \quad \text{and} \quad p(x, y) = p(x)p(y). \tag{6.46}$$

Thus, the correlation between variables with a multivariate Gaussian distribution accurately summarizes their statistical dependence. To make this more precise, consider the conditional probability for x given y:

$$p(x|y) = \frac{p(x, y)}{p(y)}, \tag{6.47}$$

where

$$p(y) = \int p(x, y)\, dx = \sqrt{\frac{\alpha\gamma - \beta^2}{2\pi\alpha}} \exp\left\{ -\left(\gamma - \frac{\beta^2}{\alpha} \right) y^2 \right\}. \tag{6.48}$$

Which implies

$$p(x|y) = \sqrt{\frac{\alpha}{2\pi}} \exp\left\{ -\alpha\left(x - \frac{\beta}{\alpha}y \right)^2 \right\}. \tag{6.49}$$

But this is a Gaussian with mean

$$\bar{x} = \frac{\beta}{\alpha}y = \frac{\sigma_x}{\sigma_y}\langle xy \rangle \cdot y. \tag{6.50}$$

In other words, the mean value of x is a linear function of y whose coefficient is proportional to the correlation.

A stationary Gaussian process can thus be modeled by a linear random process whose coefficients are the autocorrelations:

$$x_{t+1} = \sum_{i=0}^{\infty} C(i+1)x_{t-i} + n_t. \qquad (6.51)$$

Power Spectra

The power spectrum decomposes a signal into its power at each frequency. For a physical system, the power spectrum describes how the power is divided among different frequencies of oscillation. For a set of linearly coupled harmonic oscillators, for example, the power spectrum is peaked at linear combinations of the frequencies of the oscillators. Although the power spectrum is motivated by physical problems where energy and power are well defined, it is a useful quantity even in an abstract setting, where there is no connection to physics.

The Fourier transform $\hat{f}(\omega)$ of a function $f(t)$ is

$$\hat{f}(\omega) \equiv \frac{1}{\sqrt{2\pi}} \int_{-\infty}^{\infty} e^{i\omega t} f(t) \, dt. \qquad (6.52)$$

Providing this integral converges the Fourier transform can be inverted, and contains all the information about f. For a bounded signal for which the Fourier transform converges, the square of the modulus of the Fourier transform, $|\hat{f}(\omega)|^2$, describes how much energy is present at each frequency.[10]

We are often interested in signals that continue indefinitely, which we imagine to extend from $-\infty$ to ∞. In this case there is an infinite amount of energy in the signal, and the Fourier transform may not exist. This is solved by computing the Fourier transform over a finite time interval, say from 0 to T,

$$\hat{f}_T(\omega) \equiv \frac{1}{\sqrt{2\pi}} \int_{0}^{T} e^{i\omega t} f(t) \, dt, \qquad (6.53)$$

and dividing the energy by the time to compute the periodogram

$$S(\omega, T) = \frac{|\hat{f}_T(\omega)|^2}{T}. \qquad (6.54)$$

For a periodic signal $S(\omega, T)$ converges to a well defined value as $T \to \infty$. For a physical system this is "energy per unit time," or power, at each frequency ω.

For random processes convergence is more subtle. In general for a random

---

[10] Of course, for this to be the true energy, this must be put in the proper physical units.

process the limit $\lim_{T\to\infty} \hat{f}_T(\omega)$ does not exist, because the phases of the Fourier transform are random and average to zero. Because the periodogram depends on the modulus, it does not have this problem. However, for a Gaussian random process the variance of the periodogram $S(\omega, T)$ is greater than the square of the mean. The values at nearby frequencies are uncorrelated, even for large T. Thus, the periodogram does not converge in the limit as $T \to \infty$. To make it converge it is necessary to take an ensemble average first, and then take the limit as time goes to infinity. We can thus define the *power spectrum* $S(\omega)$ as:

$$S(\omega) = \lim_{T\to\infty} \langle S(\omega, T) \rangle. \qquad (6.55)$$

In practice, the power spectrum is usually computed by taking the Fourier transform of several realizations of the time series, each for the longest possible time T. The periodogram of each realization is then averaged together to get an estimate of the power spectrum. An alternative, when T can be made very large, is to obtain convergence by averaging the computed values of $S(\omega, T)$ at nearby frequencies.

The power spectrum can be defined alternatively in terms of the correlation function. The Wiener–Khinchine theorem shows that the power spectrum is the Fourier transform of the autocorrelation function. Heuristically, the argument is as follows:

$$S(\omega) = \frac{1}{2\pi} \int e^{-i\omega t} f(t) e^{i\omega t'} f(t') \, dt \, dt'. \qquad (6.56)$$

Changing variables to $\tau = t - t'$,

$$S(\omega) = \int e^{i\omega\tau} f(t) f(t-\tau) \, dt \, d\tau = \int e^{i\omega\tau} C(\tau) d\tau. \qquad (6.57)$$

This is not surprising, because if a function is nearly periodic with period T, which is what the Fourier transform measures, then a time translation of the function by T will be highly correlated with the original function, which is what the autocorrelation measures.

Through generalizations of the Wiener–Khinchine theorem it is also possible to define cross spectra for two signals $x(t)$ and $y(t)$ by Fourier transforming the cross correlation $\langle xy \rangle$. By using higher-dimensional Fourier transforms it is possible to define multifrequency spectra associated with higher moments such as $\langle x^2 y \rangle$, or multipoint autocorrelation functions such as $\langle x(t)x(t-\tau_1)x(t-\tau_2) \rangle$.

Defining the power spectrum based on the correlation function avoids the convergence problems associated with defining the power spectrum directly in terms of the Fourier transform. However, computationally this is a poor approach. Computation of the correlation function directly from its definition has problems because when the correlation function is not zero the statistical fluctuations at nearby times are correlated. The estimate therefore converges slowly. Instead, one can compute the correlation function by first computing the Fourier transform of many realizations, averaging them together to get the power spectrum, and then Fourier-transforming the power spectrum to get the correlation function. This gives a better statistical estimate.

For discretely sampled experimental data the integrations above must be replaced by sums, using the discrete Fourier transform (DFT). Discretization introduces aliasing, in which power from one frequency contributes to the power spectrum at a different frequency. This is familiar from old cowboy movies, in which the wheels of a stagecoach often appear to rotate backward even when the horses are at full gallop. This is due to the finite sampling rate of the movie camera. With a finite uniform sampling interval τ it is impossible to distinguish oscillations with period greater than τ from those with period less than τ. The critical frequency $f_c = 1/(2\tau)$ is called the *Nyquist frequency*. With a finite sampling rate the power at frequencies above f_c is reflected or *aliased* about f_c. Time intervals shorter than $\tau/2$ cannot be distinguished from those greater than $\tau/2$. Aliasing can only be eliminated by increasing the sampling rate so that the Fourier coefficients above f_c vanish.

It is also worth keeping in mind that for a finite sampling time T even a perfectly periodic signal will produce a peak of finite width in its Fourier transform. This comes about because the signal is effectively "switched on" at $t = 0$ and "switched off" at $t = T$. The process of switching the signal on and off creates a transient, and broadens the peak. (This is the underlying reason for the uncertainty principle of wave mechanics.) Of course, the best way to reduce this effect is to increase the sampling time. It can be also reduced somewhat by "windowing" the signal, or switching it on and off gradually. For a signal with many periodic components broadened, the peaks may blend together and make it difficult to distinguish a spectrum consisting of sharp lines from one with a broad background.

The power spectrum is a useful tool for the analysis of dynamical systems data. The power spectrum of a regular time series consists of sharp lines, whereas the power spectrum of a chaotic time series always has a broadband component. Of course, for a quasiperiodic time series with many frequencies, in a finite time sample it may be difficult to distinguish the peaks from a broad background. In practice, however, in experiments where it is possible to record long time series, it is usually possible to distinguish three or four frequency quasiperiodic motion from chaotic motion.

Example 6.13 The sample power spectra of the data in Fig. 6.2 are displayed in Fig. 6.8. Notice that the regularity apparent in the quasiperiodic data is easy to detect in its power spectrum.

Example 6.14 Figure 5.26 shows the power spectrum of the x Rössler attractor at several different parameter values. In each case the power spec-

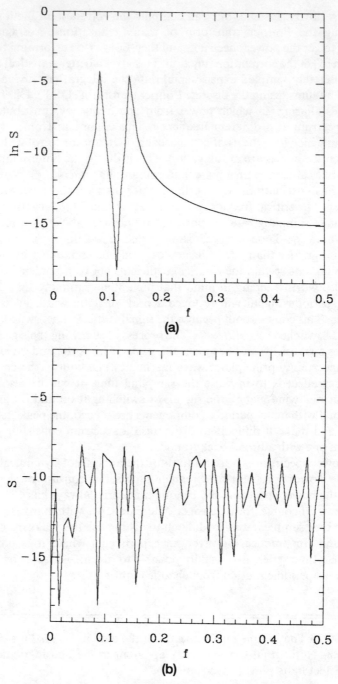

FIGURE 6.8. Ensemble averaged power spectra of the data in Fig. 6.2. (a) quasi-periodic; (b) logistic.

tra were computed using 8192 sampled points, and 50 realizations were combined in an ensemble average to produce a statistically reasonable estimate of the power spectrum.

The information contained in the power spectrum and the autotcorrelation function are equivalent, because they are just Fourier transforms of each other. The autocorrelation is a second moment of the joint probability $p(x(t), x(t - \tau))$. It is an incomplete characterization of the time series. It contains enough information to specify all the characteristics of a linear random process, but it does not contain information about the nonlinear properties. This can be very misleading. For example, the logistic map at $r = 4$ has an autocorrelation function $C(\tau) = 0$ unless $\tau = 0$. But this is a deterministic system! The autocorrelation function tells nothing about determinism, or about general statistical dependence—it only tells about *linear* statistical dependence. The autocorrelation function tells us how much of the properties of the system we could hope to capture if we modeled it as a linear random process.

There are two lessons here. The first is that if we want to model chaotic dynamical systems, we must turn to nonlinear models. This is addressed in the next chapter. The second is that we need to use other statistical measures to detect nonlinear deterministic behavior. One possibility is to use higher nonlinear moments. Another is to use other statistics; for example, the mutual information. The next section describes the information-based approach, and compares it to that based on higher-order moments.

6.5.2 Entropy and Information

The *entropy $H(p)$* of a probability density p is

$$H(p) \equiv - \int p(x) \log p(x) dx. \tag{6.58}$$

The entropy of a distribution over a discrete domain is[11]:

$$H(p) \equiv - \sum_i p_i \log p_i. \tag{6.59}$$

The units of entropy are bits, digits, or nats if the logarithm is taken base 2, 10, or e, respectively. The entropy describes the extent to which the distribution is concentrated on small sets. If the entropy is low the distribution is

---

[11] The continuum limit of the entropy for a discrete distribution is not the entropy of the corresponding continuous density. The entropy we have defined for the continuous case is sometimes called the *entropy density*. The coordinate dependence of entropy in the continuous case reflects its partition dependence in the discrete case.

concentrated at a few values of x. It may, however, be concentrated on *several* sets far from each other. Thus, the variance can be large while the entropy is small. Writing the entropy as

$$H(p) = -\langle \log p(x) \rangle, \tag{6.60}$$

we see that it is the average over something that depends explicitly on p. The entropy can thus be viewed as a *self moment* of the probability, in contrast to the ordinary moments, for example, $\langle x^2 \rangle$, which are averages over quantities that do not depend on p. The entropy and other self-moments contain a very different kind of information than the ordinary moments.

Entropy measures the degree of surprise one should feel upon learning the results of a measurement. It counts the number of possible states, weighting each by its likelihood. For a uniform distribution over two states (e.g., a fair coin toss)

$$H(p) = -\sum_{i=0,1} p_i \log_2 p_i = -2\left(\frac{1}{2} \log_2 \frac{1}{2}\right) = 1 \text{ bit.} \tag{6.61}$$

Joint and conditional entropies are defined in an analogous way based on the corresponding distributions:

$$H(x,y) = -\int p(x,y) \log(p(x,y)) dx\, dy \quad \text{and} \tag{6.62}$$

$$H(x|y) = -\int p(x|y) \log(p(x|y)) dx. \tag{6.63}$$

The negative of entropy is sometimes called information: $I(p) = -H(p)$. For our purposes the distinction between the two is semantic. Entropy is most often used to refer to measures, whereas information typically refers to probabilities.

For two random variables x and y, the *mutual information* states how much information y gives about x that is not present in x alone.

$$I(x;y) = H(x) - H(x|y). \tag{6.64}$$

$H(x)$ is the uncertainty in x; $H(x|y)$ is the uncertainty in x given y. Because knowing y cannot make x more uncertain, $H(x) \geq H(x|y)$. The mutual information is therefore nonnegative. The mutual information tells how much the uncertainty of x is decreased by knowing y.

By writing the conditional distribution in terms of the joint distribution, the mutual information can also be written

$$I(x;y) = H(x) + H(y) - H(x,y) \quad \text{or} \tag{6.65}$$

$$= I(x,y) - I(x) - I(y). \tag{6.66}$$

This makes it clear that the mutual information is a symmetric function of x

and y. The information in the joint distribution should not be confused with the mutual information, which can be viewed as a nonlinear generalization of the correlation. Indeed, the mutual information is a normalized version of the joint information, much as the correlation is a normalized covariance:

$$e^{I(x;y)} = \frac{e^{I(x,y)}}{e^{I(x)}e^{I(y)}}.$$ (6.67)

The entropy of a distribution depends on coordinates. The mutual information, in contrast, depends on the difference of two entropies, and is independent of coordinates.

Mutual information is a measure of statistical dependence. Two random variables have zero mutual information if and only if they are statistically independent. This can be seen from the following inequality:

$$I(x;y) = \int p(x,y) \log\left(\frac{p(x,y)}{p(x)p(y)}\right) \geq \int p(x,y)\left[1 - \frac{p(x)p(y)}{p(x,y)}\right]$$

$$\geq \int p(x,y) - \int p(x)p(y) = 0.$$ (6.68)

We have used the fact that

$$\log(x) \geq 1 - 1/x.$$ (6.69)

Equality holds in Eq. (6.69) only when $x = 1$, hence in Eq. (6.69) only when $p(x,y) = p(x)p(y)$. Two distributions with zero mutual information are statistically independent. Mutual information is thus a test for nonlinear independence, as opposed to correlation, which tests only for linear independence.

Example 6.15 Mutual information and χ^2. As Jaynes has pointed out, mutual information is closely related to more traditional chi-squared tests [8]. A straightforward way to test for independence is to see whether the densities $p(x,y)$ and $p(x)p(y)$ are the same. One natural way to do this is to introduce an idea of distance on probability space and measure the distance between $p(x,y)$ and $p(x)p(y)$. If this distance is zero then x and y are independent.

One measure of the distance between densities p and q is the χ^2 statistic.[12]

$$\chi^2 \equiv \int \frac{(p(x) - q(x))^2}{q(x)} \, dx.$$ (6.70)

---

[12] The χ^2 statistic is usually defined on data sets using a histogram estimate for p and q. This introduces an extra factor of N, the number of cells, into the equations.

Another possible distance measure is the Kullback information distance:

$$K(p,q) \equiv - \int q(x) \log \left(\frac{q(x)}{p(x)} \right) dx. \qquad (6.71)$$

The mutual information is the Kullback information distance between the distributions $p(x,y)$ and $p(x)p(y)$. Assuming that the two distributions p and q are nearly the same, the Kullback distance can be expanded in a Taylor series.

$$K(p,q) = - \int q \left\{ \left(\frac{p}{q} - 1 \right) - \left(\frac{p}{q} - 1 \right)^2 \Big/ 2 + \cdots \right\}$$

$$\approx \int q \left(\frac{p}{q} - 1 \right)^2 \Big/ 2 = \int \frac{(p-q)^2}{2q}. \qquad (6.72)$$

Thus, when two variables x and y are nearly statistically independent, to leading order the mutual information is proportional to the χ^2 statistic:

$$I(x;y) = K(p(x,y),p(x)p(y)) \propto \frac{1}{2}\chi^2(p(x,y),p(x)p(y)). \qquad (6.73)$$

The moments of a probability distribution, in contrast, contain very different information. The two-point autocorrelation function $\langle x(t)x(t-\tau) \rangle$ describes the linear dependence, the three-point autocorrelation $\langle x(t)x(t-\tau_1)x(t-\tau_2) \rangle$ describes quadratic dependence, and so forth. Although the behavior of any given higher-order moments describes nonlinear behavior, none of them alone characterizes the degree of determinism. None of the ordinary moments taken alone tests for statistical dependence. Although scattering experiments can measure two point correlations directly, in general it is very difficult to estimate higher-order moments from data.

An obvious nonlinear generalization of the autocorrelation function is the "auto-information" function, which is the mutual information between the probability densities for $x(t)$ and $x(t-\tau)$ [9]. Strictly speaking, for a deterministic process, this information is infinite for all τ, but as described above, the inevitable coarse-graining in the measurement process introduces an effective randomness into the process that leaves the information finite. For a chaotic process, the auto-information function decays as τ increases.

The calculated mutual information depends on the partition used in the coarse-graining, or equivalently, on the density estimation technique used. Nonetheless, we find the auto-information function to be of practical use for distinguishing uncorrelated processes from nondeterministic processes. For example, the autocorrelation function for the logistic map in Fig. 6.7b suggests that the process is not deterministic. The corresponding auto-information function in Fig. 6.9b displays the gradual loss of information about the system, showing quite clearly that there is some determinism in the process. For

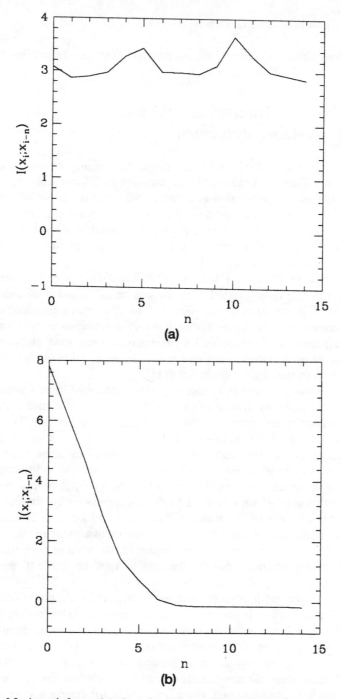

(a)

(b)

FIGURE 6.9. Auto-information functions for the data in Fig. 6.2. (a) quasiperiodic; (b) logistic. Compare these with the autocorrelation functions in Fig. 6.7. Note that the time scales have been stretched in this figure. The auto-information for the quasi-periodic case is roughly constant, as one would expect for a deterministic, nonchaotic system. The small fluctuations correspond to the fluctuations in the autocorrelation curve, though here they are due to estimation errors. For the logistic data, the auto-information clearly shows that successive iterates are not independent, although they are uncorrelated. The curve dips slightly below zero because of estimation error.

further comparison of mutual information to linear correlation functions, see Li [10].

6.6   Fractals, Dimension, and the Uncertainty Exponent

Dimension is one of the most basic properties of any geometrical object. Roughly speaking, it is related to the amount of information that must be supplied to specify the position of a point to a given degree of accuracy. For dynamics the dimension is particularly important because it provides a notion of *degrees of freedom*, the number of essential variables needed to model the dynamics. The dimension supplies the most elementary notion of the complexity of a geometric object.

The notion of dimension has many different facets. For example, one property commonly associated with dimension is "number of independent directions." The points of a three-dimensional Euclidean manifold are naturally described by three-tuples; they are not fully specified by two-tuples, and they overspecified by four-tuples. Furthermore, a three-dimensional manifold cannot be mapped continuously into a two-dimensional manifold. This leads to the notion of *topological dimension* [11].

Another set of properties commonly associated with dimension involves scaling under changes in resolution or size. For example, the volume of a cube varies as the third power of its side, whereas the "volume" of a square varies as the second. From this we deduce that a cube is three dimensional whereas a square is two dimensional. This notion of dimension is related to the pointwise dimension, defined below. It is also related to information: to specify the coordinates of a point more precisely, we must supply more information; the rate at which this must be done depends on the dimension. This leads to the concept of information dimension, which is equivalent to the pointwise dimension. We can examine the scaling of almost any quantity that varies as the resolution changes; for example, the scaling properties of the moments of a probability distribution can be used to define its generalized dimensions.

For simple geometric objects, such as the attractors characterizing regular motion, the dimension is unambiguous; any reasonable definition, based on any of the properties mentioned above, produces the same number. The dimension is equivalent to the number of degrees of freedom (i.e., the number of in dependent quantities that must be chosen to specify a state as a unique point). A fixed point has dimension zero, a limit cycle has dimension one, and a two-torus has dimension two. For a limit cycle, for example, once the phase is chosen, the motion is uniquely determined. Because we have the freedom to specify one number, we say that it has "one degree of freedom."

For more complicated geometric objects, such as chaotic attractors, the dimension is a more subtle concept that does not always agree with intuition.

The topological dimension is always an integer, but values of the other definitions mentioned here do not have to be integers. Objects with fractal properties often have fractional dimensions. In many cases definitions of dimension based on different properties yield different values. In fact, the differences between different dimensions can be used to characterize fractal structure.

Dynamical systems involve many different objects, which may have quite different dimensions associated with them. The dimension of an attractor, for example, may be significantly smaller than the dimension of the full state space. The dimension of the natural measure on an attractor may differ from that of its support. It is important to make these distinctions clear.

In the following we review several ways to define dimensions, what they mean, how they are related, and how they are used in the study of chaos.

6.6.1   Pointwise Dimension

Consider a geometric object whose dimension is unknown. One natural way to determine the unknown dimension is to examine the scaling of mass as a function of size. For example, if the mass is uniformly distributed over a smooth Euclidean manifold of dimension d, the total mass inside a ball of radius ε is proportional to ε^d, as long as ε is sufficiently small compared to the extent of the manifold. The dimension d can thus be determined by measuring the total mass, or measure, inside a series of balls with different radii. The dimension is the slope obtained by plotting the logarithm of the mass versus the logarithm of the radius.

If the mass is not uniformly distributed over the manifold, the slope may no longer agree with the dimension of the manifold. However, the scaling relationship as $\varepsilon \to 0$ can be used to *define* a dimension. For a measure μ, let $\mu(B_\varepsilon(x))$ be the measure of a ball of radius ε centered on point x. The pointwise dimension is

$$d_p(x) = \lim_{\varepsilon \to 0} \frac{\log \mu(B_\varepsilon(x))}{\log \varepsilon}. \qquad (6.74)$$

As indicated, it may depend upon the point x.

Example 6.16 The Cantor Set. A canonical example of a fractal set is the Cantor set, constructed by redistributing all the mass from the middle third of the unit interval, evenly onto the remaining two thirds, then redistributing the mass from the middle thirds of the remaining intervals, and so forth. This is usually called "removing" the middle third from the sets. The Cantor set contains no intervals and thus has Lebesgue measure zero. Consider a set of balls of radius $1/3^n$ centered at the origin. Each ball contains $1/2^n$ of the mass; thus, the pointwise dimension at the origin is given by:

$$\frac{1}{2^n} = \left(\frac{1}{3^n}\right)^d \tag{6.75}$$

$$d = \frac{\log 2}{\log 3}. \tag{6.76}$$

The Cantor set is exactly self-similar. That is, it is possible to scale a subset so that it is geometrically similar to the entire set. This is a very special property. Most fractals are not self-similar.

6.6.2   Information Dimension

The dimension is also related to the scaling relationship between the information needed to specify the position of a point and the accuracy to which the position is known. To be more precise, consider a coarse graining at resolution ε, for example, induced by a ruler whose tick marks are ε apart, or in higher dimensions, a uniform grid whose elements are ε on a side. Label each grid box by an arbitrary index i. The measure of the ith grid box B_i is $P_i = \mu(B_i)$, which can also be thought of as the "probability of occurrence" of B_i. The entropy associated with this distribution of probabilities at this level of coarse graining is:

$$H(\varepsilon) = -\sum_i P_i \log P_i. \tag{6.77}$$

The value $-\log \varepsilon$ can be thought of as the resolution of a measuring instrument. If the logarithm is taken in base two, for example, $-\log_2 \varepsilon$ is the number of bits of precision. In the limit of an arbitrarily precise instrument, the rate at which new information is acquired as a function of the resolution is given by the ratio:

$$d_I = \lim_{\varepsilon \to 0} -\frac{H(\varepsilon)}{\log \varepsilon}. \tag{6.78}$$

Roughly speaking, the information dimension is the average pointwise dimension. If we coarse-grain by covering the set with balls rather than a fixed grid, then $H(\varepsilon)$ is essentially the average of $\log \mu(B_\varepsilon(x))$, weighted according to the measure. In fact, Young has proved under very general circumstances that the pointwise dimension is equivalent to the information dimension for almost all x (with respect to the measure μ) [12].

6.6.3   Fractal Dimension

The information is not the only quantity that changes with the scale of resolution. For example, consider $N(\varepsilon)$, the number of boxes of size ε needed to cover a set. Then we can define another dimension, sometimes called the *capacity* or simply *the* fractal dimension, as:

$$d_0 = \lim_{\varepsilon \to 0} -\frac{\log N(\varepsilon)}{\log \varepsilon}. \tag{6.79}$$

This can also be viewed as an information dimension where the probability P_i is estimated by $1/N_\varepsilon$.

6.6.4   Generalized Dimensions

The entropy $H(\varepsilon)$ can be thought of as the average of the logarithm of the coarse-grained probabilities or the "logarithmic self-movent"; that is,

$$H(\varepsilon) = -\sum_i P_i \log P_i = -\langle \log P_i \rangle = -\langle \log \mu(\varepsilon) \rangle. \tag{6.80}$$

Generalized dimensions are defined based on the scaling of the other self-moments. The first-self-moment $\langle \mu(\varepsilon) \rangle$, for example, scales as

$$\langle \mu(\varepsilon) \rangle \approx \varepsilon^{d_2} \tag{6.81}$$

in the limit as $\varepsilon \to 0$. This dimension is often called the *correlation dimension* [13]. It is particularly easy to compute, and for that reason is frequently discussed in the literature [14].

Similarly, any self-moment of the coarse-grained measure can be used to define a dimension.[13] In particular, we can generalize the relation stated above for the correlation dimension, and in the limit as $\varepsilon \to 0$ write:

$$\langle \mu(\varepsilon)^{(q-1)} \rangle \approx \varepsilon^{(q-1)d_q} \tag{6.82}$$

for any $q \neq 1$. This is readily expressed in terms of the Renyi entropy:

$$H_R \equiv \frac{1}{q-1} \log\langle \mu(\varepsilon)^{(q-1)} \rangle, \tag{6.83}$$

a generalization of the Gibbs entropy. Taking the logarithm of both sides of Eq. (6.82) yields:

$$d_q = \frac{1}{q-1} \lim_{\varepsilon \to 0} \frac{\log\langle \mu(\varepsilon)^{(q-1)} \rangle}{\log \varepsilon}. \tag{6.84}$$

The value for $q = 1$ can be defined by taking the limit of the right-hand side as $q \to 1$, and corresponds to the information dimension. The value at $q = 0$ corresponds to the capacity (see problem 11).

The importance of the generalized dimensions arises because they can be used to give a statistical characterization of the diversity of scalings in frac-

---

[13] Generalized dimensions were introduced by Grassberger [15] and by Hentschel and Procaccia [16].

tals. For "simple" fractals, such as the classic Cantor set (see example 6.16), there is just one scaling. As a result, all the generalized dimensions are the same. In more typical fractals, however, such as most chaotic attractors, there are a diversity of different scalings.[14] It is easy to see how this comes about; most dynamical systems have different derivatives at different points in the flow. As a result, the folding that generates the fractal structure varies from place to place, and the fractal is not homogeneous. Enlarging different parts of the attractor as though under a microscope generates a sequence of pictures that vary from place to place. Different generalized dimensions accentuate different scalings, and so in a fractal with diverse scaling structure the generalized dimensions will usually all be different. This leads to characterization of fractals in terms of "$f(\alpha)$," the Legendre transform of the generalized dimension [17].

Dimension is a geometrical invariant. That is, it is not changed by "reasonable" coordinate transformations. In particular, the information dimension is invariant under any coordinate transformation that is smooth and invertible at all but a finite number of points. Other generalized dimensions, in contrast, are generally invariant only under transformations that are smooth and invertible everywhere [18]. Thus, for example, although the logistic map and the tent map are equivalent under a coordinate transformation, they have different generalized dimensions for $q > 1$. This has led to some discussion about whether the generalized dimensions really deserve the name "dimension," or whether they should just be called "scaling exponents." In the end, these are only words, and we will follow the common practice of calling them dimensions.

Note that although typical fractals have inhomogenous scalings, chaotic attractors have the same pointwise dimension almost everywhere. This is because the motion on an attractor is ergodic (by definition, motion on an attractor is ergodic relative to the attractor). Thus, one region of the attractor can be mapped into another by the dynamics. If the dynamical system is smooth and invertible, then this mapping can be viewed as a coordinate transformation. The invariance of the dimension implies that the dimension in each region must be the same. Note that, nonetheless, numerical calculations of the pointwise dimension may give quite different values at different points. This is because numerical calculations only sample the scalings across a finite range of ε, and in general these are not uniform. The *true* dimension is defined in the limit as $\varepsilon \to 0$, and so implicitly takes an infinite average of all the scalings centered on x.

6.6.5 Estimating Dimension from Data

There are many different methods for computing dimension. One must first choose a definition. All of them involve the scaling of a self-moment $M(\mu(\varepsilon))$,

---

[14] Fractals with inhomogeneous scalings are sometimes called "multifractals"; however, because almost all fractals are in fact multifractals, this prefix is redundant.

such as $\langle \log \mu(\varepsilon) \rangle$ or $\langle \mu(\varepsilon) \rangle$, which is often called the *correlation integral*. The numerical methods all involve picking a technique for estimating the self-moment $M(\varepsilon)$ as a function of the resolution ε, plotting $\log M(\varepsilon)$ against $\log \varepsilon$, and computing the slope, as in Fig. 6.10.

Perhaps he most popular method is due to Grassberger and Procaccia [13] (a similar method was also used earlier in the psychology literature [19]). They estimate $\langle \mu(\varepsilon) \rangle$ by counting the number of points inside balls of radius ε (a particular form of kernel density estimation). The *correlation integral* is a particularly good self-moment from a numerical point of view because it is linear, and because it can probe the measure at smaller values of ε and hence estimate $M(\varepsilon)$ across a wider dynamic range.

It is very difficult to assign error bars to an estimate of fractal dimension [20, 21]. There are statistical errors that come from having only a finite number of points. These errors get exponentially larger as the dimension increases. Worse, there are intrinsic errors: with a finite number of data points there is a lower bound on the resolution ε given roughly by the smallest interpoint distance. The computed dimension at any finite value of ε is not necessarily the same as the true dimension as $\varepsilon \to 0$. This effect is impossible to take into account because it depends on the geometrical object whose dimension is being computed. It is dangerous to draw strong conclusions from a dimension

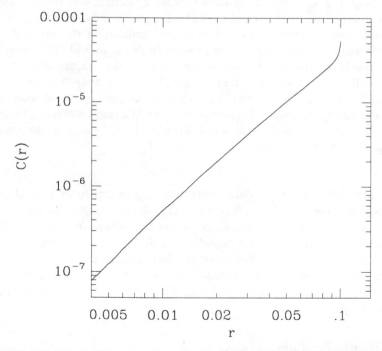

FIGURE 6.10. The $\langle \mu(\varepsilon) \rangle$ against $\log \varepsilon$ for the Rössler attractor, Fig. 5.19. The slope is roughly 2, which coincides with the graphical estimate obtained by examining the cross section.

calculation. There are no good self-consistency checks on the numbers. One should be appropriately suspicious of dimension calculations, particularly if they involve a small number of data points and the calculations yield a large value (e.g., 5 or greater).

Although dimension is defined in the limit as $\varepsilon \to 0$, in the real world things are not so simple. From a physical point of view, the scaling of the self-moments $M(\varepsilon)$ at small but finite ε is probably more relevant than that in the unobservable limit as $\varepsilon \to 0$. This is illustrated by Mandelbrot's [22] question: What is the dimension of a ball of yarn? From a great distance it is effectively a point, and appears zero dimensional; on approach it becomes a three-dimensional solid; moving closer, we discern the one-dimensional threads, which then become three-dimensional again; the threads are again composed of fibers, and so forth. These different scaling regimes would produce rather extreme oscillations in a numerical estimate of dimension. Typically when we are computing dimension we are interested in a given scaling range, but it may be very difficult to discern.

6.6.6 Embedding Dimension

It is always desirable to find the simplest model for a dynamical system. In many dynamical systems, even though the dimension of the state space is quite large, the dimension of an attractor may be quite small. We can attempt to take advantage of this by making a low-dimensional dynamical model that describes *only* motion on the attractor, by embedding the attractor in a smooth manifold and restricting the model to this manifold. The *embedding dimension* is the lowest dimension of such a manifold. We can also try to do this locally; that is, instead of embedding the entire attractor, we can just embed a piece of it. The lowest dimension in which this is possible is called the *local embedding dimension*. In general, from the Whitney embedding theorem, if the local embedding dimension is m the global embedding dimension M is bounded by the inequality:

$$2m + 1 \geq M \geq m. \tag{6.85}$$

At first sight, the embedding dimension might seem to be related to the topological dimension. In general, however, it is higher. The chaotic attractor of the Henon map at the standard parameter values, for example, has a topological dimension of one. Nonetheless, the chaotic attractor is a fractal and is not smooth—any transformation that embedded it in one dimension would not be differentiable. Consequently, both the local embedding dimension and the global embedding dimension are two, the dimension of the Henon map.

6.6.7 Fat Fractals

Fractal dimension is inadequate for describing *fat fractals*; that is, fractals with positive Lebesgue measure. To see the distinction between fat and thin fractals, modify the usual construction of the Cantor set as given in example 6.16, by deleting the central $1/3$, then $1/9$, then $1/27$, and so forth, of each successive line interval. The resulting set is topologically equivalent to the classic Cantor set, but the holes decrease in size sufficiently fast so that the resulting limit set has positive Lebesgue measure and fractal dimension one. The fact that a set has positive Lebesgue measure automatically implies that its fractal dimension is an integer. As a result, for fat sets the fractal dimension gives no information about fractal properties. An alternative is to examine the scaling of the coarse-grained measure $\mu(\varepsilon)$, where ε is the resolution; that is, the scale of the coarse graining [22–25]. For *thin* fractals, $\mu(0) = 0$, and the leading order scaling of the coarse grained measure is a simple power law in ε; that is, $\mu(\varepsilon) \approx A\varepsilon^{(D-d_f)}$, where d_f is the fractal dimension, D is the dimension of the space in which the fractal is embedded, and A is some positive constant. For fat fractals, it seems to be the case[15] that the leading order scaling in the limit as $\varepsilon \to 0$ also goes as a power law, except that it has a finite asymptote $\mu(0)$; that is,

$$\mu(\varepsilon) = \mu(0) + A\varepsilon^\beta, \tag{6.86}$$

where A and β are constants. A is unimportant, depending on the units of ε and μ. The exponent β, on the other hand, is independent of the choice of units and is an important quantity, providing a characterization of fractal properties. For fat fractals β is generally unrelated to the fractal dimension.

The fat fractal scaling exponent is closely related to the *uncertainty exponent*, originally introduced by Grebogi et al. [26]. There are many sets that have interesting fat fractal dimensions. For example,

- The chaotic invariant sets of two degree of freedom Hamiltonian systems (those with two position and two momentum variables) are often fat fractals [24].
- The parameter values where the logistic map is chaotic (as opposed to periodic) form a fat fractal in the parameter r (see Fig. 5.14). This is called *sensitive dependence to parameters* [27]. The fat fractal exponent can be used to characterize this; in this case it is roughly 0.45.
- Below the transition to chaos, the quasiperiodic parameter values between the Arnold tongues form a fat fractal. Numerical evidence and heuristic theoretical calculations suggest that the exponent has a universal value of $2/3$ (see Ref. 27).

6.6.8 Lyapunov Dimension

The dimension is related to the spectrum of Lyapunov exponents. A naive guess might be that it is the number of positive Lyapunov exponents, because this is the number of locally expanding directions. This guess is wrong. State

---

[15] Although it is possible to construct examples for which this power law scaling is violated, these appear to be pathological.

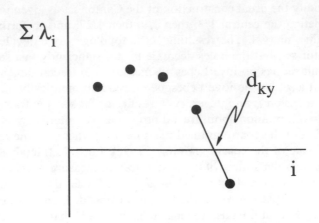

FIGURE 6.11. Kaplan–Yorke dimension.

space hypersurfaces of dimension d expand at a rate governed by the sum $\sum_{i=0}^{d} \lambda_i$. Thus, even if some of the Lyapunov exponents in this sum are negative, the dimensions associated with them still fill volume. Kaplan and Yorke have suggested the following alternative, defining the *Lyapunov dimension* as

$$d_L \equiv k + \frac{\sum_{i=1}^{k} \lambda_i}{|\lambda_{k+1}|}, \qquad (6.87)$$

where k is the largest value for which the numerator is positive [28] (see Fig. 6.11). The second term characterizes the fractional part of the dimension, if any. It is always less than one. Kaplan and Yorke conjecture that for a *typical* attractor, the Lyapunov dimension is equal to the information dimension.

There are known exceptions to the Kaplan–Yorke formula. Roughly speaking, it is valid as long as the invariant set has the structure of the Cartesian product of a Euclidean manifold and a single Cantor set. (Note that there is only one fractional part in the formula, corresponding to the single Cantor set.) A counter-example occurs if the chaotic flow has the structure of a Euclidean manifold crossed with *two* Cantor sets. Yorke argues that for chaotic attractors a single Cantor set is typical. So far there are no good counter-examples that contradict this assertion.

6.6.9 Metric Entropy

Chaotic trajectories continually generate new information, whereas predictable trajectories do not. The *metric entropy* quantifies the information flow.

Immediately after we make a measurement of a dynamical system we have a certain level of uncertainty about the state of the system. If the system is chaotic, then it will amplify this uncertainty. When we make another measurement, we reduce the uncertainty back to its original level. The metric entropy is the amount of information we gain every time we make a measurement, assuming that we have reasonably precise instruments and that we make frequent measurements.

Suppose we label the possible states of the system, as defined by a particular measuring instrument, by symbols $\{s_i\}$. Measurements at regular time intervals correspond to a string of symbols. A sequence of m measurements taken a time Δt apart on a trajectory starting from x_0 yields the symbol sequence $S_m(x_0) = s_{i_1}, s_{i_2}, \ldots, s_{i_m}$. The entropy of this string is:

$$H_m = -\sum P(S_m) \log P(S_m), \tag{6.88}$$

where the sum is over all strings of length m. The metric entropy is the rate at which the entropy increases as a function of time. To be sure we compute the true metric entropy we have to make sure we have a good measuring instrument. This can be done by finding the partition β that gives the highest rate. That is,

$$h_\mu = \sup_\beta \frac{H_m}{m\Delta t}. \tag{6.89}$$

For predictable dynamical systems, eventually new measurements provide no further new information, and the metric entropy is zero. For chaotic dynamical systems new measurements continue to provide new information, and the metric entropy is positive.

Example 6.17 Metric entropy of the binary shift map. The binary shift map produces one bit of information on every iteration. Its metric entropy is thus one bit. $P(S_m)$ is a uniform distribution on the set of all 2^m sequences of length m. H_m is thus m. Choosing units so that $\Delta t = 1$ gives $h_\mu = 1$.

For a chaotic system, on average the mutual information $I(x(0); x(t))$ decays in time, as indicated in Fig. 6.12. The metric entropy can be viewed as the initial average decay rate of the mutual information. An estimate of the metric entropy is given by the slope of auto-information curves as in Fig. 6.9 at the origin. Because we have not searched for the partition that maximizes the rate, this is not the exact metric entropy, but it is a reasonable estimate. Because the metric entropy of a nonchaotic process is zero, the curve in Fig. 6.9a should be a constant, as claimed above.

The metric entropy measures the rate at which information flows from small scales to large scales. Shaw has described the process this way [29]:

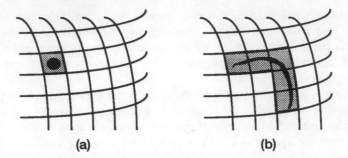

(a)   (b)

FIGURE 6.12. The decay of coarse-grained mutual information between an initial state and a later one. Although the volume of the black region remains constant or decreases from (a) to (b), the apparent volume in a coarse-grained picture increases.

Information is defined in terms of the set of possible outcomes of the measurement, a set which is fixed in advance. In order for this "pool" of information to remain of constant size, one bit of old information must be "destroyed" for each bit which is "created", or brought up from the microscales to macroscopic influence. In this way information contained in the initial conditions is systematically replaced by information *not* implicit in the initial conditions until, after a finite time or number of iterations, the state of the system is undetermined.

6.6.10   Pesin's Identity

The Lyapunov spectra and the metric entropy are related by Pesin's identity [30]:

$$h_\mu = \sum \text{ positive } \lambda_i. \qquad (6.90)$$

6.7   Dimensions, Lyapunov Exponents, and Metric Entropy in the Presence of Noise

The dimension of IID noise is infinite. This means that, strictly speaking, the dimension of a deterministic system with any dynamical noise is infinite. No matter how small the noise is, it must perturb the structure of trajectories at some scale. Because dimension is defined in the limit $\varepsilon \to 0$, this perturbation will show up in the value of the dimension. Similarly, the metric entropy of a system with noise is also infinite. Strictly speaking, this implies that the dimension and metric entropy of any real time series are always infinite. How, then, can the dimension be of any interest at all?

As already mentioned in Mandelbrot's story of the ball of yarn, in many cases we are more interested in scaling behavior at finite ε then we are in the limit as $\varepsilon \to 0$. The dimension should really be thought of as a function of ε. When we say the dimension of a real system is 3.5, we really mean that there is a scaling region over which it looks this way, fantasizing that if we could somehow get rid of all the external influences, we would see this scaling all the way down to the level of quantum fluctuations. Like any mathematical constructs used to describe the real world, the dimension and entropy are idealizations.

If the deterministic definition of the Lyapunov exponents is taken literally, the Lyapunov exponents are also infinite in the presence of noise. The definition is easily modified so that it is well behaved, however, by replacing states by ensemble averages. To see this, imagine performing an ensemble of experiments by choosing two initial conditions $\vec{x}(0)$ and $\vec{y}(0)$ so that $\|\vec{x}(0) - \vec{y}(0)\| = \varepsilon$. The leading Lyapunov exponent is:

$$\lambda_1(\vec{x}_0) = \lim_{t \to \infty} \lim_{\varepsilon \to 0} \frac{1}{t} \log\left(\frac{\|\langle \vec{y}(t) \rangle - \langle \vec{x}(t) \rangle\|}{\varepsilon}\right). \qquad (6.91)$$

This remains well defined in the presence of noise. The other Lyapunov exponents are extended similarly. The noisy Lyapunov exponents can then be used to define the metric entropy through Pesin's theorem, and the dimension through the definition of Lyapunov dimension.

Problems

6.1. For a mapping with additive dynamical and observational noise, write a formal expression for the observed state at $t + 1$ and $t + 2$ given the observed state at t. Note now the two types of noise enter differently.

6.2. Prove that the autocorrelation of the logistic map with $\mu = 4$ is $C(\tau) = \delta_{\tau,0}$. Calculate the mutual information between iterates and compare. Compute the metric entropy.

6.3. Show that $\log(x) \geq 1 - 1/x$, with equality if and only if $x = 1$.

6.4. Calculate the information in a discrete uniform distribution $p_i = 1/N$, for $i = 1, N$, and in a continuous uniform distribution $p(x) = 1/L$ for $0 < x < L$.

6.5. How much is a Gaussian worth? Find an expression for the information in a one-dimensional Gaussian distribution. Show that the information in a standard Gaussian (one with standard deviation 1, mean 0) is roughly two bits.

6.6. Consider a rectangular approximation to a delta function: $\phi(x) = 1/w$, for $|x| < w/2$, and 0 else. The density p is made up of two such peaks spaced a distance d apart: $p(x) = [\phi(x - d/2) + \phi(x + d/2)]/2$. Find its moments and its information. Notice that high information corresponds to "peakiness," but not necessarily to low variance. Show how the exponential of entropy counts the number of "states" of width w.

6.7. Show that e^H/σ is invariant under linear transformations. Argue why it is unreasonable to expect it to be invariant under nonlinear transformations.

6.8. Show that the correlation between two independent variables vanishes.

6.9. For the linearly related variables in example 6.9, examine the behavior of the correlation as the signal-to-noise ratio σ_η/σ_x becomes very large or very small compared to β. Demonstrate graphically what happens.

6.10. Generate a set of uniformly distributed random variables. For how small an N does the central limit theorem give reasonable results? (The definition of "reasonable" is up to the reader.)

6.11. Show that the generalized dimensions d_0 and d_1 correspond to the capacity and the information dimensions, respectively.

6.12. In a power spectrum, power from frequencies above the Nyquist frequency is aliased into frequencies below it. Determine where the power from frequency $f_c + \Delta f$ appears.

6.13. Compute the power spectrum of a sine wave of finite extent.

6.14. (Open-ended!) Just as the power spectrum is related to translational symmetry, it is possible to define other spectra based on other symmetries. For example, the behavior of a function under dilations indicates the degree of self-similarity it possesses. Define an analogue of the autocorrelation function that detects dilation symmetries. Find an analogue to the Wiener–Khinchine theorem—that is, find a set of basis functions for projecting a function onto such that the moduli of the projections are related to the analogue of the autocorrelation function. Is there a fast transformation analogous to the fast Fourier transform for calculating this spectrum? See [31].

References

[1] M.C. Mackey, Rev. Mod. Phys. **61**, 981 (1989); answers to frequently asked questions for the newsgroup "Sci.nonlinear", http://www.fen.bris.ac.uk/engmaths/research/nonlinear/faq.html maintained by J. Meiss.

[2] A. Kolmogorov, Eng. Mat. **2(3)** (1933).

[3] E. Slutsky, Problems of Economic Conditions **3**, 1 (1927). English translation in Econmetric **5**, 105 (1937).

[4] A.M. Yaglom, *Stationary Random Functions* (Prentice-Hall, New Jersey, 1962).

[5] V.I. Arnold, *Mathematical Methods of Classical Mechanics* (Springer-Verlag, New York, 1978).

[6] Ju. I. Kifer, USSR Izvestia **8**, 1083 (1974).

[7] M. DeGroot, *Probability and Statistics* (Addison-Wesley, Reading, MA, 1975).

[8] E.T. Jaynes, *Papers on Probability, Statistics, and Statistical Physics* (Kluwer, Boston, 1983).

[9] A.M. Fraser and H.L. Swinney, Phys. Rev. A **33**, 1134 (1986).

[10] W. Li, Technical Report, Santa Fe Institute (1989).

[11] W. Hurewicz and H. Wallman, *Dimension Theory* (Princeton University, Princeton, 1941).

[12] L. Young, Ergodic Theory and Dynamical Systems **2**, 109 (1989).

[13] P. Grassberger and I. Procaccia, Phys. Rev. Lett. **50**, 346 (1983).

[14] J. Theiler, J. Optical Soc. Am. **7**, 1055 (1990).

[15] P. Grassberger and I. Procaccia, Physica D **9**, 189 (1983).

[16] H.G.E. Hentschel and I. Procaccia, Physica D **8**, 1154 (1983).

[17] T. Tel, *Z. Naturforsch.* **43a**, 1154 (1988).

[18] E. Ott, Rev. Mod. Phys. **53**, 655 (1981).

[19] K. Fukunaga and D.R. Olsen, IEEE Trans. Comput. **C-20**, 176 (1971).

[20] J. Theiler, Ph.D. thesis, Cal. Tech. (1987).

[21] *Dimensions and Entropies in Chaotic Systems*, edited by G. Mayer–Kress (Springer-Verlag, Berlin, 1986).

[22] B. Mandelbrot, *Fractal Geometry of Nature* (Freeman, New York, 1982).

[23] J.D. Farmer, Phys. Rev. Lett., **55**, 351 (1985).

[24] D.K. Umberger and J.D. Farmer, Phys. Rev. Lett. **55**, 61 (1985).

[25] C. Grebogi, S.W. McDonald, E. Ott, and J.A. Yorke, Phys. Lett. **110A**, 1 (1985).

[26] C. Grebogi, S.W. McDonald, E. Ott, and J.A. Yorke, Phys. Rev. Lett. **99**, 415 (1983).

[27] R. E. Ecke, J.D. Farmer, and D.K. Umberger, Nonlinearity **2**, 175 (1989).

[28] J.L. Kaplan and J.A. Yorke, Ann. NY Acad. Sci. **316**, 400 (1979).

[29] R.S. Shaw, *The Dripping Faucet as a Model Dynamical System* (Aerial, Santa Cruz, CA, 1984).

[30] Ya. B. Pesin, Uspekhi Matematicheskii Nauk. **32**, 55 (1977).

[31] H.E. Moses and A.F. Quesada, J. Math. Phys. **15**, 748 (1973).

7

Modeling Chaotic Systems

Stephen G. Eubank and J. Doyne Farmer

The classic problem of dynamical systems theory is that of finding and understanding the trajectories of a given dynmical system. This section discusses the inverse problem of dynamical systems, which is that of finding and understanding the dynamics of a given trajectory. The inverse problem is complicated by a variety of factors:

- *Partial information.* We often observe only a projection of the true state. We have to reconstruct the state space from this information.
- *Measurement error.* Observations are inevitably noisy.
- *External fluctuations.* The dynamics are inevitably noisy.
- *Representation of the dynamics.* We do not know the proper form of the dynamical laws. We must choose from an infinite number of possible function representations.

These problems are usually coupled. For example, although from Takens's theorem we know that it is possible to reconstruct a state space with perfect data, the theorem is not valid in the presence of noise. The reconstruction of a state space also complicates the representation problem. For example, for a fluid flow we know that the Navier–Stokes equations provide a good description of the dynamical laws. But the Navier–Stokes equations are stated in terms of functions in three dimensions; as initial conditions they require finely sampled measurements of velocity, temperature, pressure, and so forth, over a three-dimensional grid. If we have incomplete information, such as the output of a single probe, we cannot use the Navier–Stokes equations. When we reconstruct a state space we do not know how to incorporate information from fluid dynamics into the reconstruction procedure, and so must proceed blindly.

There are two basic steps for solving the inverse problem of dynamical

---

This chapter was finished in 1991. For reasons beyond the authors' control, they were not able to revise it to bring it up to date with recent developments in the field. We are including it nonetheless because we feel it contains information that remains valuable. Reference to new developments are included at the end of the chapter. —*Editor* systems. The first step is to reconstruct the missing information. The second step is to discover the best possible approximation of the correct dynamical laws. Once this is done, the problems that we may hope to solve include:

- *Prediction.* Given the past and present, what is the future?
- *System classification.* What are the properties of the dynamics? How many degrees of freedom are there? Are the dynamics regular or chaotic?
- *Noise reduction.* Can we reduce the measurement or dynamical noise?

7.1   Chaos and Prediction

The great promise of chaos lies in the hope that randomness might become predictable. Although chaotic dynamics puts limits on long-term prediction, it implies predictability over the short term. Applications of modern non-linear data analysis techniques indicate that chaotic dynamics is quite common, and that in many cases random behavior is due to low-dimensional chaos rather than complicated dynamics involving many irreducible degrees of freedom. Until recently, however, there has been no way to exploit the presence of low-dimensional chaos to make predictions.

Chaos has caused a fundamental change in the way we think about randomness. This influence is felt strongly in physical models for random phenomena. A good example is the problem of "excess noise" in Josephson junctions. Popular models for this phenomenon [1] are now often formulated in terms of simple systems of deterministic differential equations, quite different from the statistical models that were the only option 20 years ago. Similarly, in this section we show how thinking in terms of deterministic dynamical systems, and assuming that randomness arises out of chaos rather than complexity, leads to new approaches to forecasting and nonlinear modeling.

The assumption that randomness is caused by extreme complication (i.e., the presence of many irreducible degrees of freedom), led to Kolmogorov's theory of random processes. Many people speak of random processes as though they were a fundamental *source* of randomness. This is misleading. The theory of random processes is an empirical technique for coping with inadequate information, and makes no statements about *causes* of randomness. As far as we know, the only truly fundamental source of randomness is the uncertainty principle of quantum mechanics; everything else is deterministic, at least in principle. Nonetheless, we call many phenomena such as fluid turbulence or economics random, even though they have no obvious connection to quantum mechanics. It has traditionally been assumed that the apparent randomness of these phenomena derives solely from their complexity.

We will take the practical viewpoint that randomness occurs to the extent that something cannot be predicted, which usually depends on the available information. With more data or more accurate observations, a phenomenon that previously seemed random might become more predictable, and hence less random. Randomness is in the eye of the beholder.[1] Furthermore, randomness is a matter of degree—some systems are more predictable than others.

Because chaos is defined in the context of deterministic dynamics, in some very strict sense it might be incorrect to say that chaos is random—ultimately uncertainty originates from something external to the dynamics, such as measurement error or external "noise." But sensitive dependence exaggerates uncertainty, so that small uncertainties turn into large ones. Because chaos amplifies noise exponentially any uncertainty at all is amplified to macroscopic proportions in finite time, and short-term determinism becomes long-term randomness.

Thus, the important distinction is not between chaos and randomness, but between systems with low-dimensional attractors and those with high-dimensional attractors. If a time series is produced by motion on a very high-dimensional attractor, then from a practical point of view it is impossible to gather enough information to exploit the underlying determinism. If the dynamics is modeled in a state space whose dimension is lower than that of the attractor, only a projection of the dynamics remains, and determinism is invisible—the dynamics look random. Even if the dimension of the model is large enough, the amount of data needed to make a good model for a high-dimensional attractor may be prohibitive. This problem gets exponentially worse as the dimension increases [5, 6], as is shown below.

With many degrees of freedom, the statistical approach is probably as good as any—linear models may even be optimal. But if random behavior comes from low-dimensional chaos, it is possible to make forecasts that are much better than those of linear models. Furthermore, the resulting models can give useful diagnostic information about the nature of the underlying dynamics, aiding the search for a description in terms of first principles.

7.2   State Space Reconstruction

When we observe a dynamical system, we often receive only partial information about its behavior. In this case, the first problem we must confront is that of reconstructing a description of the dynamics in terms of states that make it as deterministic as possible.

Consider a time series $\{v(t_i)\}, i = 0, 1, \ldots, N$. Assume for the moment that v is a scalar, although the extension to the case in which it is a vector is straightforward. Typically $\{v(t_i)\}$ is a projection of dynamics in a higher-dimensional state space.

A typical example occurs in fluid flow experiments, which in principle can be modeled accurately by deterministic partial differential equations. How-

---

[1] Another perhaps more fundamental notion of randomness is due to Kolmogorov and Chaitin [2]. Whether or not chaotic systems are random in this sense is controversial [3, 4].

ever, in practice this may not be useful unless the data is in the correct form. For example, suppose a single probe measures a given component of the velocity at a fixed point in space. This data is simply inadequate to provide initial conditions for the Navier–Stokes equations. To build a model from the data at hand one is forced to reconstruct a state space from a single time series. There are several ways to do this.

7.2.1   Derivative Coordinates

One method for reconstructing a state space is based on derivatives [7]. State space coordinates x_i can be assigned as:

$$x_1 = v(t)$$

$$x_2 = \frac{dv}{dt}$$

$$x_3 = \frac{d^2v}{dt^2}$$

$$\vdots$$

(7.1)

In practice, time series are only sampled at finite intervals, and the derivatives must be approximated with a numerical scheme. For a time series sampled at a uniform time interval Δt, using a crude method for approximating the derivative at time t_i gives the three-dimensional state:

$$\vec{x}(t_i) = \left(v(t_i), \frac{v(t_i) - v(t_{i-1})}{\Delta t}, \frac{v(t_i) - 2v(t_{i-1}) + v(t_{i-2})}{(\Delta t)^2} \right).$$

(7.2)

For clean data derivatives often give a nice embedding, as long as the dimension is not too large. But because numerical differentiation amplifies high-frequency noise, high-dimensional embeddings are impractical with this method [5, 7]. For finely sampled data this problem can be reduced somewhat by smoothing.

7.2.2   Delay Coordinates

An alternate method creates state vectors $\vec{x}(t)$ by assigning coordinates:

$$x_1(t) = v(t),$$

$$x_2(t) = v(t - \tau),$$

$$\vdots$$

$$x_d(t) = v(t - (d - 1)\tau),$$

(7.3)

where τ is a delay time [7, 8]. If the dynamics takes place on an attractor of dimension D, then a necessary condition for determinism is $d \geq D$.

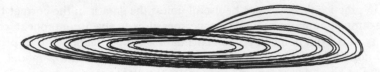

FIGURE 7.1. Delay reconstruction of the Rössler attractor in Fig. 5.19.

For noise-free data, Takens has shown that if r is the dimension of a manifold containing the attractor, almost any delay or derivative embedding in $2r + 1$ dimensions or less will preserve topological properties of the attractor [8]. More specifically, the embedding will be a diffeomorphism—a differentiable mapping with a differentiable inverse—from the true state space to the delay space.

A delay reconstruction of the Rössler attractor is shown in Fig. 7.1. Note the obvious similarity between the attractor in the true state space (Fig. 5.19) and this one.

It is not clear how to generalize the embedding theorem to the noisy case. It is often assumed that for small noise levels, something similar to Takens's result holds. This is a nontrivial assumption, because noise is by definition infinite dimensional, and even a small coupling of a finite dimensional system to an infinite dimensional one might be expected to drastically alter embedding characteristics.

In principle, τ is arbitrary as long as x is not periodic, in which case τ must not be rationally related to the period. In practice, if τ is too small the coordinates become singular, so that $x_j \approx x_{j+1}$ and the embedded trajectories lie on a line in the reconstructed space. If τ is too big, chaos makes x_1 and x_d causally disconnected. Suggestions for appropriate values of τ include the first minimum in the autocorrelation function and the first minimum in the auto-information function for v. The idea behind these suggestions is to ensure that each new coordinate adds new information. Taken together, in the presence of noise the bounds on reasonable values of τ imply an effective upper bound on the embedding dimension d. In practice d is often chosen by trial and error, starting with a low value and increasing it, searching for optimal results. A more systematic procedure based on mutual information has been explored by Fraser and Swinney [9].

The use of delay coordinates to reconstruct a state space is not original to dynamical systems theory. It goes at least as far back as Yule, who in 1927 made a model for sunspot activity based on a linear combination of past values [10]. This idea is also implicit in Kolmogorov's definition of a random process. The important contribution from dynamical systems theory is the demonstration that reconstruction preserves geometrical *invariants* of the dynamics, such as attractor dimension, metric entropy, and the positive Lyapunov exponents.

7.2.3 Broomhead and King Coordinates

Broomhead and King [11] have suggested an alternative approach to reconstruction. They apply the Karhunen and Loeve principal value decomposition to the delay coordinate representation, and produce embeddings that seem to have the nice properties of derivatives, but without the numerical problems. The simplest way to implement their procedure is to compute the covariance matrix $\langle v(t_i)v(t_{i+j})\rangle_t$ for some set of delays j and compute its eigenvectors and eigenvalues α_i. The eigenvalues α_i are the average root–mean–square projection of the d-dimensional delay coordinate time series onto the eigenvectors. Ordering them according to size, the first eigenvector has the maximum possible projection, the second has the largest possible projection for any fixed vector orthogonal to the first, and so on. The numerical calculations of Broomhead and King demonstrate that under good circumstances α_i falls off exponentially with i, until it reaches a floor determined by the noise level.

Gibson [12] has demonstrated that the eigenvectors of the Broomhead and King procedure are approximations to Legendre polynomials over the lag window. This implies that Broomhead and King coordinates are in fact very similar to derivative coordinates, smoothed over the lag window. Furthermore, the fall off steepens as the sampling time Δt decreases, allowing fewer significant values to rise above the noise floor.

7.2.4 Reconstruction as Optimal Encoding

It is useful to compare the embedding problem to the problem of data compression. A good embedding is one that compresses all the information about the future contained in the past into as few variables as possible. This can be accomplished systematically by decomposing the information in a delayed variable into three parts: one that is irrelevant to the future; a second that is redundant with information given by other degrees of freedom in the reconstructed state space; and a third that contains new information about the future. The irrelevant information is thrown away. The redundant information is "averaged over" to give a better estimate of those degrees of freedom that have already been included in the reconstructed space. Finally, the new information is incorporated into a new degree of freedom. Explicit algorithms for doing this are under development [13].

7.3   Modeling Chaotic Dynamics

7.3.1   Choosing an Appropriate Model

Once the data has been embedded in a state space, the next task is to model the dynamics. There are several approaches [14–19]. The simplest assumes that the dynamics can be written as a map in the form

$$\vec{x}(t+T) = f_T(\vec{x}(t)). \tag{7.4}$$

The problem is to estimate f_T. We will call this estimate $\hat{f}_T$, and approximate the dynamics by a map of the form

$$\hat{x}(t, T) = \hat{f}_T(\vec{x}(t)), \tag{7.5}$$

where $\hat{x}(t, T)$ represents an estimate of the state $\vec{x}(t+T)$ based upon $\vec{x}(t)$.

Chaotic dynamics does not occur unless f is nonlinear, so to approximate chaotic dynamics the model $\hat{f}_T$ must be nonlinear. There are an infinite number of ways to represent nonlinear functions, and finding good nonlinear approximations is a difficult problem. Picking a particular representation is an ad hoc choice, which may or may not be a good one. Prior information may suggest a good representation, but in the absence of any theoretical understanding modelers must make arbitrary choices.

To make a good approximation for a function f, a representation must be able to conform to its variations. If f is well behaved, any complete representation will provide a good approximation, as long as it has enough free parameters, and there is enough data to stably specify them. However, if f is complicated, there is no guarantee that any given representation will approximate it efficiently.

The dependence on representation can be reduced by *local approximation*. The basic idea is to break up the domain of f into local neighborhoods and fit different parameters or *charts* in each neighborhood. When f is smooth the neighborhoods can be small enough that f does not vary sharply in any of them, making the constraints of a particular representation less important.

Local approximation usually produces better fits for a given number of data points than global approximation, particularly for large data sets. Most global representations reach a point of diminishing returns where adding more parameters or more data only gives a marginal improvement in accuracy. Higher order terms may oscillate, and actually cause the behavior to get worse. Past a certain point, adding more local neighborhoods is usually more efficient than adding more parameters and going to higher order. Local approximation makes it possible to use a given functional representation efficiently. The key is to choose a local neighborhood size such that each neighborhood has just enough points to make the local parameter fits stable in the sense that adding more points does not make significant improvements.

7.3.2   Order of Approximation

One measure of the quality of a model is its ability to make predictions. Let $\hat{x}(t, T)$ be a prediction of the value $\vec{x}(t+T)$, based only on data available at time t or before. The error E of the prediction is:

$$E = \hat{x}(t, T) - \vec{x}(t+T). \tag{7.6}$$

The geometric mean is a more useful indicator of average error in a set of predictions than the arithmetic mean. For a series of n measurements this is

$$\bar{E} = \frac{(\prod_{i=1}^{n} |E|)^{1/n}}{(\prod_{i=1}^{n} |\vec{x}(t) - \bar{x}|)^{1/n}}, \tag{7.7}$$

where $\bar{x}$ is the mean value of $\vec{x}$. The term in the denominator is a normal-
ization factor, chosen so that constant prediction of the mean value, $\hat{x}(t, T) = \bar{x}$ gives a normalized error of unity. The normalized error is zero when all the
predictions are perfect.

Properties of the modeling method sometimes give information about the
expected error. Approximation schemes can be classified according to the
order of the derivatives that errors depend on. For example, suppose the charts
are polynomials of degree m, but the true dynamics is a function f of higher
degree. In the ideal case the error depends on the $(m + 1)^{st}$ derivative of f.
This implies that the errors are proportional to ε^{m+1}, where ε is the spacing
between data points. The average spacing between N points uniformly dis-
tributed over a D dimensional space is $\varepsilon \approx N^{-1/D}$. Calling q the *order of
approximation*, one finds[2]:

$$\bar{E} \sim N^{-q/D}, \tag{7.8}$$

where in this case $q = m + 1$.

Achieving the ideal case where $q = m + 1$ is difficult for large q, because in
general fitting a polynomial of degree m does not produce a fit that is accurate
to order $m + 1$. For example, suppose the number of data points is equal to
the number of free parameters. The approximation may go through each
point precisely, but in between it may oscillate wildly, producing an extremely
inaccurate approximation. This may limit the accuracy even when more
neighbors are used.

To avoid this confusion we will use Eq. (7.8) to *define* the order of local
approximation, taking the limit as $N \to \infty$, and letting D be the information
dimension of the underlying measure of the data points:

$$q \equiv -\lim_{N \to \infty} \frac{D \log \bar{E}}{\log N}. \tag{7.9}$$

In general q may depend on D, f, the specific neighborhoods used, and other
factors.

A trivial example of local approximation is *first order*, or *nearest neighbor*
approximation. This amounts to simply looking through the data set for the
nearest neighbor, and predicting that the current state will do what the
neighbor did a time T later. We approximate $x(t + T)$ by $\hat{x}(t, T) = x(t' + T)$,
where $x(t')$ is the nearest neighbor of $x(t)$. For example, to predict tomorrow's weather, search the historical record to find the weather pattern most
similar to that of today and predict that tomorrow's weather pattern will be
the same as the neighboring pattern one day later.[3] First order approxima-

---

[2] We use the symbol "$\sim$" to mean "scales as"; that is, $z \sim y(x)$ implies $z = Cy(x)$, where C
includes all dependencies on variables other than x.

[3] This was attempted by E.N. Lorenz, who examined roughly 4000 weather maps [20]. The results
were not very successful because it was difficult to find good nearest neighbors, apparently
because of the high dimensionality of weather.

tion can sometimes be improved by finding more neighbors and averaging their predictions, for example, by weighting according to distance from the current state.

An approach that is usually superior is *local linear* or *second order* approximation. For the neighborhood $\{\vec{x}(t')\}$, fit a linear polynomial to the pairs $(\vec{x}(t'), \vec{x}(t' + T))$. When the number of nearest neighbors $M = d + 1$ and the simplex formed by the neighbors encloses $\vec{x}(t)$, this is equivalent to linear interpolation. If the data is noisy, the chart may be more stable when the number of neighbors is greater than the minimum value. Again, this procedure can be improved somewhat by weighting the contributions of the neighboring points according to their distance from the current state. Linear approximation has the nice property that the number of free parameters and consequently the neighborhood size grows slowly with the embedding dimension.

It is also possible to use local polynomial approximation of any order. However, using a higher-order polynomial does not mean that a higher order of approximation is obtained. The number of parameters needed for a polynomial increases roughly as d^m, where m is the order of the polynomial and d is the dimension. The number of data points in the neighborhood must be greater than or equal to the number of free parameters; typically, to get good results it must be greater. Thus, using a higher-order polynomial requires a larger neighborhood. In high dimensions the gain in accuracy by using a higher polynomial is rapidly offset by the increase in the size of the neighborhood. Until a proper theory is developed, determining the order of approximation of a given method must be done by trial and error.

7.3.3   Scaling of Errors

Naturally, for a chaotic system errors depend strongly on the extrapolation time. The rate at which errors grow depends on the way predictions are made. There are two choices:

- *iterative forecasts* fit a model for $T = 1$ and iterate to make predictions for $T = 2, 3, \ldots$
- *direct forecasts* fit a new model for each individual T.

On the surface, direct forecasting might seem more accurate, because each model is "tailored" for the time it is supposed to predict, and there is no accumulation of errors due to iteration. In fact, if the model is sufficiently accurate, the opposite is true. Approximation errors for iterative forecasting grow roughly according to the largest Lyapunov exponent $\lambda_{\max}$, whereas for direct forecasts the errors grow as $q\lambda_{\max}$. More precisely, in ref. [18] it is demonstrated that in the limit as first $N \to \infty$ and then $T \to \infty$, for direct forecasting the errors vary as:

$$\bar{E} \sim N^{-q/D} e^{q\lambda_{\max}T}, \tag{7.10}$$

whereas for iterative forecasting they grow as[4]:

$$\bar{E} \sim N^{-q/D} e^{\lambda_{max} T}. \tag{7.11}$$

The superiority of iterative estimates comes from the fact that they make use of the regular structure of the higher iterates. The time series are generated by iterating a dynamical system, and so iterative approximations are more natural. The power of the iterative procedure is reminiscent of Barnsley's methods for constructing complicated fractals from the recursive application of simple affine mappings [21].

Example 7.1 Error scaling in nonlinear models of the sine map. The validity of these estimates for a simple example is demonstrated in Fig. 7.2, where we

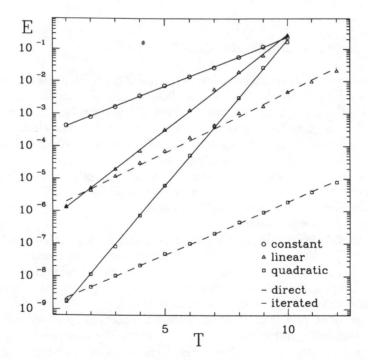

FIGURE 7.2. Error scaling for a nonlinear model of the sine map, Eq. (7.12). Letting $q = m + 1$, where m is the order of the polynomial, as expected from Eqs. (7.10) and (7.11), the logarithm of the error for direct forecasts grows roughly according to $q\lambda$, whereas for iterative forecasts it grows according to λ, independent of the order of approximation. λ is the Lyapunov exponent, which in this case is one bit per iteration.

---

[4] For a continuous time series N has an ambiguity because of the sampling time. It is not possible to forecast arbitrarily well by just sampling arbitrarily finely. As long as the time series is sampled reasonably finely, the dependence is on the number of characteristic times present in the time series, rather than the number of data points.

show the approximation error as a function of the extrapolation time T. Forecasts are made using local polynomial approximation, building the data base from a 5000 point time series generated by the sine map:

$$x_{t+1} = \sin(\pi x_t). \tag{7.12}$$

We make 500 forecasts, and plot the average error as a function of the extrapolation time T. The error grows roughly according to Eq. (7.10) for direct approximation and according to Eq. (7.11) for iterative approximation.

The above results are for scaling in the limit as $\vec{E} \to 0$. In situations where the model is not good enough to achieve this limit, these scaling laws cannot be expected to hold.

7.4   System Characterization

The problem of system characterization is that of trying to deduce properties of the dynamics from the data. The traditional approach to this question in time series analysis is to fit a linear model to the data, and determine the optimal order (dimension) and the optimal parameters of the model. This is obviously unsatisfactory in the presence of nonlinear behavior. A simple generalization is to fit the best nonlinear model. However, the result in and of itself is usually not very illuminating. Without any prior knowledge, any model that we fit is likely to be ad hoc. We are more likely to be interested in such questions as the following: How nonlinear is the model? How many degrees of freedom does it have?

Dynamical systems theory provides a language for properly posing these questions, which become: What is the dimension of the dynamics? How chaotic is dynamics, or in other words, what is the metric entropy, or the largest Lyapunov exponent? It thus gives us a *model-independent* language for posing the problems of systems characterization.

The problem with this is that algorithms for computing dimension, Lyapunov exponents, and metric entropy are notoriously unreliable. They produce a number. But as the self-consistency checks on this number are usually rather weak, it is difficult to know a priori how much to trust the number. Also, there are many subtleties involved in computing dimension, and many incorrect conclusions have been reached through naive application of these methods.

Forecasting provides one possible test for self-consistency. When the forecasts of a nonlinear model are significantly better than those of a linear model, there is good reason to believe the nonlinear model is better. The great advantage of this is that it is very unlikely that a long series of good forecasts happen at random; we may make one or two lucky guesses, but we are unlikely to make a long series of them. This can be made precise by applying standard tests for statistical significance.

Furthermore, when the order of approximation is known, the scaling properties of Eq. (7.11) can be used to compute the dimension and the largest Lyapunov exponent. Varying the number of data points and looking at the scaling in the prediction error gives an estimate of the dimension, whereas varying the extrapolation time T gives the exponent.

7.5 Noise Reduction

Noise reduction is a vague term for a class of problems in which we wish to decompose a signal into a part that we want and a part that we do not want. To be a little more precise, suppose we observe a noisy time series $\{y_t\}$, where $t = 1, 2, \ldots, N$. Assume that it can be decomposed as

$$y_t = x_t + n_t, \tag{7.13}$$

where x_t is the "desired" part, or the *signal*, and n_t is the "noise." We want to get rid of the noise. We can only observe the noisy signal y_t, but we wish to recover the pure signal x_t.

What we call "noise" is usually subjective, and to reduce it we must have a criterion for distinguishing n_t from x_t. For example, if the noise is predominantly at high frequencies, and the signal predominantly at low frequencies, we can reduce the noise by simply applying a low-pass filter. However, if the noise and the signal have similar frequencies, spectral methods are ineffective and more sophisticated criteria are required.

Recent methods for noise reduction [18, 22–24] assume that the distinguishing feature of the signal $\{x_t\}$ is that it is a trajectory of an invertible dynamical system f; that is,

$$x_{t+1} = f(x_t). \tag{7.14}$$

The noise $\{n_t\}$ might also be a trajectory of a dynamical system; if so, we will assume that it is substantially different from f. Often in cases of interest this means that the dimension of the noise is much greater than that of the signal. When noise is described as "random," this typically could be stated more precisely by saying that it is high dimensional.

The noise reduction problem can be broken into two parts: learning the dynamical system f, and reducing the noise once it is known. There are three cases:

1. We are given f.
2. We must learn an approximation to f, but we have available some "clean" data with known values of x_t.
3. We must learn f directly from the noisy data y_t.

In all three of these cases we are ultimately faced with problem (1), that of reducing the noise once we know f or have an approximation to it.

The best approach to noise reduction depends on the goal. For example, we may not care about recovering each value of the signal x_t exactly—we may be interested only in recovering a statistical property, such as the power spectrum. Statistical noise reduction is intrinsically simpler than detailed noise reduction.

7.5.1   Shadowing

The shadowing problem is that of finding a purely deterministic orbit $x_{t+1} = f(x_t)$ that stays close to y_t, or in other words, a deterministic orbit that *shadows* the noisy orbit.

The problems of shadowing and noise reduction are closely related. On one hand, shadowing is not a noise reduction problem per se, because the shadowing orbit x is an artificial construction. On the other hand, if we pretend that the shadowing orbit x is a signal, then the "effective noise" $\tilde{n}_t$ can be defined trivially so that

$$y_t = x_t + \tilde{n}_t. \tag{7.15}$$

Conversely, if we pretend that observational noise is dynamical noise, then the true orbit x is automatically a shadowing orbit. Although it may not be the *best possible* shadowing orbit, it should be at least close. This suggests that an approximation to the true orbit is given by an optimal shadowing orbit. Under certain conditions that remain to be made more precise, these two should asymptotically coincide.

In the dynamical systems literature the shadowing problem was originally motivated by concerns about chaos in noisy environments. Chaotic dynamics strongly amplify the smallest fluctuations, so that a noisy orbit quickly diverges from a purely deterministic orbit with the same initial conditions. Thus, in physical experiments, where noise is always present, the detailed behavior of a trajectory is quite different from that of a purely deterministic trajectory with the same starting value. This raises the following question: In the presence of chaos, can purely deterministic dynamical equations be valid models of physical systems? Similarly, in computer experiments where there is chaos, because of roundoff error simulated trajectories may deviate significantly from those of the equations they are supposed to simulate. Do they nonetheless correctly reproduce their statistical properties?

Even though noise radically alters the details of individual orbits when the dynamics are chaotic, for small noise levels we can still hope that the *qualitative* properties of noisy orbits and purely deterministic orbits are similar. For instance, suppose there is *some* purely deterministic orbit, albeit with a different initial condition, that stays close to the noisy orbit. Providing this orbit is a *typical* orbit of the dynamical system, then we are assured that noisy orbits will have the same statistical properties for length scales larger than that of the noise.[5]

Anosov and Bowen [25, 26] presented a solution to this problem for the special case of everywhere hyperbolic (Anosov) dynamical systems. Loosely speaking, *everywhere hyperbolic* means that at each point the dynamics can be factored into directions where the motion is either exponentially expanding or exponentially contracting. Everywhere hyperbolic dynamical systems do not have homoclinic tangencies; that is, the angle between stable and unstable manifolds is bounded away from zero.

Assume additive dynamical noise. Also assume that f is everywhere hyperbolic, and the noise is bounded; that is, $\|n_t\| < \varepsilon$, where $\varepsilon > 0$. Anosov and Bowen proved that for every $\delta > 0$ there is an ε such that every noisy orbit y_t of f is shadowed by an orbit x_t with $\|y_t - x_t\| < \delta$ for all t.

The Anosov–Bowen construction depends on the fact that the stable and unstable manifolds are never parallel where they intersect. Although it was originally believed that Anosov systems were generic, we now know that quite the opposite is true; most dynamical systems of interest have homoclinic tangencies. Furthermore, in many cases they display *sensitive dependence on parameters* [27, 28], so that arbitrarily near any parameter value with a chaotic attractor there is a parameter value with a stable periodic attractor. In this case the noise may induce bifurcations, so that even though an orbit is periodic in the deterministic limit, with the addition of some noise it becomes chaotic [29]. In this case there is no shadowing orbit, and thus the shadowing lemma is not true in general.[6]

Hammel et al. [30] applied the Anosov–Bowen construction to two-dimensional maps that were not hyperbolic and found, at low noise levels, good deterministic shadowing orbits for finite segments of noisy trajectories. The length of these segments is limited by the presence of homoclinic tangencies. As the length of the trajectory segment goes up, so does the probability of finding a homoclinic tangency of a given severity; that is, where the angle is smaller than a given value. This problem becomes more severe as the noise level increases.

The Anosov–Bowen construction is cumbersome and difficult to apply in more than two dimensions, and does not provide a practical tool for solving

---

[5] The presence of a shadowing orbit per se does not imply that the statistical properties are the same; the shadowing orbit might be atypical. For example, consider the binary shift map $x_{t+1} = 2x_t \bmod 1$. Iteration on a computer asymptotically results in $x_t = 0$ as $t \to \infty$ under most common round-off algorithms. $x_t = 0$ is a valid deterministic orbit, but it is highly atypical, with misleading statistical properties.

[6] The Anosov–Bowen shadowing lemma is not true in general, but if we modify the original problem to allow small changes of parameters in f, then there may still exist a shadowing orbit, for a "nearby" f. Thus, there may be a modification of the theorem along these lines that applies to the general case of nonhyperbolic flow.

experimental noise reduction problems. Using related ideas, several groups have recently developed practical methods for noise reduction.

The basic principle behind improved methods for noise reduction can be viewed as nonlinear generalization of the simple moving average smoothing technique, as illustrated in Fig. 7.3. In a simple moving average the points of a time series around a given point are simply averaged together. Averaging acts as a low-pass filter, which removes the high-frequency components. The noise reduction obtainable from this technique is severely limited, however, because averaging over too large an interval causes distortion of the signal. The problem is that simple averaging fails to take the dynamics of the signal into account. Nonlinear methods improve on the usual moving average noise filter by using a model of the dynamics to transport points from different times to a common time, and then averaging them together. As a result many more points can be averaged together without distortion of the signal, improving the noise reduction.

When points are transported from the past to the present, errors may be amplified. This amplification is usually direction dependent. In a chaotic

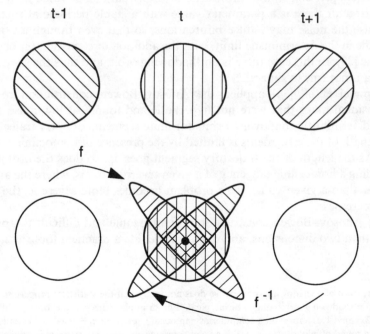

FIGURE 7.3. Nonlinear smoothing. The circles represent noisy measurements of a deterministic trajectory at three different times. Transporting successive measurements to the same time distorts the associated noise probability distributions. A correctly weighted average of the transported points produces a better estimate of the true value y_t. The estimates can be further improved by iterating this process.

dynamical system, on average the errors grow exponentially along the unstable manifold, and contract exponentially along the stable manifold. The opposite is true of points transported from the future to the present. Thus, points from the past are accurate along the stable manifold, whereas those from the future are accurate along the unstable manifold, as illustrated in Fig. 7.3. For good noise reduction it is essential to take this into account.

One method does this by assigning weighting matrices to the contribution of each point according to the derivatives of the dynamical system [31]. To derive the proper weighting matrices, the noise fluctuations are assumed to be IID Gaussians. A local linearization of the dynamics induces rules for pulling back and pushing forward probability distributions to different points in time. This produces weighting matrices that properly take into account the amplification of errors by the stable and unstable manifolds. For more details, see ref. [31].

7.5.2   Optimal Solution of Shadowing Problem with Euclidean Norm

An alternate method is based on shadowing [24]. The original Anosov–Bowen construction is concerned with the worst-case distance between points on the noisy orbit y and points on the shadowing orbit x. This amounts to using the L_1 or sup norm to measure the distance between two trajectories. Instead, we use a root-mean-square or Euclidean norm. The distance between two trajectory segments x_t and y_t is thus defined as

$$D(x, y) = \sqrt{\frac{1}{N} \sum_{t=1}^{N} \|y_t - x_t\|^2}, \qquad (7.16)$$

where $\|y_t - x_t\|$ is the Euclidean distance between the vectors x_t and y_t. The Euclidean norm is used in two senses, both to measure the distance between the two points x_t and y_t, and to measure the distance between two trajectories x and y.

Minimizing D is equivalent to minimizing D^2. For a given noisy trajectory segment y, we wish to find a trajectory segment x that minimizes $D^2(x, y)$, subject to the constraint that x is *deterministic with respect to* f, that is,

$$x_{t+1} = f(x_t) \qquad (7.17)$$

for $t = 1, 2, \ldots, N - 1$. This problem is straightforward to solve with the method of Lagrange multipliers. It is equivalent to minimizing

$$S = \sum_{t=1}^{N} \|y_t - x_t\|^2 + 2 \sum_{t=1}^{N-1} [f(x_t) - x_{t+1}]^{\dagger} \lambda_t, \qquad (7.18)$$

where $\dagger$ denotes the transpose and $\{\lambda_t\}$ are the Lagrange multipliers. The factor of two in front of the sum is a convenience that simplifies subsequent expressions.

Differentiating with respect to the unknowns and searching for an extremum gives the following system of equations:

$$\frac{\partial S}{\partial x_t} = 0 = -(y_t - x_t) + f'^{\dagger}(x_t)\lambda_t - \lambda_{t-1}$$

$$\frac{\partial S}{\partial \lambda_t} = 0 = f(x_t) - x_{t+1},$$

(7.19)

where $\lambda_t = 0$ if $t < 1$ or $t > N - 1$. $f'(x_t) = df/dx(x_t)$ is the Jacobian matrix of f at x_t. In the first equation t runs from 1 to N, whereas in the second t runs from 1 to $N - 1$. Typically f is nonlinear and this system of equations has no closed form solution.

Because Eq. (7.19) is nonlinear, numerical approximation must be used to solve it in general. Any of a large variety of search algorithms can be used, for example, Newton's method. A careful analysis shows that such a search requires inverting a matrix M. The structure of M complicates the search. To get the largest possible noise reduction we want to make N as large as possible. However, when the underlying dynamics are chaotic M becomes ill conditioned. Furthermore, approximate homoclinic tangencies cause it to be nearly rank deficient. Both of these factors make M difficult to invert when N is large.

7.5.3 Numerical Results

The details of implementing this algorithm are beyond the scope of this paper. They are described in more detail in ref. [24]. This section presents a few numerical results on two problems.

Noise was removed from a sample data set using an iterative procedure: approximate f, apply the noise reduction algorithm, and then approximate f again using the smoothed data, repeating until the results cease to improve. If a sufficiently large approximation neighborhood is used, the learned representation of f averages over the behavior of neighboring points, providing an initial noise reduction. Application of the noise reduction algorithm averages together points from other times as well, including those from distant parts of the state space, thus amplifying the initial noise reduction.

Fig. 7.4 demonstrates the result of noise reduction when the dynamics must be learned directly from the noisy time series. Fig. 7.4a shows a phase plot of a portion of the 5000 points in the noisy time series. Local quadratic fits were made over neighborhoods of 50 points to approximate the dynamics. The procedure of learning the dynamics and reducing the noise was iterated three times to produce the noise-reduced phase plot shown in Fig. 7.4b. For comparison, the original uncontaminated phase plot is shown in Fig. 7.4c. There are discrepancies between (b) and (c): there are some stray points caused by

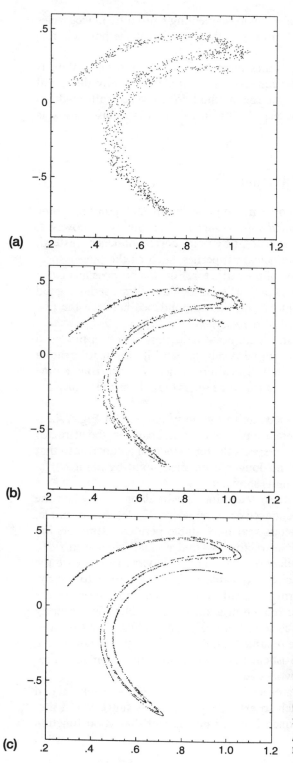

(a)

(b)

(c)

FIGURE 7.4. (a) A phase plot showing 1500 points of a noisy time series, obtained by adding uniformly distributed noise of magnitude 0.02 to each component of the data shown in (c). (b) The result of applying the noise reduction method of Farmer and Sidorowich to the data in (a). See text for details of the technique. (c) A plot of 1500 successive iterates of the Ikeda map with $\mu = 0.7$. This is the "true" data for (a), before adding noise.

homoclinic tangencies; there is also some segmenting caused by inaccuracies in the approximation algorithm. However, the majority of the points correspond in detail.

Reducing noise when the dynamics must be learned from noisy data is considerably harder than when clean data is available. Typically the initial noise levels must be lower to get the method to converge, although as shown in Fig. 7.4 it can be successful with initial signal-to-noise ratios as low as 50.

7.5.4   Statistical Noise Reduction

We have addressed the problem of removing noise from each point in a time series. This might be called *detailed noise reduction*. In some cases, however, we may not care whether there is a point-by-point correspondence. Instead, we may be interested only in statistical properties, such as the power spectrum, the dimension, or the shape of the attractor. For this purpose any clean time series with the same statistical properties as the "true" time series is good enough, even if there is no point-by-point correspondence between the two time series. We call this *statistical noise reduction*.

A very simple technique for statistical noise reduction is to fit a model to the noisy data, pick an arbitrary initial condition, and iterate it to create a new time series. In many cases the statistical properties of the surrogate time series are much closer to those of the true time series than those of the original noisy data.

The result of applying this procedure to the noisy data set of Fig. 7.4a is shown in Fig. 7.5. Although there are noticeable discrepancies, the attractor is reproduced fairly well. In some respects the reproduction is superior to that of Fig. 7.4b; however, there is no longer a detailed point-by-point correspondence to the set of "true" points shown in Fig. 7.4c.

This approach to statistical noise reduction can be dangerous when the dynamics displays sensitive dependence on parameters [27, 28]. For example, in some cases the set of parameters that have stable periodic attractors are dense within those with chaotic attractors, so that a small change in f can cause a change from chaotic behavior to periodic behavior. In this case the errors accumulated from the noise may cause an approximation to the dynamics whose statistical properties are radically different from those of the true time series. Detailed noise reduction avoids this problem by ensuring that the shadowing orbit stays close to the noisy orbit. Thus, application of detailed noise reduction is more reliable than the simple method above, even when the final goal is only statistical noise reduction. But when statistical noise reduction works, it often works very well.

The statistical noise reduction procedure described above is closely related to the "bootstrapping" approach to extending time series suggested in previous work [14, 18]. This approach is based on the fact that good function

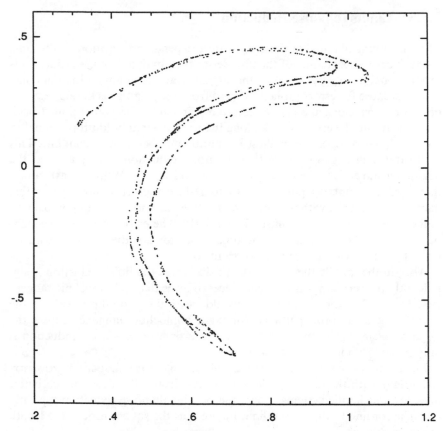

FIGURE 7.5. The result of applying the simple statistical noise reduction technique described in section 7.5.4 to the data of Fig. 7.4a. The overall features are quite similar, but in contrast to the reconstruction of Fig. 7.4b, there is no detailed correspondence in the points of the time series.

approximation schemes may give information about the underlying functional form that is accurate on length scales smaller than the typical separation between data points. Fitting a model and iterating can be used to artificially extend a data set, a procedure that can be very useful, for example, in computing fractal dimension. Casdagli [14] has demonstrated this method by approximating the invariant measures of several attractors based on only a small amount of data.

7.5.5 Limits to Noise Reduction

The maximum obtainable noise reduction depends on position in the time series. Near the beginning of the time series the obtainable noise reduction is proportional to $(\Lambda_s)^t$, where Λ_s is the largest Lyapunov number less than one. This is because for errors pushed forward from the past this is the slowest rate of contraction along the stable manifold. Similarly, for points pulled back from the future the errors shrink along the unstable manifold proportional to $(\Lambda_u)^{(N-t)}$, where Λ_u is the smallest Lyapunov number greater than one. This is illustrated in Fig. 7.6. Note that the noise reduction is very poor at the beginning and ends of the time series. (Refer to Fig. 7.3.) Without past values, there are no points to push forward to reduce the error along the stable manifold. Similarly, without future values there are no points to pull back to reduce the errors along the unstable manifold. The error improves exponentially moving from the end of the time series toward the middle, at a rate given by the relevant Lyapunov exponents of the map.

Also, in the middle there are large peaks where the noise reduction is significantly poorer than it is at other places. This is due to homoclinic tangencies. In Fig. 7.3, for example, this would mean the two ellipsoids have the roughly the same principal axis. An exact homoclinic tangency causes the same problem that occurs at the ends of the time series—noise reduction is only possible along one direction.

For a chaotic system the maximum obtainable noise reduction grows *exponentially* with the length of the time series. Ironically, this is much better than for nonchaotic systems, where the obtainable noise reduction typically grows according to the central limit theorem as the square root of the length of the time series.

The bad aspect for chaotic systems is that it is not possible to reduce noise significantly at the ends of the time series. This implies that there is a strong limit to how much noise reduction can be used to improve prediction. For the purposes of prediction, the future of the present state is unknown. Noise reduction can be used to reduce errors along the stable manifold, but not along the unstable manifold. Although noise reduction can be used to get a better approximation for f, it cannot reduce all the error on the present state.

For regular systems, in contrast, the obtainable noise reduction is more or less independent of the position in the time series. For regular systems noise reduction *can* be used to give significant improvements in predictions, but for chaotic systems it cannot. Also, regular systems do not suffer from the problems caused by homoclinic tangencies.

Continuous flows always have at least one Lyapunov number equal to one. This implies that we cannot reduce noise significantly in directions along trajectories. This is not usually a problem, however, because we are typically concerned with reproducing the form of a trajectory, and not with the detailed position of the samples. In high-dimensional systems, Lyapunov numbers close to one could cause a problem; in the directions where this

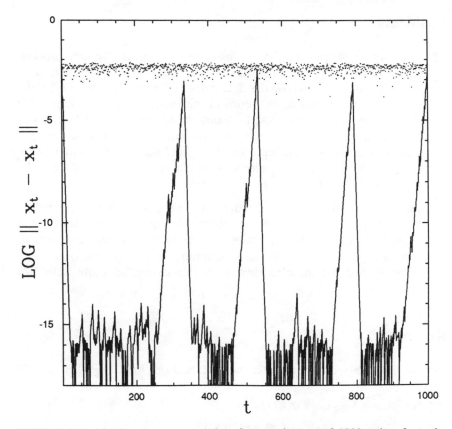

FIGURE 7.6. Absolute error versus time for a trajectory of 1000 points from the Hénon map. The initial noise level of .005 is indicated by scattered points. Here the map was assumed to be known, so the answers are much more accurate than those shown earlier. The solid curve is the noise level after two iterates of the optimal shadowing technique.

occurs the obtainable noise reduction is similar to that of regular dynamical systems.

In most practical applications where we must learn f, the real limit to noise reduction comes from our ability to approximate f, which in turn is limited by statistics. As demonstrated in [18], it becomes difficult to approximate f accurately when the dimension of the attractor or limit set is large. Thus, these noise reduction techniques are not likely to be effective for high-dimensional systems.

For more recent discussions of modeling chaotic systems, including prediction, control, noise reduction, and other topics, see [32].

Problems

7.1. A linear autoregressive *filter* is essentially a linear random process, whose "noisy inputs" are the signal to be filtered. The correlation function of the filter depends on the eigenvalue of the matrix that specifies the linear system. By picking the terms in this matrix properly it is possible to alter the spectral response; for example, to make a filter that passes low frequencies but does not pass high frequencies. Consider using a chaotic time series as the input to a filter, and show that this may cause the (Kaplan–Yorke) dimensionality of the output of the filter to be higher than that of the original time series.

7.2. Suppose the nonlinear noise reduction described earlier is applied to a two-dimensional map. Moving from the beginning of a time series toward the middle, which Lyapunov exponent describes the limit to nonlinear noise reduction? What about moving from the end of the time series toward the middle? How does this change in more than two dimensions?

References

[1] B.A. Huberman, J.P. Crutchfield, and N.H. Packard, Appl. Phys. Lett. **37**, 750 (1980).

[2] G.J. Chaitin, Sci. Am. **232**, 47 (1975).

[3] S. Wolfram, Phys. Rev. Lett. **55**, 449 (1985).

[4] J. Ford, Phys. Today **36**, 40 (1983).

[5] H. Froehling, J.P. Crutchfield, J.D. Farmer, N.H. Packard, and R.S. Shaw, Physica D **3**, 605 (1981).

[6] J. Guckenheimer, Nature **298**, 358 (1982).

[7] N.H. Packard, J.P. Crutchfield, J.D. Farmer, and R.S. Shaw, Phys. Rev. Lett. **45**, 712 (1980).

[8] F. Takens, in *Dynamical Systems and Turbulence*, edited by D. Rand and L.-S. Young (Springer-Verlag, Berlin, 1981).

[9] A.M. Fraser and H.L. Swinney, Phys. Rev. A **33**, 1134 (1986).

[10] G.U. Yule, Philos. Trans. Roy. Soc. London A **226**, 267 (1927).

[11] D.S. Broomhead and G.P. King, Physica D **20**, 217 (1987).

[12] J.F. Gibson, J.D. Farmer, M. Casdagli, and S. Eubank, Physica D **57**, 1 (1992).

[13] M. Casdagli, S. Eubank, J.D. Farmer, and J. Gibson, Physica D **51**, 52 (1991).

[14] M. Casdagli, Physica D **35**, 335 (1989).

[15] J. Cremers and A. Hübler, Z. Naturforsch. **42a**, 797 (1987).

[16] J.P. Crutchfield and B.S. McNamara, Complex Systems **1**, 417 (1987).

[17] J.D. Farmer and J.J. Sidorowich, Phys. Rev. Lett. **59**, 845 (1987).

[18] J.D. Farmer and J.J. Sidorowich, in *Evolution, Learning and Cognition*, edited by Y.C. Lee (World Scientific, Singapore, 1988).

[19] A. Lapedes and R. Farber, Technical Report LA-UR-87-2662, Los Alamos National Laboratory (1987).

[20] E.N. Lorenz, J. Atmos. Sci. **26**, 636 (1969).

[21] M.F. Barnsley and S. Demko, in *Chaotic Dynamics and Fractals*, edited by M.F. Barnsley (Academic, New York, 1986).

[22] E.J. Kostelich and J.A. Yorke, Phys. Rev. A **38**, 1649 (1988).

[23] S.M. Hammel, Technical Report, Naval Surface Warfare Center, Silver Spring, MD 1989.

[24] J.D. Farmer and J.J. Sidorowich, Physica D **47**, 373 (1991).

[25] D.V. Anosov, Proc. Steklov Inst. Math. **90**, 45 (1967).

[26] R. Bowen, J. Diff. Eq. **18**, 92 (1975).

[27] J.D. Farmer, Phys. Rev. Lett. **55**, 351 (1985).

[28] E.N. Lorenz, Telus **6**, 1 (1964).

[29] B.A. Huberman, J. Crutchfield, J.D. Farmer, Phys. Rep. **92**, 45 (1982).

[30] S.M. Hammel, J.A. Yorke, and C. Grebogi, J. Complexity **3**, 136 (1987).

[31] J.D. Farmer and J.J. Sidorowich, in *Dynamic Patterns in Complex Systems*, edited by J.A.S. Kelso, A.J. Mandell, and M.F. Shlesinger (World Scientific, Singapore, 1988).

[32] H.D.I. Abarbanel, *Analysis of Observed Chaotic Data* (Springer-Verlag, New York, 1995); H.D.I. Abarbanel, R. Brown, J.L. Sidorowich, L.Sh. Tsimring, Rev. Mod. Phys. **65**, 1131 (1993); *Nonlinear Modeling and Forecasting*, edited by M. Casdagli and S. Eubank (Addison-Wesley, Reading, MA, 1992); D. Kaplan and L. Glass, *Understanding Nonlinear Dynamics* (Springer-Verlag, New York, 1995); E. Ott, *Chaos in Dynamical Systems* (Cambridge University, Cambridge, 1993); S. Strogatz, *Nonlinear Dynamics and Chaos* (Addison-Wesley, Reading, MA, 1994). See also the extensive bibliography of chaos and nonlinear dynamics maintained at http://www.uni-mainz.de/FB/Physik/Chaos/chaosbib.html, and the useful links to these topics that can be found at http://cnls-www.lanl.gov/nbt/intro.html.

III

Pattern Formation and
Disorderly Growth

8

Phenomenology of Growth

Leonard M. Sander

8.1 Aggregation: Patterns and Fractals Far from Equilibrium

In these lectures we will try to give an idea of a new picture of rapid growth of *extended structures* (crystals, films, colloids, etc.), which has arisen in recent years by using simple (but nonlinear) models that often produce fractals. These models give a good deal of insight into the fact that fractals are often observed in nature in situations that are very far from equilibrium.

Much of the activity in this area was inspired by the work of Witten and Sander [1] on one of the models that we will discuss below, the diffusion-limited aggregation (DLA) process. In recent years several reviews have treated these subjects from various points of view. The student is particularly encouraged to consult the recent book of Vicsek [2] and the review article of Sander [3], and the forthcoming book of Meakin [3].

When a physical system grows, matter is added from outside. Let us consider a simple type of growth, *irreversible aggregation*; that is, imagine the formation of a large structure by the repeated random sticking of subunits, for example, atoms or molecules, each obeying the same dynamics. After the units stick, we assume that they have no subsequent motion, except to move with the aggregated structure. We will ask whether we can predict anything useful about the aggregate when it has become very large compared to the individual constituents. Aggregation is an extreme idealization of irreversible growth.

A bit of thought will reveal that there are two subclasses of this type of aggregation that are useful to distinguish. On the one hand, subunits (we will usually call them particles whether they are atoms, molecules, or even larger bits of matter) can be added one at a time during the formation of the aggregate. This is referred to as "particle aggregation." On the other hand, we can consider an atmosphere of particles in random motion that stick together to form dimers when they encounter one another. In many common situations, the dimers then continue to wander and combine with other dimers (and with the remaining particles) to form larger clusters, and so on. This general class

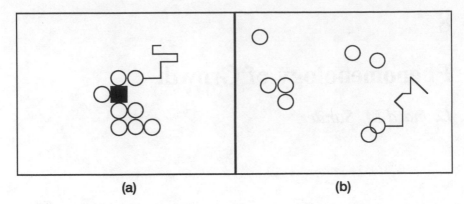

FIGURE 8.1. (a) Particle aggregation; (b) Cluster aggregation.

of process is called "cluster aggregation." A familiar example is the formation of a colloid or an aerosol such as a smoke, see Fig. 8.1. We will discuss both of these processes below.

Irreversible aggregation is quite different from the formation of objects
near equilibrium. When an equilibrium crystal grows, atoms aggregate, but
then they undergo relaxation and rearrangement to find the lowest energy
configuration. Growth is slow compared to relaxation rates. The internal
structure and external shape of an object formed by slow growth are describable by ordinary equilibrium statistical mechanics. In the case of crystallization, the equilibrium shape is usually relatively simple.

As we drive the system away from equilibrium, we often find a new regime
of morphology, that of pattern formation. In this case competition between
the ordering processes (that dominate the equilibrium) and the growth sometimes gives rise to intricate patterns, which can persist even for large objects.
Examples of these are the beautiful feathery shapes of dendritic crystals (e.g.,
snowflakes). For this case the symmetry of the equilibrium crystal is retained,
and it controls the shape of the structure.

Even farther from equilibrium a new regime can appear: disorderly, irreversible growth. In our idealization of irreversibility in terms of aggregation,
once a particle has added to the aggregate, it remains attached in that place
without any subsequent relaxation or detachment; growth is (infinitely) fast
compared to relaxation. Of course, in nature, we are never in either extreme
limit, but it is useful to study them both. In fact, as we will see, processes in
nature often take place sufficiently far from equilibrium so that the aggregation approximation is useful. In this case the shape of the cluster encodes its
history: we will always be dealing with situations involving very long-term
memory. This is what makes this subject entirely different from equilibrium
physics.

In this regime it is remarkable (and the key to much of the progress we have
made) that even though there is usually no hint of the symmetry of the equilibrium structure left, the growths need not be merely amorphous, but exhibit

scaling. The fact that this can occur was first demonstrated experimentally for a cluster aggregate of metal smoke particles. Theoretical study was initiated by Witten and Sander [1]. The fractal geometry discussed in Part I is vital for the analysis of growth in this regime.

8.2 Natural Systems

In this section we will describe, in general terms, the physical processes of growth for a few typical systems. These will be discussed in detail in the rest of this section. We have chosen a few examples from the wealth of possibilities in the literature to illustrate the general points we will make. In all the cases we discuss, the microscopic growth process has been realized in a model that sometimes makes fractals.

8.2.1 Ballistic Growth

Thin films are important in many industrial technologies. The most common way to grow a thin film is by the process of vapor deposition: a chamber is prepared with a furnace or other source of particles, which are allowed to fall on a surface (the substrate) and build up the film.

To visualize this process, materials scientists (see Leamy et al. [4]) have devised a simple model. In Fig. 8.2a we show how the process works. An atom drops in a straight line from far away and attaches (aggregates, in fact) on the substrate or on the other atoms. The key to the nonlinearity and complicated nature of the process is that atoms sometimes stick to the *sides* of other atoms. In Fig. 8.2b we show a simplified version of the model. In this case the atoms are constrained (artificially, but it is a very useful approximation) to live on the points of a regular array, a lattice. Aggregating atoms

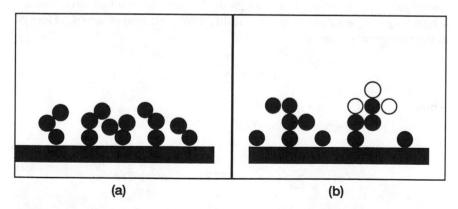

(a)   (b)

FIGURE 8.2. (a) Off-lattice ballistic aggregation. (b) Ballistic aggregation on a lattice. The white circles are some possible growth sites.

are added to either the top of a column, or to the side of the adjacent column, whichever is higher.

The interest in this process is the fact that the upper surface produced here is rough, with a roughness that grows with the thickness of the sample. In fact, as we will see in the next chapter, the roughness is a quantity with scaling properties: the average thickness of the surface grows as a power law of the height. The upper surface is a self-affine fractal. A large literature exists on this process, and on applications to real film growth. The interested student should consult the books of Family and Viscek [4], Barabasi and Stanley [4], and the review article of Halpin-Healy and Zhang [4].

The geometry of growth described above is the closest to the physical process being considered; namely, the growth of a deposit from a substrate. However we can equally well consider (and we will find it useful to do so) the growth of a cluster from a center, say a point in our evaporation chamber. In this case we can take the trajectories to be random in direction. For large clusters the roughness of the surface is similar to that of the deposits considered above.

Example 8.1 The ballistic aggregation process is entirely nontrivial because of the presence of coupling between the columns. Show that without the coupling the lattice process is trivial, and the roughness of the surface grows as $\sqrt{H}$, where H is the average height. The most convenient definition of roughness, σ, which we will use throughout, is the root-mean square deviation of the height from the average:

$$\sigma^2 = \langle (h_i - H)^2 \rangle = \langle h_i^2 \rangle - H^2. \tag{8.1}$$

This reduces to a classic problem in combinatorial mathematics, the solution to which can be obtained from the famous Poisson distribution. We will give a simplified solution. Because each column is independent, n_i, the number of particles in column i, is the sum of independent random variables, q_t, which take on the value 1 with probability $1/N$, and zero otherwise. Here N is the number of columns:

$$n_i = \sum_{t=1}^{T} q_t.$$

Thus:

$$\langle n_i \rangle = T \langle q \rangle = T/N$$

$$H = aT/N$$

$$\langle n_i^2 \rangle = \sum_{ts} \langle q_t q_s \rangle = T(T-1)/N^2 + T/N$$

$$\sigma^2 = a^2 \{ \langle n_i^2 \rangle - \langle n_i \rangle^2 \} \approx aH. \tag{8.2}$$

In the second to last line we have used the fact that $\langle q_t q_s \rangle = 1/N^2$, if $t \neq s$, and $1/N$ if $t = s$. We assume that $N \gg 1$.

8.2.2   Diffusion-Limited Growth

In many physical cases, growth is not controlled by the arrival of particles in a straight line. For example, it often occurs that the mean free path of the particles is short and they arrive after many steps of a random walk. That is, the growth is controlled by diffusion. We will give below some physical examples of the sort of process that we are concerned with.

A model that reproduces much of the physical nature of the growth in this case was introduced by Witten and Sander [1]. It is called diffusion-limited aggregation (DLA). In this case, a substrate is prepared as before, and particles are allowed to wander in from far away, one at a time, by a random walk and attach to the substrate or the other particles, whichever they encounter first. Once more, we can also think about the growth of a cluster from a point. The two cases are illustrated in Fig. 8.3. As we will see, DLA deposits and clusters are not merely rough, as ballistic deposits are; they are complex, disorderly fractals. A small example is given in Fig. 8.3. In nature, we often see fractals that are of this general type, and we will try in what follows to show what features of the model are physically realistic and why it might be similar to what happens in nature.

Before we introduce the specific physical systems we will attempt to describe with the DLA model, it is useful to introduce a set of equations that describe

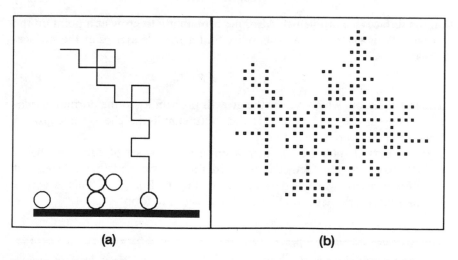

(a)   (b)

FIGURE 8.3. (a) Diffusion-limited aggregation. The track of a particle is shown. (b) A diffusion-limited aggregation cluster.

the process. As we will see, the first two physical examples we will introduce are more or less well described by the model as it stands. The other two examples have the same mathematical description in the limit of large sizes, but do not involve particles aggregating. The DLA model still serves to solve the macroscopic equations in this case.

The random walker may be described by a function, $u(\mathbf{r}, t)$, which gives the probability to find the particle at a given position at a given time. The function, u, satisfies a well-known equation [5]:

$$u(\mathbf{r}, t + \tau) - u(\mathbf{r}, t) = \frac{1}{z} \sum_{\delta} [u(\mathbf{r} + \delta, t) - u(\mathbf{r}, t)]. \tag{8.3}$$

This equation simply says that the walker has an equal probability to jump from any z neighboring sites located at $\mathbf{r} + \delta$ to the site in question at any time. This is the discrete version of the continuum diffusion equation:

$$\partial u / \partial t = \eta \nabla^2 u, \tag{8.4}$$

where the diffusion constant, η, is given by $\delta^2 / z\tau$. In the case of a single walker at a time we can neglect the right-hand side of the equation because the behavior we are interested in is independent of the time it takes to attach a single particle: the average over realizations of the random walk depends only on position, not on time. Thus:

$$\nabla^2 u = 0. \tag{8.5a}$$

The boundary conditions for the equation are that there be a finite flux onto the growing aggregate from infinity, and that the particles be absorbed on the surface. This last condition is equivalent to the condition

$$u|_s = 0 \tag{8.5b}$$

on the surface (see problems). Also, the probability to grow at a point on the surface, P_s, is given by the probability that a particle arrives at the surface. Thus:

$$P_s \sim u(\mathbf{r} + \delta) \sim \partial u / \partial n|_s. \tag{8.5c}$$

The last equation, which relates the growth probability to the normal derivative on the surface, is to be used in the continuum limit. The average growth velocity, $v|_s$, is proportional to $\partial u / \partial n|_s$.

In the next chapter we will demonstrate that this set of equations has a growth instability. In fact, it is believed that no simple, time-independent structure is stable when it develops according to Eq. (8.5). This is why a complex disorderly structure like a DLA cluster can form.

Example 8.2 Show that for a DLA cluster growing in three dimensions the growth rate of the number of particles is proportional to the mean radius, provided the probability is kept fixed far away.

This result is important in showing that certain bounds exist on the fractal dimension of DLA clusters. To show it, take a large sphere, radius R, which is far away from the cluster, and a small sphere whose radius ρ is typical for the pattern. Now the solution to Eq. (8.5) is easy to write between the spheres:

$$u(r) = A + B/r.$$

Suppose that u attains some value u_∞, at R, and we require it to be 0 on the small sphere (this defines the "typical radius," ρ). Then, matching coefficients,

$$u = u_\infty [1 - \rho/r]. \tag{8.6}$$

Now the total amount of matter added, dM/dt, is the total over the surface of $P|_s$:

$$dM/dt \sim \pi\rho^2 \partial u/\partial n|_s \sim \rho.$$

Crystallization

In certain cases, crystal growth has almost exactly the same description as DLA, as we will now see. When a crystal grows far from equilibrium there are many physical effects that must be considered. For a careful discussion, the student should consult one of several excellent recent review articles and the book edited by Godreche [6].

In many cases, the bottleneck for growth is the arrival of matter from outside, say in the solidification of a crystal from a dilute solution. (The more commonly studied case of limitation by the diffusion of latent heat is mathematically identical). The concentration of matter is certainly given by Eq. (8.4), the diffusion equation. We can, in this case, estimate the size of the term $\partial u/\partial t$ by noting that if there is a typical velocity of growth, v, then $\partial u/\partial t \sim v\partial u/\partial x$. Now:

$$|\nabla^2 u| \approx (1/L_d)|\partial u/\partial x|, \tag{8.7}$$

where $L_d = \eta/v$, the diffusion length, sets the scale for the diffusion field. In practical cases L_d is sometimes very much larger than the other scales in the problem, so that the right-hand member of Eq. (8.7) can be neglected, and we are back to (8.5a).

The boundary condition on the concentration at a flat surface of the growing crystal is given by the equilibrium concentration at that point, which we can take to be our zero for u. Far from the growing crystal u is fixed at some positive value (the solution is supersaturated). However, there is an important complication that arises from classical thermodynamics; because the pressure inside a curved surface is different from that outside,

$$\Delta P = P_{\text{in}} - P_{\text{out}} = \gamma\kappa. \tag{8.8}$$

Here γ is the surface tension and κ the curvature. The equilibrium condition is

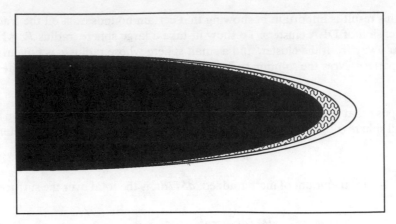

FIGURE 8.4. A rough sketch of a dendrite showing stages of growth. Real dendritic crystals have side branches in addition to the parabolic shape sketched.

shifted proportional to ΔP. Then the boundary condition is

$$u|_s \sim \gamma \kappa. \tag{8.9}$$

This equation is different from Eq. (8.5b), and controls the formation of orderly patterns when they exist. We will argue below that this condition, which means that very curved surfaces act as if the supersaturation is smaller, and grow slower, acts as a cutoff quite analogous to the finite particle size in the DLA model. However, the student should be warned that this point (that the two sorts of cutoff act in a similar manner) is quite controversial, and not accepted by many experts in the field.

The final boundary condition says that matter arrives at the interface by diffusion. The diffusion current is ∇u and thus its normal component at the surface is proportional to the growth rate there. Thus, the growth velocity can be written:

$$v \sim \partial u / \partial n|_s. \tag{8.10}$$

This is a version of Eq. (8.5c).

The usual interpretation of the system (8.7), (8.9), and (8.10) is as a set of *deterministic* equations to be solved for a steady-state shape given some suitable initial conditions. An example of a steady-state solution is a *dendrite*, sketched in Fig. 8.4, which is often observed in nature, for example as one of the branches of a snowflake. In this case there is a roughly parabolic tip that advances in time without changing shape. The DLA model produces something quite different, an ensemble of disorderly growths; however, the two cases arise from essentially the same physics. The relationship between the two points of view will be explored in chapter 9 (section 9.2). The feature that we have left out, and whose importance will be elucidated below, is *anisotropy*.

Electrochemical Deposition

Electrochemical deposition of metals is a very old subject with an extensive literature [7]. In the particular version we are interested in (see Fig. 8.5) metal is deposited on a cathode from an electrolytic solution. In this case we are dealing with an almost literal realization of the DLA model provided the electric field in the solution is not important for the motion of the ions over most of the bulk. This is the case in the experiments of Brady and Ball [8], who considered the electrodeposition of copper from $CuSO_4$ and found fractal deposits. To ensure that the electric force is not important in this case they added a second component to the solution to screen the fields; the ions diffuse freely and are described by the same equations as in the last section. We will discuss this experiment in detail in the next chapter. The resulting deposits are of the DLA type.

In another set of experiments [9–11], it appears that the driving force for the motion of the ions *is* the electric field. The ions follow field lines. However, in this case we are once more in a regime that can be treated by the same model equations (Eqs. 8.5) because we can reinterpret the u to mean the electrostatic potential. Equation (8.5a) is the classic Laplace equation for a grounded conductor (8.5b), which grows when on ohmic current arrives at the surface. The current, by Ohm's law, is proportional to $\partial u/\partial n$ (i.e., the electric field) at the surface. The process is DLA-like (and can produce patterns very like DLA clusters) but is really more similar to the process described in the section on dielectric breakdown, than to diffusion-limited electrodeposition. Here, again, there is the possibility of unstable growth giving rise to DLA-type fractals, but also stable, dendritic crystals, depending on the experimental conditions.

Since the first generation of experiments in this area many groups have

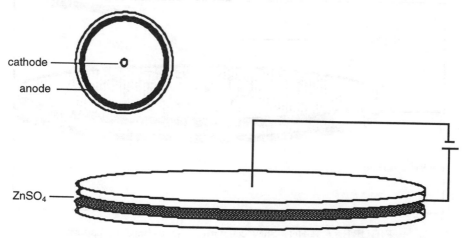

FIGURE 8.5. An electrochemical cell of the type used by Grier et al. [11].

taken a careful look at the microscopic processes that underlie the pattern formation in electrochemical cells. For an example of the recent literature, the student should consult article Kuhn and Argoul [11].

Viscous Fingering

Viscous fingering is a less familiar example of a diffusion-limited process, but it is remarkably similar to the examples cited above, and has great technological importance as well. It refers to the unstable motion of the interface of two fluids, one more viscous than the other, when the less viscous one is injected into the more viscous [12]. In technology, the fingering instability is a problem for oil recovery processes in which air or water is injected into an oil field to drive the oil to a distant well. Clearly, if the pattern of the interface is complex, it is difficult to know where to put the well.

The phenomenon occurs not only in flow in a porous medium such as an oil-bearing rock, but also for flow between two plates, a so-called Hele–Shaw cell. It is illustrated in Fig. 8.6. The pattern of the relatively inviscid injected fluid plays the role of the crystal in the sections above, or the DLA cluster.

To relate this process to what we have discussed so far, we need only point out that flow in porous media or a Hele–Shaw cell is described by an empirical rule called D'Arcy's law [12], which relates the fluid velocity, v, to the gradient of the pressure, u, in the viscous fluid:

$$v = -K\nabla u. \tag{8.11}$$

The proportionality constant is given by $K = k/\mu$, where k is called the permeability and μ is the viscosity. For a Hele–Shaw cell it can be shown that $k = b^2/12$, where b is the distance between the plates.

We note that most fluids are almost incompressible. This implies that

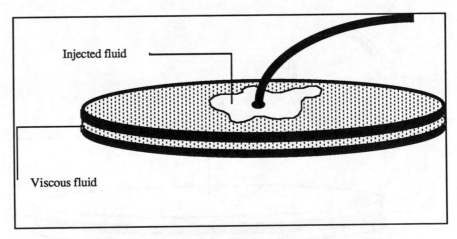

FIGURE 8.6. A Hele–Shaw cell with a viscous fingering pattern.

$\nabla \cdot \mathbf{v} = \nabla^2 u = 0$. We take the zero of pressure to be that in the inviscid fluid (because it has small viscosity, its pressure there is approximately constant). Thus $u|_s = \gamma\kappa$, because, from Eq. (8.8) there is a pressure drop due to curvature. And D'Arcy's law itself, applied at the interface between the two fluids, gives the third ingredient needed; namely, exactly Eq. (8.10) in the form $v|_s = -K\partial u/\partial n$.

Thus we have another example of the kind of diffusion-limited growth that should be describable using DLA. Experiments on large cells with the geometry shown in Fig. 8.6 show fractal behavior over a large range of scales. Experiments in long straight channels display deterministic growth of a single "finger."

Dielectric Breakdown

Our final detailed example of a DLA-type process is the breakdown of a dielectric material when it is exposed to a very large electric field. Niemeyer et al. [13] showed that the pattern formed in some cases is exactly that of DLA.

Suppose we have a material with two electrodes with a large potential between them. Let u be the electrostatic potential between them. We have the Laplace equation for u, Eq. (8.5a), at once. The breakdown channel where the material is highly ionized has a large conductivity, and thus a constant potential, Eq. (8.5b). And, it is reasonable to suppose (though may depend on the material in question) that the probability of breakdown is proportional to the electric field, ∇u, as in Eq. (8.5c).

Other Patterns of the DLA Type

The DLA process is remarkable in that it shows up, at least in approximate form, in many different physical contexts. For example, when bacteria grow on agar in a petri dish they normally form a colony that is roughly round and whose surface is rather rough. In fact, they are similar in form to the ballistic aggregates that we described previously. We will discuss this more below. However, if the bacteria are *starved* their growth is limited by the diffusion to the colony of the nutrient material on the agar from outside. This leads to remarkable effects, and in many cases to living colonies in the form of DLA clusters. For this literature the student should consult Vicsek's book [2].

Another example (of many possible) is the formation of islands on a thin film during vapor deposition. In this process (which is very important industrially) atoms can land on a surface and diffuse around until they nucleate islands, then the islands grow until a layer is formed, and then the process starts again. This is called layer-by-layer growth, and is rather common. The remarkable thing is the shape of the islands: because their growth is limited by diffusion they have a sprawling, fractal shape unless *edge diffusion* is so rapid, the atoms run around the edge of the island and smooth it out as it grows. See [4b] and [13a].

8.2.3   Growth of Colloids and Aerosols

When an atmosphere of particles can diffuse and combine, forming dimers, trimers, and so forth, we have a different sort of physics than that which we have dealt with so far. In colloidal solutions or the formation of smokes and other aerosols, the important process is the *aggregation of aggregates*. This has long been recognized in the community of chemists who deal with such processes, and is based on the classic work of Smoluchowski [14].

In recent years a model [15] has been invented to deal with this process, and the model produces fractal aggregates. It bears the name of cluster–cluster aggregation. In the model one begins with a large collection of particles, each of which is allowed to diffuse until it encounters another. Then the resulting cluster continues to move (perhaps with a smaller diffusion coefficient, though this is not very important) until large clusters result. In this case, as for DLA, the clusters are self-similar fractals.

Problems

8.1. It is relatively easy to write a computer program for the ballistic aggregation model for a one-dimensional substrate. Do so on a substrate of base 100, and let your film grow to a mean height of 200. Plot at each step the mean height and σ. It will be useful for you to show that the rule for ballistic aggregation is the following: if i denotes the column chosen at the time step in question, then:

$$n_i(t + \tau) = \text{Max}[n_{i(t)+1}, n_{i\pm1}(t)].$$

8.2. Consider a random walk in one dimension and suppose a "wall" at $x = 0$ absorbs walkers that arrive. Show that in this case Eq. (8.3) must be solved with boundary condition $u(0, t) = 0$.

8.3. Show that if a concentration field is described by a diffusion equation, then the current density is $-\eta\nabla u$.

8.3. Show that L_d has the units of length.

8.4. What is the analog of the diffusion length for Hele–Shaw flow? (Hint: in order that the Laplace-like equation turn into a diffusion equation, the fluid cannot be completely incompressible).

References

[1] T.A. Witten and L.M. Sander, Phys. Rev. Lett. **47**, 1400 (1981); T.A. Witten and L.M. Sander, Phys. Rev. B **27**, 5686 (1983).

[2] T. Vicsek, *Fractal Growth Phenomena* (World Scientific, Singapore, 1992).

[3] L. Sander, Nature **322**, 789 (1986); P. Meakin, *Fractal Scaling and Growth Far From Equilibrium* (Cambridge University, Cambridge, in press).

[4] H. Leamy, G. Gilmer, and A. Dirks, in *Current Topics in Materials Science*, Vol. 6 (North-Holland, Amsterdam, 1980); F. Family and T. Vicsek, *The Dynamics of Growing Interfaces* (World Scientific, Singapore, 1992); A. Barabasi and H.E. Stanley, *Fractal Concepts in Surface Growth* (Cambridge University, Cambridge, 1995); T. Halpin-Healy and Y.-C. Zhang, *Kinetic Phenomena*, Phys. Rep. **254**, 215 (1995).

[5] L.E. Reichl, *A Modern Course in Statistical Physics* (University of Texas, Austin, 1980).

[6] J.S. Langer, Rev. Mod. Phys. **52**, 1 (1980); D. Kessler, J. Koplik, and H. Levine, Adv. in Phys. **37**, 255 (1988); G. Godreche, Solids Far from Equilibrium. (Cambridge University, Cambridge, 1991).

[7] J. Bockris and A. Reddy, *Modern Electrochemistry* (Plenum, New York, 1970).

[8] R. Brady and R.C. Ball, Nature (London) **309**, 225 (1984).

[9] M. Matsushita, M. Sano, Y. Hayakawa, H. Honjo, and Y. Sawada, Phys. Rev. Lett. **52**, 286 (1984).

[10] A. Dougherty, P.D. Kaplan, and J.P. Gollub, Phys. Rev. Lett. **58**, 1652 (1987).

[11] D. Grier, E. Ben-Jacob, R. Clarke, and L.M. Sander, Phys. Rev. Lett. **56**, 1264 (1986); A. Kuhn and F. Argoul, Phys. Rev. E **49**, 4298 (1994).

[12] D. Bensimon, L.P. Kadanoff, S. Liang, B. Shraiman, and C. Tang, Rev. Mod. Phys. **58**, 977 (1986).

[13] L. Niemeyer, L. Pietronero, and H. Weismann, Phys. Rev. Lett. **52**, 1033 (1984); L.M. Sander, "The Diffusive Instability in Growth: DLA, Dendritic Growth, and Solidification," Symposium of Fractal Aspects of Materials, edited by F. Family et al. Proceedings: Materials Research Society Pittsburgh (1994).

[14] M. Smoluchowski, Z. Phys. Chem. **92**, 129 (1915).

[15] P. Meakin, Phys. Rev. Lett. **51**, 1119 (1983); M. Kolb, R. Botet, and R. Jullien, Phys. Rev. Lett. **51**, 119 (1983).

9

Models and Applications

Leonard M. Sander

In the previous chapter we introduced the fundamental models that we will analyze in more detail here. Because many simulations of these models have been performed, we will give a review of the current situation. In addition, for the particular case of diffusion-limited growth, we will present a possible view of the troublesome question of the relationship between orderly and disorderly patterns. We will briefly review the experimental situation.

9.1 Ballistic Growth

9.1.1 Simulations and Scaling

When the ballistic growth model of the previous chapter was first introduced, it was not realized that the surfaces produced were self-affine fractals in some regimes. This property is important because it means that scaling should occur in the model for many scales of length. If that is so, the short-range details of the potential between the atoms and the surface, the exact form of the potential, and so forth, should not be important for long-range properties. As in the case of equilibrium systems near critical points or dynamical systems, only the general features of the growth should affect the scaling. The model should be *universal*. For a review of this property see [1].

We first point out a simple property: ballistic aggregates are not fractal in the usual sense. Careful numerical simulations verify this point [1, 2]. It is most easily proved using Ball and Witten's argument [3]. Consider a ballistic cluster growing in a radial geometry. Suppose there is a fixed flux of particles from far away at random angles. Clearly the growth rate of the mass will be proportional to the surface area. In d dimensions this is

$$dM/dt \sim \rho^{d-1}, \qquad (9.1)$$

where ρ is a mean radius of the cluster. Now

$$dM/dt = \frac{dM}{d\rho}\frac{d\rho}{dt} \sim \rho^{D-1}v. \qquad (9.2)$$

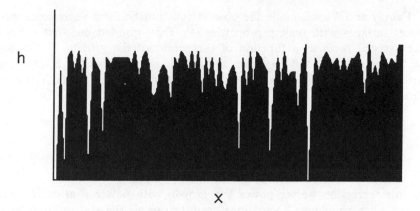

h

X

FIGURE 9.1. A ballistic aggregation deposit.

The speed, v, is the rate of increase of the mean radius, and D is the fractal dimension (see Part I). Now we can surely assume that v is bounded. Thus we find:

$$\rho^{d-D} \leq \text{const.} \tag{9.3}$$

The only way for this to happen is for $d \leq D$. But because the fractal dimension cannot exceed the dimension of space, $d = D$. Thus, the bulk of a cluster grown is solid and the density achieves a constant value. The surface is rough, but its thickness, σ, must grow more slowly than the mean height, H. An example of one of these rough surfaces is given in Fig. 9.1.

Example 9.1 Give an argument to show that ballistic aggregates are not fractal in the deposition geometry of Fig. 8.2.

The simplest argument is indirect. Suppose the deposit is fractal. Then its mean density per unit height decreases without bound:

$$dm/dh \sim h^{-k}. \tag{9.4}$$

For this to happen, we must never "fill in" the deposit. Suppose that the deposit has essentially reached its limiting density below a certain height, H_1, and let it grow to H_2. If $H_{1,2}$ are large enough, then the number of particles between them (of order $[H_2 - H_1]H_1^{-k}$) can be made as small as we like; in particular, smaller than $[L/a]^{d-1}$, where L is the width of the $d - 1$ dimensional substrate. But then there are "holes" in the layer, and we will fill in below H_1; the probability of falling below H_1 is just the relative size of the holes, and thus the probability of an appreciable change in dm/dt below is as large as we like, a contradiction. This argument fails for diffusing particles, as we will see.

Family and Vicsek made the observation that ballistic aggregation produces surfaces with scaling properties [4]. They pointed out that σ has a very special form as a function of the height of the growth. If the simulation starts from a flat surface, the roughness increases for a while and then saturates at a time that depends on the width of the deposit. The exact form is:

$$\sigma = L^{\alpha} f(H/L^z)$$

$$= \left\{ \begin{array}{ll} H^{\beta}; & H \ll L^z \\ L^{\alpha}; & H \gg L^z \end{array} \right\}. \tag{9.5}$$

In this expression we see power law growth with power β at early times. Later, when roughness fluctuations begin to sample the system boundaries, the roughness attains a steady-state in the characteristic fashion of a self-affine fractal. The scaling function, f, interpolates between the regimes. For the roughness to be independent of L at early times, when the "sideways" growth has not had time to reach the boundaries, and the surface is made up of independently fluctuating pieces, we must postulate that:

$$f(y) \sim y^{\beta}; \qquad \beta z = \alpha. \tag{9.6}$$

The scaling exponents α, z (or α, β) characterize the scaling. The surface is not fractal and the roughness grows more slowly than the height or the width. Thus $\alpha, \beta < 1$. We will see that $z > 1$.

Example 9.2 It is interesting to motivate the assumption (9.5) by looking at a simpler model, due to Edwards and Wilkinson [5] and studied in the present context by Sander [6]. It is a *linear* version of ballistic aggregation in which we replace the nonlinear function, Max, in ballistic aggregation (cf. problem 8.1) by the average, and linearize with respect to the noise. That is, consider a growth process in which the rule for the chosen column is:

$$h(i, t + \tau) = Ave[h(i, t) + a, h(i \pm 1, t)]. \tag{9.7}$$

Show that Eq. (9.6) holds with $\alpha = 1/2$ and $\beta = 1/4$ in two dimensions.

This is most easily done by noting that we can rewrite Eq. (9.7) defining a noise variable q_i, as in Example 8.1, which takes value 1 for column i, chosen at random, and zero for the rest. Then for all i

$$h(i, t + \tau) = q_i Ave[h(i, t) + a, h(i \pm 1, t)] + (1 - q_i) h(i, t)$$

$$h(i, t + \tau) - h(i, t) = q_i Ave[-2h(i, t) + a, h(i \pm 1, t)]$$

$$= \{q_i \nabla_i^2 h(i, t) + q_i^a\}/3$$

$$\approx k\nabla_i^2 h(i, t) + ka + \zeta. \tag{9.8}$$

Here, $\nabla_i^2 h(i, t) = h(i + 1, t) + h(i - 1, t) - 2h(i, t)$ is the lattice Laplacian, $k = \langle q_i \rangle / 3$, and $\zeta = a[q_i - \langle q_i \rangle]/3$ is a fluctuating noise term. We have neglected the term in $\zeta \nabla_i^2 h$. Equation (9.8) is the lattice version of a diffusion equation driven by noise: if we put $h - kat = s$ and coarse-grain in space, then

$$\partial s / \partial t \approx [ka^2 / \tau] \nabla^2 s(x, t) + a\zeta(x; t). \qquad (9.9)$$

This equation can be solved by a Fourier transformation:

$$s(K, t) = \exp(-\eta K^2 t) \int_0^t a\zeta[K; T] \exp[\eta K^2 T] \, dT. \qquad (9.10)$$

Here, $\eta = ka^2 / \tau$. If we coarse grain in time, by the central limit theorem ζ becomes a white noise:

$$\langle \zeta(x; t)\zeta(y; u) \rangle = X\delta(x - y)\delta(t - u). \qquad (9.11)$$

Now:

$$\sigma^2 = \langle s(0, t)^2 \rangle = \int \langle s(K, t)s(-K, t) \rangle \, dK$$

$$\sim \int_{1/L} \frac{1 - \exp[-2\eta K^2 t]}{2\eta K^2} \, dK. \qquad (9.12)$$

The limit on the integration accounts for the finite size of the sample. A change of variables $KL = p$, (see problems) yields Eq. (9.5) with $\alpha = 1/2$ and $z = 2$. (i.e. $\beta = 1/4$). In three dimensions we must replace dK by KdK. In this case $\alpha = \beta = 0$ (i.e., the surface thickens only logarithmically in time).

It is reasonable, and it can be numerically verified, that the surfaces in question satisfy a stronger assumption. Suppose we think of the roughness growing by fluctuations of the deposit of particles at different places. From the above, there is a kind of "diffusion" of roughness that moves $|\delta \mathbf{x}|^z$ in a "time" H. (H can be thought of as a time because the density is constant.) Now suppose that this same idea can be applied locally. We can look at the growth of the mean square difference between the height fluctuations, $s(\mathbf{x}, t) = h(\mathbf{x}, t) - H(t)$, at different points. It is natural to assume the following:

$$\{\langle (s(\mathbf{x} + \delta \mathbf{x}, t + \tau) - s(\mathbf{x}, t))^2 \rangle\}^{1/2} \sim |\delta \mathbf{x}|^\alpha G(\tau / |\delta \mathbf{x}|^z). \qquad (9.13)$$

The scaling function $G(y)$ must be taken to be constant for small argument y (so that the self-affine behavior is preserved within the range of the spread of the fluctuations) and approach y^β for large y, exactly as above.

The scaling behavior of the surface is thus given by the exponents α, β. There is a considerable literature which involves numerical simulations of the model in various versions to find these exponents. In two dimensions $\alpha = 1/2$, $\beta = 1/3$; these are known to be exact. The simplest derivation is given in Meakin et al. [2]. Note that β is smaller than that for the trivial model of uncoupled columns in the previous chapter (1/2, with α undefined), and larger than for the linearized model above.

In three dimensions $\alpha \approx 0.35$ and $\beta \approx 0.21$ [7]. A sketch of the kind of data analysis necessary to find these exponents is given in Fig. 9.2. The exponents seem, as expected, to be independent of a large number of details of the model.

In fact, there is very strong evidence for universality; consider a completely different model, the Eden model [7], which has the same scaling law with the same exponents. This is a model for the growth of a cell colony such as those mentioned in the previous chapter. At each stage of the growth every perimeter site has an equal probability to grow, rather than needed to grow by addition from outside. This is an example of reaction-limited growth; in a materials science context we could be thinking of a process of growth for which the probability of sticking of a molecule is small. The molecule then explores many sites, and its probability to stick at each is more or less the same. Or we could be thinking of a cell colony. Here the probability of fission of any cell at the perimeter of the colony is the same, provided they are all well nourished.

9.1.2   Continuum Models

Because many models have the same scaling on the large scale, it is reasonable to try to exploit this by trying for a large-scale theory directly. We will explain how this has been done in this section.

In example 9.2 we have given a linearized version of the model in a continuum description. It is reasonable to assume that the scaling behavior should still be given correctly after coarse-graining, passing to the continuum limit.

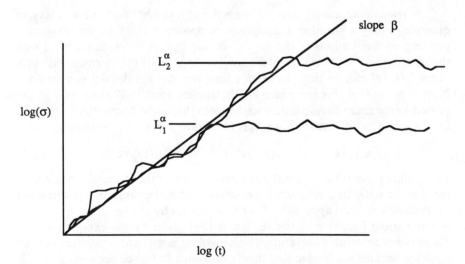

FIGURE 9.2. Sketch of the numerical method for finding the exponents α, β by a log–log fit to σ as a function of time.

However, the problem with the model of example 9.2 is that it is linear and in this respect it does not represent the physics of the problem correctly. Kardar, Parisi, and Zhang [8] have shown that the slmplest nonlinear continuum version of Eq. (9.9) does have many features in common with the lattice models. They proposed:

$$\partial s/\partial t = \eta \nabla^2 s(x,t) + \lambda |\nabla s|^2 + \zeta(x;t) \tag{9.14}$$

as the simplest generalization consistent with the symmetry of the problem. The new term, $\lambda |\nabla s|^2$, builds into the physics the fact that in ballistic aggregation, and in many other versions of growth, slopes grow at a different rate from flat surfaces. It is surprising that the addition of this simple feature should be adequate to restore scaling in the same form as the lattice model, but all our current evidence is that it is so. The first term admits the possibility of smoothing processes; it is the only term in the linear model, of course, and roughly represents the effects of processes leading to equilibrium.

Meakin et al. [2] pointed out that it is an immediate result of Eq. (9.14) that the exponents α, β are not independent. In fact,

$$2/\alpha = 1 + 1/\beta, \quad \text{or} \quad \alpha + z = 2. \tag{9.15}$$

The derivation is relatively simple: see for example refs. [1] or [8]. All known lattice simulations satisfy this rule, except, of course for the linear model, which has $\lambda = 0$.

There is a good deal of current research on the implications of Eq. (9.14) for dimensions above two. [In two dimensions the exact exponents $\alpha = 1/2$, $\beta = 1/3$ can be found directly from Eq. (9.14)]. Much of the difficulty seems to arise from the fact that the model appears to show a *phase transition* from linear (as in example 9.1) behavior to nonlinear as a function of the parameters of the model. Halpin–Healy [9] speculated that this should be true for dimensions above three. Yan, Kessler, and Sander [10] seemed to find numerically that the transition exists in three *and* in two dimensions. However, this is quite controversial and other authors have found no transition in two dimensions [10].

In the case of real thin films the purpose of much of the technology is to produce smooth deposits. That is, relaxation processes are encouraged to make layer-by-layer crystal growth possible. It is not clear where this kind of growth fits in to our picture. In fact, the scaling behavior described here, which is on a fairly sound theoretical basis, lacks experimental confirmation. For a detailed discussion of this, see ref. [1].

9.2 Diffusion-Limited Growth

9.2.1 Simulations and Scaling

DLA clusters, in contrast to ballistic clusters, seem to be self-similar fractals with a fractal dimension that depends on the dimension, d, of the space in which the simulation is done. The best current results are that $D = 1.7$ for

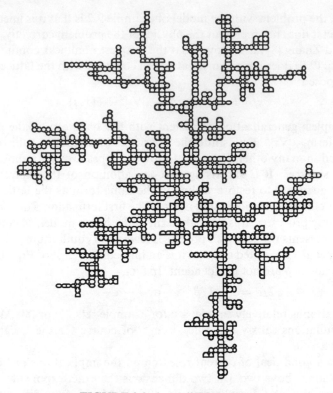

FIGURE 9.3. A small DLA cluster.

$d = 2$ and $D = 2.5$ for $d = 3$ [11]. A rather small DLA cluster is given as Fig. 9.3. Current simulations have succeeded in making very large clusters: The current record is in the millions of particles [12].

There are two popular methods to measure fractal scaling: one is to look for power laws in the correlation function, and the other to look at the increase of a typical radius with M. For a fractal (see Part I) $\rho \sim M^D$.

We should point out that the reasoning above that showed that ballistic aggregates are not fractal can be applied to DLA. However, the result is different: we will find that DLA can be fractal. We return to the argument of Ball and Witten [3]. Consider a DLA cluster growing in a radial geometry with a fixed probability to find a random walker far away. Extending the argument of example 8.2 to d dimensions we have:

$$dM/dt \sim \rho^{d-2}, \tag{9.16}$$

where ρ is, again, a mean radius of the cluster. Now from Eq. (9.2):

$$\rho^{d-2} \sim \rho^{D-1}v. \tag{9.17}$$

The speed, v, is bounded. Thus we find:

$$\rho^{d-D-1} \leq \text{const.} \qquad (9.18)$$

The only way for this to happen is for

$$d - 1 \leq D \leq d. \qquad (9.19)$$

This bound on the fractal dimension is satisfied by all known simulations.

The qualitative interpretation of this result, the reason that the "holes" in the DLA aggregate of Fig. 9.3 do not fill in, is a kind of *screening*. That is, when a random walker wanders into a DLA cluster, it cannot penetrate very far into a hole because it is likely to hit the sides in its wandering.

Example 9.3 Estimate the screening effect with the help of Eq. (8.5) and show that the length over which screening occurs is smaller than ρ. Replace the density with an average.

In this case we are to allow particles to be absorbed by the aggregate, and consider the average effect. We radially average the density of a cluster such as that of Fig. 9.3, and replace it by a function $f(r)$. Because $M \sim \rho^D$ the density near the surface, $f(\rho) \sim \rho^{D-d}$. Now the percentage absorption of random walkers should be proportional to f. We have then:

$$\partial^2 u / \partial w^2 \sim fu,$$

where w is measured from the surface of the deposit. Now the solution to this is $u \sim \exp[-w/c]$, where the absorption length $c \sim \rho^{(d-D)/2}$, which becomes small compared to ρ as the cluster grows.

This result, although correct, is misleading if taken too literally. The length, c, is significant only if we neglect the huge fluctuations of the density of the cluster around its periphery.

There have been a large number of simulations that have looked at different features of *universality* of the model, which we expect for the same reasons here as in the previous section. For example, the original work of Witten and Sander [13] and Meakin [11] showed that the fractal scaling properties were independent of the *sticking probability*. That is, if we allow a particle to sometimes bounce off the aggregate before sticking it thickens the branches but does not change the overall scaling.

The pattern is independent of initial conditions in the sense that an originally smooth nucleus of whatever type quickly develops into an object with fractal correlations. For example, the student can amuse him or herself by altering the program in the appendix to start the growth on a square. The square quickly grows fractal "hair."

FIGURE 9.4. The Mullins–Sekerka instability. In the left hand picture equipotential lines are shown. Because they are crowded above the bump, it gets sharper.

For small aggregates the pattern is independent of the lattice on which the growth takes place. For large aggregates the pattern *does* depend on the lattice [13]. In fact, for large aggregates the pattern deforms and takes on a shape, for which the individual branches are rather like the dendritic shape in Fig. 9.4, aligned with the lattice axes. The lattice anisotropy apparently acts to distort the cluster. Another anisotropy effect was given by Ball et al. [14], who showed that if the sticking probability is anisotropic on a square lattice, so that it is more likely to stick right and left than up or down, then the cluster distorts very quickly into an elongated shape where one radius increases as $M^{1/3}$ and the other as $M^{2/3}$. The universality of the model is not preserved in the presence of anisotropy.

However, the model is well represented by the continuum version embodied in Eq. (8.5). Numerical simulations of these equations [15] (using a relaxation method to solve the Laplace equation) lead to DLA-type fractals. In fact, the same authors have measured a real dielectric breakdown pattern (a Lichtenberg figure) and found fractal behavior of the DLA type.

Niemeyer et al. [15] also gave an interesting extension of the model to allow the growth probability to be given by:

$$P_s \sim [\partial u/\partial n|_s]^{\kappa}, \tag{9.20}$$

where the growth exponent, κ, represents possible nonlinearities in breakdown, and also gives rise to fractals whose fractal dimension depends on κ. In fact, $D(\kappa = 0) = d$, and D approaches 1 for large κ.

9.2.2 The Mullins–Sekerka Instability

We have seen that the DLA model gives rise to complex objects. A dendritic crystal, such as a snowflake, which is a different object but also complex, arises in slightly different conditions from similar physics. It should be evident that some common feature makes diffusion-limited growth likely to produce complex objects.

The fundamental reason for the varied shapes produced is that diffusion-limited growth is afflicted with a growth instability [16]. We will indicate how this is shown by doing the classic linear stability analysis of Mullins and Sekerka [17]. The situation in the nonlinear regimes is discussed in the next section.

We ask then, a simple question: why do Eq. (8.5) not simply describe a smooth surface that advances in time? We can ask the same question about the averaged, deterministic, version of the equations where Eq. (8.5c) becomes $v \sim \partial u/\partial n|_s$. In fact, if we start with a flat surface it does advance in time (with $v \sim t^{-1/2}$), but this solution is not stable: any slight distortion leads to unstable growth.

The calculation of the spectrum of the instability is given in the standard references [16]. It is sufficient for our purposes to reason qualitatively. Suppose we start with a flat surface with a small bump (see Fig. 9.4). Consider the electrostatic interpretation of Eq. (8.5), which is given in the previous chapter. We are asked to find the potential, u, near a grounded conductor with a bump, and then advance the different parts of the surface at a speed proportional to $\partial u/\partial n|_s$, the surface electric field. Now the field is largest near the bump, as is known in elementary electrostatics: this is the principle of the lightning rod. Thus, the bump develops in time faster than the rest of the surface and gets even sharper. To stabilize the growth, other effects such as surface tension in fluid flow or crystal growth, or finite particl size in DLA, are necessary. And, clearly, these effects cannot stabilize large-scale instabilities. On some scale the surface must deform.

In the nonlinear regime the instabilities proliferate and mix; that is, different bumps interact. If, as we will discuss in the next lecture, the result is sometimes fractal growth, we can now see qualitatively why the growth speed becomes *multifractal* as discussed in Part I. The amplifying effect of a bump on an electric field means that each sharp tip in Fig. 9.3 will have a very singular growth. The collection of different singularities will be multifractal.

9.2.3   Orderly and Disorderly Growth

The calculation of what actually happens in the case of diffusive growth in the orderly case has been one of the triumphs of recent years in mathematical physics. This work is reviewed in ref. [18]. The theory is very intricate, and we will review only the barest outlines as they relate to the disorderly case.

Consider first the case of a viscous finger growing in a channel. It has long been known [19] that there could be a single finger of stable shape that moves uniformly down the channel. However, the Saffman–Taylor solution in the absence of surface tension does not determine the percentage of the channel filled. In this case it has been proved that that a single finger of similar shape is stable provided the surface tension is taken into account, and the surface tension determines the percentage of the channel which is filled.

By contrast, radial viscous fingering (such as shown in Fig. 9.5) is intrinsically unstable, and no steady state seems possible. It seems that the shape is described by DLA over a range of scales. This has been proved experimentally by several groups, notably Rauseo et al. [20] and Couder [21].

For the case of crystallization with a finite diffusion length, the situation is mathematically far more complex, but qualitatively similar. That is, there is a

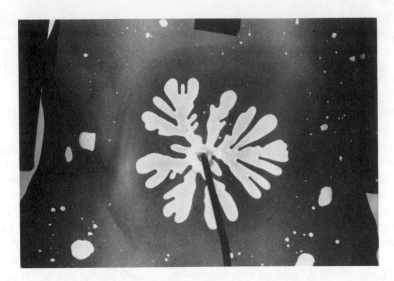

FIGURE 9.5. A fractal viscous fingering pattern.

parabolic solution that neglects surface tension, but with an undetermined velocity. A stable solution with well-defined velocity is found (presumably a real dendrite) provided another effect is taken into account, namely the *anisotropy* of the surface tension. This small effect provides the necessary stabilization to have a steady-state solution with fixed velocity.

When, then, do we see DLA-like behavior? This is a very difficult question, but we believer that at least the following qualitative picture is consistent with all the evidence we have. First, we must be in a situation with diffusion length that is long compared to the size of the deposit. Otherwise the diffusion length provides a scale beyond which fractal correlations are screened (see problems). That is, L_d is the length beyond which the screening effect of different branches is remembered. In a lattice model this can be demonstrated by doing a DLA simulation with many particles at a time [22]. The mean distance between particles, which plays the role of L_d in this case [23], is the distance over which a particle "forgets" that it is unable to penetrate. In the absence of other effects, with finite L_d we expect fractal behavior up to L_d and homogeneous patterns on larger scales. There are other finite lengths in other diffusion-limited growth problems, such as directional solidification, which are not relevant to our discussion here [18].

Also, some kind of anisotropy is necessary to stabilize diffusive growth: in the two cases we have discussed it is provided by either the walls of the channel or the surface tension. If it is large enough to overcome the intrinsic noise that is always present in any experiment, then the relevant steady state will be attained. But suppose the crystal is polycrystalline for some reason, with no long-range correlation for the crystalline axes [24], which is analogous to the

at a large cluster is likely to encounter another of comparable size, and not
any individual particles. This is in fact the case observed, most often, but to
e when it occurs, we have to be a bit more precise about the way in which
e simulations are done.

In the discussion on the previous chapter, we noted that we must continue
let aggregates diffuse. We must decide how fast they are to diffuse; namely,
at the diffusion coefficient of an aggregate is to be. A physical choice is to
e Stokes' law friction for the diffusion; in this case the retarding force (and
us the diffusion coefficient, by the Einstein relation) is proportional to the
verse radius of the cluster. Thus:

$$\eta_M \sim M^\gamma, \tag{9.21}$$

th $\gamma = -1/D$. In the simulations it is usual to take γ to be a free parameter.
fact, the fractal dimensions are independent of γ (and equal to the values
ove) provided γ is less than 1. However, if it exceeds 1, that is, large clusters
ve much faster than small ones (an unphysical assumption, but with amus-
 consequences), one large cluster begins to "eat" the smaller ones who
er grow. Thus we are back to particle aggregation of the DLA type.
here are two variations of the model that one can study. First, we can
k at ballistic cluster–cluster aggregation. In this case we *still* produce frac-
, because of the exclusion effect mentioned above, but the fractal dimen-
 is higher. We have $D(2) = 1.55$.
nd one can consider a physically motived process "reaction-limited"
regation. Suppose that the probability of sticking is small at each try.
e again we expect that after many tries the clusters will penetrate more
 for the diffusive case. In the limit of a large number of tries in order to
k, $D(2)$ approaches 1.53.

pendix: A DLA Program

is a program written in TrueBasic™ for the Macintosh. It produced the
in Fig. 9.2. There are two important tricks implemented here. The ran-
 walker is allowed to take large steps when it is outside the aggregate. It
ot encounter any matter, so it can take a step as large as the distance to
earest point, but in a random direction. Also, the walker need not start
way from the aggregate. It can start at a random point on a circle of size
 that just encloses the cluster. The first trick was invented by P. Meakin,
he second by M.E. Sander.

```
ram dla ! DLA program incorporating large steps
blob(-90 to 90, -90 to 90), dl(4), dm(4)
dl(1)=0 ! steps on a square lattice
dl(2)=0
dl(3)=1
```

case of fluid flow in the radial geometry. Then, whatever the initial source of
instability, it will be amplified and a DLA-like pattern will develop.

9.2.4   Electrochemical Deposition: A Case Study

In recent years, a number of groups have taken up the study of electrochemical deposition as a model system for studying fractals and dendrites
(i.e., the relationship between orderly and disorderly diffusive growth). Several groups, including ours at the University of Michigan in Ann Arbor, have
studied the transitions between various sorts of patterns in the formation of
zinc deposits [25, 26]. For other, more recent work, see ref. [26] and references
therein.

Our cells (Fig. 8.5) consist of plates of plastic spaced by about 0.1 mm
between which a film of electrolyte is confined. The deposits typically are
made on a small cathode in the center of the cell and the anode is in the form
of a ring a few centimeters in diameter.

As we pointed out above, it may at first glance seem that electrochemical
growth cannot be treated as diffusion-limited because in real electrochemical
cells the ions rarely move under the influence of a density gradient alone. If
the electric fields in the system are not screened out, there is a drift current in
addition to the diffusion current. In particular, the electrical characteristics of
our apparatus is ohmic, indicating that there are definitely electric fields in the
solution. In fact, this is quite misleading, because in the quasistatic limit *both*
the electrostatic potential and the density obey the same equation, the Laplace equation.

We find that by changing growth conditions we can pass from fractal to
dendritic growth; see Fig. 9.6. In the fractal regime (which is the case of very
slow growth in our case) the patterns have been carefully measured and have
a fractal dimension close to that of the DLA simulations, about 1.7. The
dendritic regime occurs for high voltages and concentrations (i.e., for fast
growth). It seems clear, from the discussion above, that these zinc leaves are
being stabilized by some sort of anisotropy. Its origin in the physics and electrochemistry of metal surfaces is presently under investigation.

However, and this was the surprise in the original work [25, 26], for intermediate growth rates we find a pattern that is not seen in viscous fingering
experiments. It has many tip-splittings like a fractal, but its overall outline is
stably round in a round cell, and its average density is constant (see Fig. 9.7).
We call this the dense radial pattern. How does this fit into the overall
picture?

The main thing that puzzled us and other workers in the field was that this
pattern seems to be immune to the Mullins–Sekerka instability. The outline is
round, and appears to become rounder as the pattern grows. But, in terms of
the considerations above, its seems that a round shape should be unstable.
One of the filaments should get ahead of the others by a fluctuation, and then
grow unstably because it is then in a region where the voltage is higher and

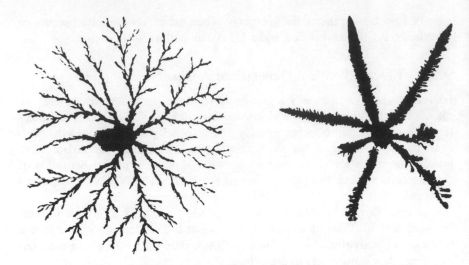

FIGURE 9.6. Fractal and dendritic patterns of zinc electrodeposits.

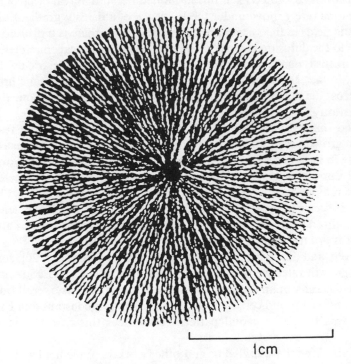

1cm

FIGURE 9.7. The dense radial pattern.

the electric field at the surface is bigger. This does not
Instead, the pattern typically grows rounder as it gets large

Our explanation of the effect [27] is that the metal dep
ciable resistance compared to the electrolyte. If we mod
disk, then we may associate with the disk an effective resist
this may be a rather large number, even if the metal that
conductor, because the pattern is composed of a number o
(in our case, about 0.05 mm in diameter). In fact, we can
by measuring the dependence of the total resistance of the
occupied by metal. We find that the ratio of ρ_d to ρ_e, t
electrolyte, is about 0.1. This is within a factor of 2 of wh
the geometry of the pattern.

Now the boundary condition on the potential that repl

$$u_s = \sigma \rho_d r_s \ln(r_s/r_o)v.$$

This is nothing more than Ohm's law for a disk of metal
inner (cathode) radius r_o, using the conservation of m
current.

We now performed a linear stability analysis of Ec
(9.20). We found that the growth rate of a bump on t
(stable growth) if the bump is wide enough for ρ_d/ρ
0.05.

There is a subtlety in this analysis that we must no
(8), that the current inside the deposit flows radially, a
measured by the distance from the center. If the depo
disk, this would not be valid, because the current c
inside as well as outside. However, the deposit is not a
is justified. All of our measurements of dense radial p
sistent with the picture that we have given, namely I
lized by finite resistance.

9.3   Cluster–Cluster Aggregation

The cluster–cluster aggregation model is quite differ
gation and from DLA. Fractal clusters are produ
fractal dimension is much smaller than that of a DL
evident: the screening effect that we have mentione
much better for cluster aggregates. As hard as it is to
a DLA cluster, it is that much harder to introduce a

Numerical results are in accord with this statemen
measured for cluster–cluster aggregates [28] are /
(compare DLA with 1.7, 2.5, respectively).

However, the statement assumes that the devel
such that at any time the cluster size distribution i

```
let dl(4)=-1
let dm(1)=-1
let dm(2)=1
let dm(3)=0
let dm(4)=0
let rr=30 !This controls the ultimate size - must not
be greater than array dim
let nblob=0
for i=-rr to rr
for j=-rr to rr
let blob(i,j)=0
next j
next i
let blob(0,0)=10 ! blob=10 or more means occupied
for k=1 to 4
let blob (dl(k),dm(k))=1 ! blob=1 to 4 means a neighbor
of aggregate
next k
let rmax=2
let r2=rmax*rmax
let r2big=4*r2
randomize
call start (rmax,l,m)
do while rmax <rr
  let x=int(4*rnd)+1 !take a step
  let l=l+dl(x)
  let m=m+dm(x)
  let rsq=l*l+m*m
  if rsq <r2 then !see if attachment is possible, if so
grow and start over
    if blob(l,m)>0 then
      let blob(l,m)=blob(l,m)+10
      for k=1 to 4
        let ll=l+dl(k)
        let mm=m+dm(k)
        let blob(ll,mm)=blob(ll,mm)+1
      next k
      let nblob=nblob+1
      let rtry=sqr(rsq)+2
      if rtry>rmax then
        let rmax=rtry
        let r2=rmax*rmax
        let r2big=4*r2
      end if
      print nblob;
```

```
      call start(rmax,l,m)
    end if
  else
      if rsq<r2big then !between rmax and 2*rmax take big
steps
        let step=sqr(rsq)-rmax
        call start(step,ll,mm)
        let l=l+ll
        let m=m+mm
        let rsq=l*l+m*m
      end if
      if rsq>r2big then ! Throw away
      call start(rmax,l,m)
    end if
  end if
loop
call plot(blob,rr)
end
sub start(rmax,l,m)
let th=2*3.14159*rnd
let l=int(rmax*cos(th)+0.5)
let m=int(rmax*sin(th)+0.5)
end sub
sub plot(blob(,),rr)
clear
let rad=0.5
let asrat=1.5
let xrr=asrat*rr
set window -xrr,xrr,-rr,rr
for i=-rr to rr
for j=-rr to rr
if blob(i,j)>9 then
  box circle i-rad,i+rad,j-rad,j+rad
end if
next j
next i
end sub
```

Problems

9.1. Show that Eq. (9.13) implies Eq. (9.5).

9.2. Show that Eq. (9.7) implies Eq. (9.5) by taking $h(\mathbf{x}, 0) = 0$ for all $\mathbf{x}$; integrate over $\delta \mathbf{x}$.

9.3. Write a program to make DLA clusters on a square lattice in deposition geometry. Use a fairly small lattice (say 10×20), and to simplify things always launch the random walker from the top of the lattice. Use periodic boundary conditions. For some hints, see the appendix to this chapter, which is a radial DLA program written in TrueBasic™.

9.4. Extend example 8.2 to derive Eq. (9.16).

9.5. Does the ballistic aggregation model have a growth instability? Draw pictures such as Fig. 9.4, and show that the model is marginal; bumps neither grow or shrink.

9.6. Justify the interpretation of L_d as the range of propagation of information in the diffusion field by considering diffusive growth of a plane parallel to the x-axis at constant velocity v. Go into the frame moving with v, and show that $\partial u/\partial t = \eta \nabla^2 u$ becomes

$$1 u \xi \xi + u \xi / L_d = 0,$$

where the new variable $\xi = y - vt$. Thus u decays as $\exp[-\xi/L_d]$ in front of the moving surface.

References

[1] F. Family and T. Vicsek, *Dynamics of Growing Interfaces* (World Scientific, Singapore, 1992); A. Barabási, and H.E. Stanley, *Fractal Concepts in Surface Growth* (Cambridge University, Cambridge, 1995).

[2] P. Meakin, R. Ball, P. Ramanlal, and L. Sander, Phys. Rev. A **34**, 5091 (1986).

[3] R.C. Ball, and T.A. Witten, Phys. Rev. A **29**, 2966 (1984).

[4] F. Family and T. Vicsek, J. Phys. A **18**, L75 (1985).

[5] S.F. Edwards, and D.R. Wilkinson, Proc. Roy. Soc. London **A381** 17 (1982).

[6] L. Sander, Proceedings of Symposium on Multiple Scattering of Waves and Random Rough Surfaces, (University Park, PA, July, 1985). Published in *Multiple Scattering of Waves in Random Media and Random Rough Surfaces*, edited by V.V. Varadan and V.K. Varaden (Pennsylvania State University, University Park, 1986).

[7] D.E. Wolf, and J. Kertesz, Europhys. Lett. **4**, 651 (1987); J. Kim and J. Kosterlitz, Phys. Rev. Lett. **62**, 2289 (1989); M. Eden, Proceedings of the Fourth Berkeley Symposium on Mathematical Statistics and Probability **4**, 22.3 (1961).

[8] M. Kardar, G. Parisi, and Y. Zhang, Phys. Rev. Lett. **56**, 889 (1986); J. Krug and H. Spohn. in *Solids Far from Equilibrium*, edited by G. Godreche (Cambridge University, Cambridge, 1991).

[9] T. Halpin-Healy, Phys. Rev. Lett. **63**, 442 (1989).

[10] H. Yan, D. Kessler, and L.M. Sander, Phys. Rev. Lett. **64**, 926 (1990); L. Tang, T. Natterman, and B. Forrest, Phys. Rev. Lett. **65**, 2422 (1990); L. Tang, B.M. Forrest, and D. Wolf, Phys. Rev. A **45**, 7162 (1992).

[11] S. Tolman and P. Meakin, Phys. Rev. A **40**, 428 (1989); P. Meakin, Phys. Rev. A **26**, 1495 (1983).

[12] P. Ossadnik, Physica A **195**, 319 (1993).

[13] T.A. Witten and L.M. Sander, Phys. Rev. B **27**, 5686 (1983), P. Meakin, R. Ball, P. Ramanlal, and L. Sander, Phys. Rev. A **34**, 5091 (1986).

[14] R.C. Ball, R. Brady, G. Rossi, and B. Thompson, Phys. Rev. Lett. **55**, 1406 (1985).

[15] L. Niemeyer, L. Pietronero, and H. Weismann, Phys. Rev. Lett. **52**, 1033 (1984).

[16] J.S. Langer, Rev. Mod. Phys. **52**, 1 (1980); See the article by C. Caroli, B. Caroli, and B. Roulet in G. Godreche, *Solids Far From Equilibrium* (Cambridge Univesity, Cambridge, 1991); E. Louis, O. Pla, L.M. Sander, et al. Mod. Phys. Lett. B **8**, 1739 (1994).

[17] W.W. Mullins and R. Sekerka, J. Appl. Phys. **28**, 333 (1957).

[18] D. Kessler, J. Koplik, and H. Levine, Adv. Phys. **37**, 255 (1988).

[19] P. Saffman and G.I. Taylor, Proc. Roy. Soc. London, A **245**, 312 (1958).

[20] S. Rauseo, P. Barnes, and J. Maher, Phys. Rev. A **35**, 5686 (1986).

[21] Y. Couder, in *Random Fluctuations and Pattern Growth*, edited by H.E. Stanley and N. Ostrowsky, (Boston, Kluwer, 1988).

[22] R. Voss, Phys. Rev, B **30**, 334 (1984).

[23] M. Nauenberg, R. Richter, and L. Sander, Phys. Rev., B **28**, 1649 (1983).

[24] G. Radnoczi, T. Vicsek, L. Sander, and D. Grier, Phys. Rev. A **35**, 4012 (1987).

[25] Y. Sawada, A. Dougherty, and J.P. Gollub, Phys. Rev. Lett. **56**, 1260 (1986).

[26] D. Grier, E. Ben-Jacob, R. Clarke, and L.M. Sander, Phys. Rev. Lett. **56**, 1264 (1986); A. Kuhn and F. Argoul, Phys. Rev. E **49**, 4298 (1994).

[27] D. Grier, D. Kessler, and L.M. Sander, Phys. Rev. Lett. **59**, 2315 (1987); D. Grier and D. Mueth, Phys. Rev. E **48**, 3841 (1993).

[28] P. Meakin, in *Phase Transitions and Critical Phenomena*, Vol. 12, edited by C. Domb and J. Lebowitz (Academic, New York, 1988).

IV

Solitons

10

Integrable Systems

Lui Lam

10.1 Introduction

Solitons are robust, localized traveling waves of permanent form. They are found everywhere and have sizes of 10^{-7}–10^{10} cm. They exist in the sky as density waves in spiral galaxies, and the giant Red Spot in the atmosphere of Jupiter [1]; they exist in the ocean [2] as waves bombarding oil wells; they exist in much smaller natural and laboratory systems such as plasmas [3], molecular systems [4], laser pulses propagating in solids [5], superfluid ^{3}He [6], superconducting Josephson junctions [7], magnetic systems [8], structural phase transitions [9], liquid crystals [10], polymers [11, 12], and fluid flows [3, 13], as well as elementary particles [14]. They may even have something to do with the newly discovered high T_c superconductors [15]. So, what is a soliton?

Solitons are special nonsingular solutions of some nonlinear partial differential equations (PDEs): (i) They are spatially localized; (ii) a single soliton is a traveling wave (i.e., it is a wave of permanent form); (iii) they are stable; (iv) when a single soliton collides with another one, both of them retain their identities after collision—like the elastic collision of two particles (Fig. 10.1).

However, in many systems the fourth property cannot hold. It turns out that this nice but stringent elastic-collision property is intimately related to a specific property of the system, which is called integrability [16] (see section 10.3). We therefore differentiate two kinds of systems; namely, integrable systems and nonintegrable systems, and discuss them separately.

Note that many mathematicians insist that the name "soliton" should be reserved for those wave solutions that possess simultaneously all four properties listed above; this is not true for most physicists. As pramatists, physicists have to deal with the real world. There are simply too many real physical systems in nature that are nonintegrable (see chapter 11). For these systems, the concept of soliton even without property iv is found to be so useful and fruitful that one cannot afford not to use it. As we shall see later, the word "soliton" is used so loosely these days that sometimes not even properties ii and iii are retained.

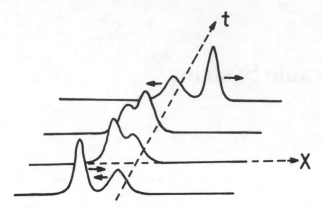

FIGURE 10.1. Collision of two solitons. After collision, the two solitons separate from each other with original shapes and velocities as before collision, but with a phase shift.

In Part II, we see examples that the time evolution of a system of a few degrees of freedom is represented by ordinary different equations (ODEs). The system may be chaotic in time. When this number of degrees of freedom N is infinite, the discrete index i $(= 1, \ldots, N)$ representing each degree of freedom may be replaced by a continuous variable x in the continuum limit, and we have a nonlinear PDE instead of a system of ODEs. As expected, this nonlinear PDE could give very complicated chaos in both space and time. The surprising thing is that, in special cases, some very regular and coherent solutions, such as solitons, can exist.

As an important branch of nonlinear science [17, 18], the study of solitons is much more well developed than the others. Modern advances in soliton study have been made continuously since 1965, the year the word "soliton" was first introduced by Zabusky and Kruskal [19]. In contrast, for example, the field of chaos did not become really active and alive until 1978, when universality of chaos was discovered by Feigenbaum [20]. In recent years, we have seen interwoven development of solitons with other branches of non-linear physics such as the coexistence of solitons and chaos, and the study of solitons using cellular automata (see section 11.9).

Here is a brief guide to the literature. For beginners, the short article "The Birth of a Paradigm" by Scott [21] is highly recommended. Bullough [22] gives derivations of many standard soliton equations and plenty of computer results. Reference [23] by Bishop et al. is a very readable essay on the relevance of solitons in condensed matter physics, while [24] is a compact introduction to the mathematical aspects of solitons in integrable systems.

Of the many textbooks on solitons, we recommend *Waves Called Solitons* by Remoissenet [25] for its easy reading and emphasis on physical applications, and *Solitons and Nonlinear Wave Equations* by Dodd et al. [26] for its wide scope of coverage. The three other books—*Solitons: Mathematical Methods for Physicists* by Eilenberger [27], *Elements of Soliton Theory* by Lamb [28], and *Solitons in Mathematics and Physics* by Newell [29]—may also be consulted. More specialized topics are covered in [3–5, 10, 16, 30, 31], and, of course, there are many conference proceedings, too numerous to be listed here.

10.2   Origin and History of Solitons

For a linear wave with dispersion, for example

$$\theta_t - \theta_{xxx} = 0, \tag{10.1}$$

where $\theta = \theta(x, t)$ with $\theta_t \equiv \partial\theta/\partial t$, and so forth, the solution is given by $\theta_k = \theta_o \exp[i(kx - \omega t)]$ with $\omega = k^3$; θ_o is a constant. The phase velocity $\omega/k = k^2$ depends on k. Each component θ_k in the linear superposition $\theta = \sum_k a_k \theta_k$ will therefore travel with a different velocity—a phenomenon called *dispersion*. In this case a pulse constructed as a linear superposition of θ_k actually spreads out as it propagates.

On the other hand, for a nonlinear equation without dispersion such as

$$\theta_t + \theta\theta_x = 0, \tag{10.2}$$

one has the formal solution, $\theta = f(x - ct)$ with "velocity" $c = \theta$. Different points of a pulse then travel with different velocities proportional to their heights (Fig. 10.2), resulting in a squeeze of the pulse width as it travels.

For some special nonlinear equations with dispersion, it is then possible that this squeeze of the pulse width due to nonlinearity is balanced exactly by the expansion of the width due to dispersion, leading to a traveling wave of permanent shape, and hence a soliton. In fact, the Kowteweg–de Vries (KdV) equation (with $\alpha = $ const)

$$\theta_t + \alpha\theta\theta_x + \theta_{xxx} = 0, \tag{10.3}$$

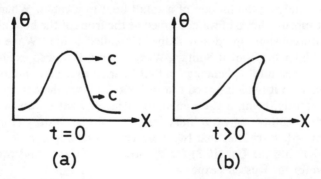

FIGURE 10.2. Solution of Eq. (10.2) at $t > 0$ (b), given an initial profile in (a). $c = c(\theta)$.

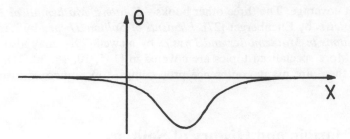

FIGURE 10.3. A single soliton of the KdV equation.

which may be viewed as a combination of Eqs. (10.1) and (10.2), is such an equation. The single soliton solution is given by

$$\theta(x, t) = (12/\alpha)a^2 \, \text{sech}^2[a(x - 4a^2t - x_o)], \qquad (10.4)$$

which is bell-shaped and localized in space for fixed t (Fig. 10.3); a and x_o are arbitrary constants. It is of the form $\theta = \theta(x - ct)$, a traveling wave. Note that the amplitude $(12/\alpha)a^2$, the wave width $1/a$, and the velocity $4a^2$ are all related to each other—a characteristic property of solitons, which separates them from traveling wave solutions of *linear* equations where all three quantities are usually independent of each other. [Note that for KdV solitons, the "tall" one must be "slim" and runs faster. Similarly, even though the sine-Gordon solitons have the same amplitude, the "slim" one always runs faster than the "fat" one (see section 10.4.3). However, this fact need not be true for solitons from other equations (see section 11.3.3).]

In many cases, a very simple but useful way to visualize the orgin and occurrence of solitons is to first assume a traveling wave solution for the nonlinear PDE and turn it into an ODE, in which a soliton solution can then be understood from the simple picture of a "particle rolling in a hill." (See sections 10.4.3 and 11.1.)

Historically, soliton was first observed by John Scott Russell [32] in 1834. Russell was studying the motion of a small boat in a canal. When the boat suddenly stopped, a lump of water formed at the front of the boat and moved forward with constant speed and shape. He called it the "Wave of Translation" and later the "Great Solitary Wave." Russell was so excited by this "singular and beautiful" phenomenon that he tried to explain many things in the universe with it (which turned out to be wrong; see the end of [30]). But more importantly, being a good engineer, Russell went on and did experiments, recreating these "great solitary waves" in his laboratory (Fig. 10.4). It was not until 61 years later that two theorists, Korteweg and de Vries, succeeded in deriving the Eq. (10.3) for shallow water waves and provided an explanation for the Russell's experiments.

The recent intense interest in solitons occurred after the numerical work of Zabusky and Kruskal [19] on the KdV equation, which in turn was inspired

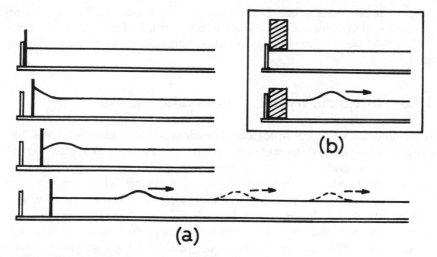

FIGURE 10.4. Two ways of generating solitons in a tank of shallow water, (a) and (b), in the experiments of Russell (1844).

by the work of Fermi, Pasta, and Ulam (see Fig. 10.5 and problem 10.1). They coined the word "soliton" to represent the numerically discovered particle-like property of these solitary waves; namely, the pairwise elastic-collision property. This brings us to the discussion of property iv described in section 10.1.

(Note that an *analytic* expression representing the elastic collision of two solitons in the sine-Gordon equation was actually derived by Perring and Skyrme [33] in 1962—3 years before the *numerical* discovery of a similar

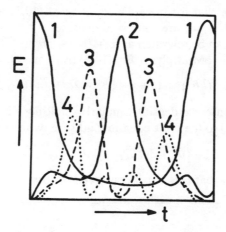

FIGURE 10.5. The FPU problem—distribution of energy E as a function of time t in different modes in a (1-D) nonlinear lattice.

property for the KdV equation in [19]—but went unnoticed. This is a history paralleling that in chaos, in which the significance of the work of Lorenz [34] in 1963 was not appreciated until, in this case unfortunately, 15 years later.)

10.3   Integrability and Conservation Laws

A Hamiltonian system of N degrees of freedom is said to be integrable if there exist, in principle, N different functions that are constants of motion [35]. (If a system is integrable it cannot be chaotic [36].) Note that integrability is not the same as the ability to compute the solution in terms of known functions, which may depend on one's skill and luck. Known trivial examples of integrable systems include the systems with $N = 1$ for which the energy is constant, and systems with $N = 2$ for which the energy and the total momentum are conserved. The not-so-trivial examples with $N > 2$ include the rigid body rotating around the fixed center of mass, the axially symmetric top, and Kovalevskaya's asymmetric top [35].

For a PDE the equivalence of constants-of-motion are conservation laws. By a conservation law we mean an equation of the form

$$I_t + J_x = 0 \tag{10.5}$$

such that $J \to$ const when $|x| \to \infty$. By Eq. (10.5) one has $\int_{-\infty}^{\infty} dx\, I =$ const, a constant of motion; I and J are called the conserved density and the flux, respectively. In fact, for the KdV equation there exist an infinite of conservation laws.

Example 10.1 Show that for the KdV equation of (10.3) with $\alpha = 6, I_1 \equiv \theta$, $I_2 \equiv \theta^2$, and $I_3 \equiv \theta^3 + \frac{1}{2}\theta_x$ are conserved densities.

Equation (10.3) can be rewritten as $\theta_t + (\theta_{xx} - 3\theta^2)_x = 0$, implying that I_1 is a conserved density. Multiplying Eq. (10.3) by θ, one obtains

$$\left(\tfrac{1}{2}\theta^2\right)_t + \left(\theta\theta_{xx} - \tfrac{1}{2}\theta_x - 2\theta^3\right)_x = 0,$$

and hence I_2 is a conserved density. Similarly, multiply Eq. (10.3) by $3\theta^2$ and then apply $\theta_x(\partial/\partial x)$ to the same equation; add the two to obtain

$$\left(\theta^3 + \tfrac{1}{2}\theta_x^2\right)_t + \left(-\tfrac{9}{2}\theta^4 + 3\theta^2\theta_{xx} - 6\theta\theta_x^2 + \theta_x\theta_{xxx} - \tfrac{1}{2}\theta_{xx}^2\right)_x = 0.$$

We see immediately that I_3 is also a conserved density.

For the KdV equation the first three conserved densities are relatively easy to guess; it is not so for the fourth (problem 10.2) and higher ones. Fortunately, a systematic method of finding them is available using something called the Gardner transformation [24].

The importance of the existence of an infinite number of conservation laws is that they are *believed* to be essential for the elastic-collision property of solitons to be established. The solitons have to and can maintain their identities after collision because of the many constraints required by the infinite number of conservation laws.

10.4   Soliton Equations and Their Solutions

In this section some prototype soliton equations and their soliton solutions are given. Single solitons as well as multisolitons are discussed. All these solitons possess the elastic-collision property. The derivations of these equations in representative physical systems are given in section 10.6.

10.4.1   Korteweg–de Vries Equation

The KdV equation and the corresponding single soliton are given in Eqs. (10.3) and (10.4), respectively. There is no "breather" soliton (see section 10.4.3) for this equation. The equation describes phenomena with weak nonlinearity and weak dispersion, including waves in shallow water, ion-acoustic and magnetohydrodynamic waves in plasmas, and phonon packets in nonlinear crystals.

10.4.2   Nonlinear Schrödinger Equation

The nonlinear Schrödinger (NLS) equation is given by

$$i\theta_t + \theta_{xx} + \alpha\theta|\theta|^2 = 0, \tag{10.6}$$

where $\theta(x, t)$ is complex. The single soliton solution assumes the form

$$\theta = \theta_0 \, \text{sech}[(\alpha/2)^{1/2}\theta_0(x - at)] \, \exp[i(a/2)(x - bt)], \tag{10.7}$$

where a and b are the velocities of the "envelope" and the "carrier," respectively. Breather solitons do exist. The NLS equation describes phenomena with weak nonlinearity and strong dispersion, such as waves in deep water, self-focusing of laser in dielectrics, propagation of signals in optical fibers, 1D Heisenberg magnets, and vortices in fluid flow.

10.4.3   Sine–Gordon Equation

The sine–Gordon (sG) equation is given by

$$\theta_{xx} - \theta_{tt} = \sin \theta. \tag{10.8}$$

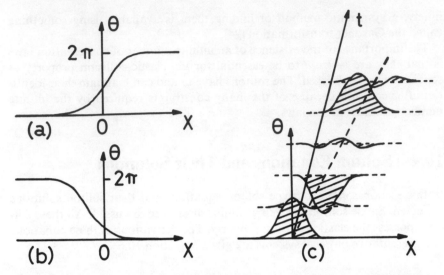

FIGURE 10.6. Solitons of the sG equation. (a) Kink; (b) Antikink; and (c) Breather.

There exist three basic types of solitons (Fig. 10.6).

kink:

$$\theta = 4 \tan^{-1}\{\exp[(x - ct - x_0)/(1 - c^2)^{1/2}]\}, \tag{10.9}$$

antikink:

$$\theta = 4 \tan^{-1}\{\exp[-(x - ct - x_0)/(1 - c^2)^{1/2}]\}, \tag{10.10}$$

and *breather*:

$$\theta = 4 \tan^{-1}\{(\tan a) \sin[(\cos a)(t - t_0)] \operatorname{sech}[(\sin a)(x - x_0)]\}, \tag{10.11}$$

where c (< 1), x_0, t_0, and a are constants. A breather is not a traveling wave, and may be considered as the bound state of a kink–antikink pair.

A 2-soliton solution,

$$\theta = 4 \tan^{-1}\left\{\frac{c \sinh[x/(1 - c^2)^{1/2}]}{\cosh[ct/(1 - c^2)^{1/2}]}\right\}, \tag{10.12}$$

also exists, which describes the collison of two single solitons (a kink–antikink pair). (See problem 10.3.) Note that multisoliton, as well as the breather, are not traveling waves. We therefore see that only simple, single solitons are traveling waves.

A kink (or antikink) has energy $E = 8(1 - c^2)^{-1/2}$. Consequently, it takes a finite energy to excite a kink.

A simple mechanical analogue picture can be used to understand these solitons. Assuming a traveling wave, $\theta = \theta(\tau)$ with $\tau \equiv x - ct$, one may transform Eq. (10.8) into an ODE,

$$(1 - c^2)\theta_{\tau\tau} = \sin\theta = -\partial V/\partial\theta, \qquad (10.13)$$

where $V(\theta) = 1 + \cos\theta$ (Fig. 10.7a). For $c < 1$, taking θ as the "displacement" and τ as the "time," Eq. (10.13) describes the motion of a single particle of mass m ($\equiv 1 - c^2$) in a periodic potential V. The kink solution of Eq. (10.9) corresponds to the particle moving from hilltop A to the adjacent hilltop B (with "velocity" $\theta_\tau = 0$ at A and B). In the phase space, this corresponds to the separatrix between 0 and 2π, denoted as curve 1 in Fig. 10.7b. Similarly, the antikink of Eq. (10.10) corresponds to the reverse motion from B to A in Fig. 10.7a and the separatrix 2 in Fig. 10.7b. Energy conservation guarantees that the particle starting from A with zero initial velocity can reach B and stop there, irrespective of the magnitude of the mass m. Therefore, the velocity c of the wave is arbitrary. (For $c > 1$, $m = c^2 - 1$ and $V = 1 - \cos\theta$. The hilltops A and B are now at $\theta = \pi$ and 3π, respectively, and one still has soliton solutions.) It is easy to imagine that the particle may move from A to B, and then to C or a few more hilltops before it stops on one of them. This would correspond to a multisoliton (Fig. 10.8).

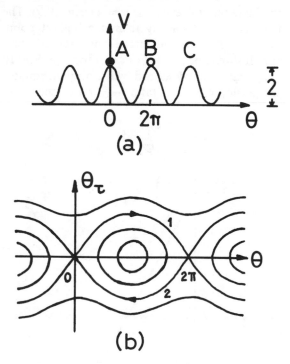

(a)

(b)

FIGURE 10.7. Motion of a particle in a potential $V = 1 + \cos\theta$ (a), and the corresponding phase space diagram (b). In (b) the trajectories 1 and 2 in $[0, 2\pi]$ correspond to the kink and antikink solitons, respectively.

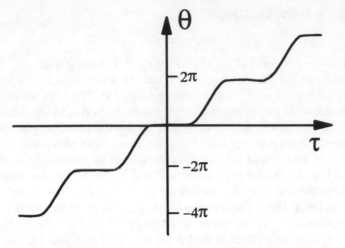

FIGURE 10.8. A multisoliton of the sG equation. The number of kinks may also be finite in number.

Equation (10.13) may also be understood as the equation of motion for a simple pendulum. Here the stable state is $\theta = \pi$ (Fig. 10.9). The kink soliton corresponds to the pendulum starting from the highest point ($\theta = 0$) and returning to it ($\theta = 2\pi$) after one full circle. This may also be inferred from the phase diagram. It should be pointed out that the V in Fig. 10.7a is not the potential of the physical system described by the original equation, Eq. (10.8). Although the two asympotic states (at $\tau \to \pm\infty$) of the kink are unstable

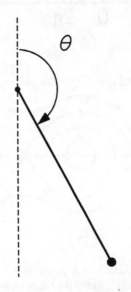

FIGURE 10.9. A simple pendulum.

states of the pendulum, they can nevertheless be stable states of the dynamical system described by Eq. (10.8). Note that the simple pendulum is not the system described by Eq. (10.8), and there is no contradiction involved here. (Solitons connecting stable to unstable states can be found in section 11.3.)

The sG equation has been used to describe crystal dislocation, walls in magnetic systems, disclinations in liquid crystals, magnetic fluxes on a Josephson line [31], and so forth. Furthermore, because the sG equation is Lorentz invariant, it is much studied by particle physicists as a model field theory [14] with the hope that elementary particles may eventually be interpreted as solitons. Interestingly, in this regard, the sG solitons with $c > 1$ will then correspond to particles traveling faster than light—the tachyons.

Exact soliton solutions to the two-dimensional (2D) version of the sG equation [with θ_{yy} added to the left-hand side of Eq. (10.8)] are available [37]. (Similarly, a nontrivial 2D generalization of the NLS of Eq. (10.6) is found to be integrable and solvable [38].)

10.4.4 Kadomtsev–Petviashvili Equation

Integrable soliton equations in 2D are much rarer than those in 1D. An important example of the former was given by Kadomtsev and Petviashvili [39] (KP), which appeared first in the stability study of the KdV solitons to transverse perturbations. The KP equation,

$$(\theta_t + 6\theta\theta_x + \theta_{xxx})_x + 3\theta_{yy} = 0, \tag{10.14}$$

has single soliton solution,

$$\theta(x, y, t) = 2a^2 \operatorname{sech}^2\{a[x + by - (3b^2 + 4a^2)t + x_0]\}, \tag{10.15}$$

which travels in an arbitrary direction in the (x, y) plane, as well as multisolitons. A 2-soliton is shown in Fig. 10.10, which resembles some real nonlinear waves observed in the shallow water off the Oregon coast (see, e.g., p. 34 of [40]).

When the sign of the last term in Eq. (10.14) is reversed, one obtains the so-called KP2 equation,

$$(\theta_t + 6\theta\theta_x + \theta_{xxx})_x - 3\theta_{yy} = 0, \tag{10.16}$$

which has the soliton solution,

$$\theta(x, y, t) = 4\frac{(a^2y^2 - X^2 + a^{-2})}{(a^2y^2 + X^2 + a^{-2})^2}, \tag{10.17}$$

where $X \equiv x + a^{-1} - 3a^2t$. However, these solitons are unstable [3].

The KP equation describes water surface waves and ion-acoustic waves in a plasma. As is the case with the KdV equation, only overtaking (but not head-on) collisions are described by the KP equation.

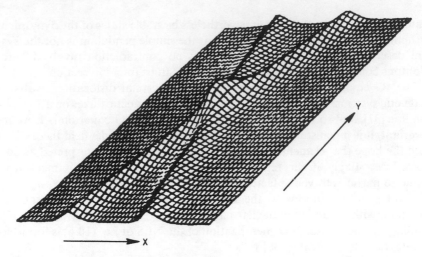

FIGURE 10.10. A snapshot of a 2-soliton of the KP equation [40].

10.5   Methods of Solution

Given a nonlinear PDE, there is no general way of knowing whether it has
soliton solutions or not, or how the soliton solutions can be found. The following are some methods that were developed and applied successfully to
particular cases.

10.5.1   Inverse Scattering Method

The inverse scattering method (ISM) was invented by Gardner et al. [41] in
1967 to solve the KdV equation. In 1968, Lax [42] gave a deeper and more
general formulation of the ISM. The NLS equation was solved by Zakharov
and Shabat [43] in 1971 using the ISM; in 1973 the sG equation was solved by
Ablowitz et al. [44].

Given a nonlinear equation $\theta_t = N(\theta)$, $\theta = \theta(x, t)$, the ISM consists of
three parts:

(i) Find operators L and B (which depend on the solution θ) such that

$$iL_t = BL - LB. \tag{10.18}$$

(ii) For the scattering operator L, the eigenvalue equation is given by

$$L\psi = \lambda\psi. \tag{10.19}$$

(iii) The time dependence of the scattering wave is given by

$$i\psi_t = B\psi. \tag{10.20}$$

Instead of solving directly the nonlinear problem of finding $\theta(x, t)$ from the
given initial condition $\theta(x, 0)$, the ISM proceeds in three steps involving

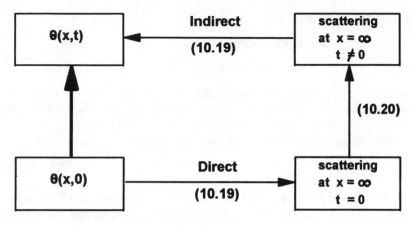

FIGURE 10.11. Sketch of the inverse scattering method.

the solution of linear integral equations (Fig. 10.11). The major difficulty of the ISM is that there is no systematic method of finding the operators L and B (if they exist at all).

As an example let us consider the KdV equation. One finds

$$L = -\partial^2/\partial x^2 + \theta(x, t) \tag{10.21}$$

and

$$B = -4i\partial^3/\partial x^3 + 3i(\theta\partial/\partial x + \partial\theta/\partial x). \tag{10.22}$$

The bound states of L correspond to solitons. For the sG equation, L and B are 2×2 matrices [44].

10.5.2 Bäcklund Transformation

Given $d\theta = P\,dx + Q\,dt$, there correspond two first-order differential equations

$$\theta_x = P, \qquad \theta_t = Q. \tag{10.23}$$

Integrability requires $P_t = Q_x$. The Bäcklund transformation (BT) method also consists of three steps [24].

(i) Find P and Q (as functions of θ).
(ii) Given solution θ_0.
(iii) Use $\theta_{1,x} = P(\theta_1, \theta_0)$ and $\theta_{1,t} = Q(\theta_1, \theta_0)$ to find new solution θ_1.

Example 10.2 In the case of the sG equation, Eq. (10.8) may be rewritten as $\theta_{\xi\eta} = \sin\theta$ with the transformations $\xi = \frac{1}{2}(x + t)$ and $\eta = \frac{1}{2}(x - t)$. The BT for this case is found to be

$$\theta_{1,\xi} = 2a \sin[\tfrac{1}{2}(\theta_1 + \theta_0)] + \theta_{0,\xi}$$
$$\theta_{1,\eta} = (2/a) \sin[\tfrac{1}{2}(\theta_1 - \theta_0)] - \theta_{0,\eta}, \tag{10.24}$$

where a is a constant. Assuming $\theta_0 = 0$ (i) one obtains $\theta_1 = 4 \tan^{-1}[\exp(a\xi + \eta/a)]$ from Eq. (10.24), and (ii) the soliton solutions of Eqs. (10.9) and (10.10) are recovered by taking $c = (1 - a^2)/(1 + a^2)$.

The solutions can be obtained by direct substitutions.

In short, the BT method enables one to find a new solution from a given one. When applied repeatively, the BT can give the breather and the N-soliton solutions of the sG equation. It can also be applied to the KdV and other equations. The difficulty is in finding the functions P and Q.

10.5.3   Hirota Method

The Hirota method [30] changes the independent variables and the original equation into the form $F(D_x^m D_t^n) = 0$. Analytic solutions are then obtained by perturbation. Here,

$$D_x^m D_t^n (a \cdot b) \equiv \left[\left(\frac{\partial}{\partial x} - \frac{\partial}{\partial x'} \right)^m \left(\frac{\partial}{\partial t} - \frac{\partial}{\partial t'} \right)^n a(x, t) b(x', t') \right]_{\substack{x=x' \\ t=t'}}. \tag{10.25}$$

Let us consider the sG equation as an example. Let $\theta(x, t) = 4 \tan^{-1}[g(x, t)/f(x, t)]$. By Eq. (10.8) one obtains

$$(D_x^2 - D_t^2)(f \cdot g) = fg$$
$$(D_x^2 - D_t^2)(f \cdot f - g \cdot g) = 0. \tag{10.26}$$

Obviously, $g = 0$ and $f = 1$ is a solution. Let

$$f = 1 + \varepsilon^2 f^{(1)} + \varepsilon^4 f^{(2)} + \cdots$$
$$g = \varepsilon g^{(1)} + \varepsilon^3 g^{(2)} + \cdots. \tag{10.27}$$

Substituting Eq. (10.27) into Eq. (10.26) and equating the coefficients of the ε terms, one obtains the linear equation

$$g_{xx}^{(1)} - g_{tt}^{(1)} = g^{(1)}. \tag{10.28}$$

The solution is given by $g^{(1)} = \exp(kx - \omega t + \delta)$, $k^2 = \omega^2 - 1$; or equivalently, $\theta = 4 \tan^{-1}[\exp(kx - \omega t + \delta)]$, which is the same as Eq. (10.9).

10.5.4   Numerical Method

When analytic solutions cannot be found one can always solve the nonlinear PDE by numerical methods. Sometimes the numerical solutions may even lead to a correct guess of their analytic forms. References [26, 45] give a general discussion on numerical methods for solitons of one and higher spatial dimensions.

Note that numerical methods do not distinguish between linear and nonlinear equations, and for that matter, integrable and nonintegrable systems.

10.6 Physical Soliton Systems

In this section the physical origins of the first three soliton equations of section 10.4 are discussed. Corresponding to each soliton equation there are many different physical systems, one of which is presented here for its simplicity. Note that for the same physical system, under different conditions more than one soliton equation may be obtained [28].

In the literature, the kink (bell)-shaped solitons are called topological (nontopological) solitons. In the former (latter) the two asymptotic states at $\pm\infty$ are different from (the same with) each other. Also, for historical reasons, the same type of soliton may be called by different names in different physical systems. For example, the kink solitons assume the name of "walls" in magnetic systems and liquid crystals, "fluxons" in long Joshephon junctions, and "discommensurations" in crystals with competing interactions.

10.6.1 Shallow Water Waves

For a shape-preserved dispersionless linear wave propagating with velocity c in one dimension along the axis x, the wave can be described by ξ ($\equiv x - ct$). In the laboratory frame in which this is observed, let the height of the wave be $h(x, t)$. It is obvious that in a reference frame moving with the same velocity c, the height of the wave $h(\xi, \tau)$ would then be independent of τ; that is,

$$h_\tau = 0. \tag{10.29}$$

Here, $\tau \equiv t$. Equation (10.29) is equivalent to

$$h_t + ch_x = 0. \tag{10.30}$$

As shown in section 10.2, nonlinearity introduces the dependence of h into c. To first approximation, $c = c_0 + ah$, where a is a constant. In the moving frame, this changes Eq. (10.30) to

$$h_\tau + ahh_\xi = 0. \tag{10.31}$$

On the other hand, dispersion changes the dispersion relation from $c = \omega/k = c_0$ to $c = c_0 + bk^2$, or $\omega = c_0 k + bk^3$, where b is a constant. Because k^3 corresponds to the third derivative with respect to x or ξ, this corresponds to adding a term of $b\partial^3/\partial\xi^3$ to Eq. (10.31). The combined effect of nonlinearity and dispersion results in the equation

$$h_\tau + ahh_\xi + bh_{\xi\xi\xi} = 0. \tag{10.32}$$

In terms of the new variables $h' \equiv (a/\alpha)b^{-1/3}h$ and $\xi' \equiv b^{-1/3}\xi$ (and then dropping the primes), Eq. (10.32) becomes the KdV equation of Eq. (10.3).

The heuristic "derivation" above [5] can be made more rigorous by applying the perturbation method to the equation of motion of a nonviscous fluid [28].

10.6.2   Dislocations in Crystals

The Frenkel–Kontorova model of dislocations in crystals is shown in Fig. 10.12. The atoms in the upper chain have neighboring harmonic interaction with each other. Atoms in the lower chain are assumed to be fixed, and provide a periodic potential $V(\phi_n) = A[1 - \cos(2\pi\phi_n/a)]$ for the upper chain. Here a is the lattice constant. As shown in Fig. 10.12, when the upper chain is extended there are only 5 atoms in 6 lattice cells and we have vacancy dislocation. On the contrary, if the upper chain is contracted there will be interstitial dislocation.

The equation of motion for atoms of the upper chain is given by

$$md^2\phi_n/dt^2 = -dV/d\phi_n + k[(\phi_{n+1} - \phi_n) - (\phi_n - \phi_{n-1})]. \qquad (10.33)$$

If ϕ_n varies slowly with n, one may take the continuum limit, $n \to x/a$ and $\phi_n \to \phi(x)$ (and $\Sigma_n \to a^{-1} \int dx$), and obtain

$$m\phi_{tt} = -2\pi(A/a) \sin(2\pi\phi/a) + ka^2\phi_{xx}. \qquad (10.34)$$

Through the transformations, $\theta \equiv 2\pi\phi/a$, $\xi \equiv x/\lambda$, $\eta \equiv t/\tau$, $\lambda \equiv a(ak/2\pi A)^{1/2}$, and $\tau \equiv (ma/2\pi A)^{1/2}$, Eq. (10.34) can be rewritten as $\theta_{\xi\xi} - \theta_{\eta\eta} = \sin \theta$, the sG equation.

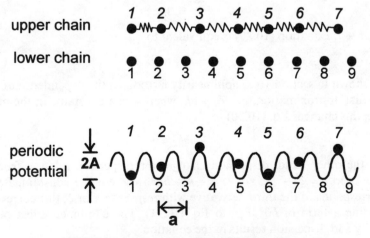

FIGURE 10.12. Model of crystal dislocations.

10.6.3   Self-Focusing of Light

In a medium with inhomogeneous refractive index light does not travel in straight lines. For example, mirage observed by travelers in a desert is a consequence of this phenomenon. This inhomogeneity of the refractive index n may be caused by thermal effect if n depends sensitively on temperature. For a laser beam of finite cross-section propagating in a medium, the intensity of the beam varies in space, resulting in inhomogeneity in temperature and hence inhomogeneity in the refractive index. The beam may then shrink in size—a phenomenon called self-focusing.

The equation describing self-focusing of light may be derived as follows [22]. For a medium with inversion symmetry, the polarization is given by

$$\mathbf{P} = \alpha \mathbf{E} + \alpha_{NL}|E|^2 \mathbf{E}. \tag{10.35}$$

Three of the Maxwell equations for a neutral, homogeneous medium are given by

$$\nabla \times \mathbf{E} = -c^{-1}\partial \mathbf{B}/\partial t \tag{10.36}$$

$$\nabla \times \mathbf{B} = c^{-1}\partial \mathbf{D}/\partial t \tag{10.37}$$

$$\nabla \cdot \mathbf{E} = 0, \tag{10.38}$$

where $\mathbf{D} = \mathbf{E} + 4\pi\mathbf{P}$. Combining these three equations and using the identity, $\nabla \times (\nabla \times \mathbf{E}) = \nabla(\nabla \cdot \mathbf{E}) - \nabla^2\mathbf{E}$, one obtains

$$\nabla^2\mathbf{E} - c^{-2}\partial^2\mathbf{E}/\partial t^2 = (4\pi/c^2)\partial^2\mathbf{P}/\partial t^2. \tag{10.39}$$

Putting Eq. (10.35) into Eq. (10.39) one has

$$(\partial^2 - c^{-2}\partial^2/\partial t^2)E = (4\pi\alpha/c^2)\partial^2 E/\partial t^2 + (4\pi\alpha_{NL}/c^2)\partial^2(|E|^2E)/\partial t^2. \tag{10.40}$$

Let $E = \varepsilon(x, y, z, t)\exp[i(\omega t - kz)] + cc$, where z is the propagation direction of the beam and cc denotes complex conjugate. The complex amplitude ε is assumed to be slowly varying.

By Eq. (10.40) one has

$$(\omega^2/c^2 - k^2 + 4\pi\alpha\omega^2/c^2)\varepsilon - 2ik\varepsilon_z + (2i\omega/c^2)(1 + 4\pi\alpha)\varepsilon_t$$
$$+ \varepsilon_{xx} + \varepsilon_{yy} + 12\pi(\omega^2/c^2)\alpha_{NL}|\varepsilon|^2\varepsilon = 0. \tag{10.41}$$

In the linear regime, the cubic term $|\varepsilon|^2\varepsilon$ is dropped and $\varepsilon = $ const; Eq. (10.41) gives the linear dispersion relation

$$\omega^2/c^2 - k^2 + 4\pi\alpha\omega^2/c^2 = 0. \tag{10.42}$$

Note that in Eq. (10.41), only first harmonics are kept. Now in the weakly nonlinear regime, that is, when Eq. (10.42) is still valid, Eq. (10.41) simplifies to

$$\varepsilon_{xx} + \varepsilon_{yy} + 12\pi(\omega^2/c^2)\alpha_{NL}|\varepsilon|^2\varepsilon - 2ik\varepsilon_z + (2i\omega/c^2)(1 + 4\pi\alpha)\varepsilon_t = 0. \tag{10.43}$$

In the steady state, $\varepsilon_t = 0$, and after rescaling, Eq. (10.43) reduces to

$$\varepsilon_{xx} + \varepsilon_{yy} + 2|\varepsilon|^2\varepsilon - i\varepsilon_z = 0. \qquad (10.44)$$

This is the equation that describes optical self-focusing, which is equivalent to the NLS equation if one-dimensional spatial variation of ε is assumed. In such a case ($\varepsilon_{yy} = 0$, say) one has

$$-i\varepsilon_z + \varepsilon_{xx} + 2|\varepsilon|^2\varepsilon = 0, \qquad (10.45)$$

which is the NLS equation of Eq. (10.6), with $-z$ here playing the role of "time."

10.7   Conclusions

The integrable soliton equations (KdV, NLS, and sG) discussed in this chapter can be derived from suitable Hamiltonians [28]. This Hamiltonian structure is related to the existence of an infinite number of conservation laws. Although a general criterion in identifying integrable soliton equations is still lacking, there is some progress being made in the use of the so-called Painlevé test [24, 35]. See [16] for details.

All real physical systems are subjected to perturbations such as thermal agitation or noise. An integrable system under perpurbation will, in general, becomes nonintegrable. We are thus required to consider nonintegrable systems when comparison between theory and experiment is desired, if not for other reasons. And, as will be discussed in chapter 11, there are other reasons.

Problems

10.1. For a linear chain of coupled atoms interacting with nearest-neighbor quadratic interaction, the equation of motion is given by

$$m\ddot{y}_i = k(y_{i+1} - 2y_i + y_{i-1}) + k\alpha[(y_{i+1} - y_i)^2 - (y_i - y_{i-1})^2],$$

where $y_i = y_i(t)$.

(i) By Taylor expanding $y_{i\pm1}$ to fourth order in a, where a is the lattice constant, show that the equation of motion is replaced by

$$y_{t't'} = y_{x'x'} + \varepsilon y_{x'} y_{x'x'} + \beta y_{x'x'x'x'} + 0(\varepsilon a^2, a^4),$$

where $\varepsilon \equiv 2a\alpha$, $\beta \equiv a^2/12$, $t' \equiv \omega t$, $\omega \equiv (k/m)^{1/2}$, and $x' \equiv x/a$ (with $x \equiv ia$).

(ii) Show that by the change of variables, $T \equiv \varepsilon t/2$ and $X \equiv x - t$, the above equation becomes

$$v_T + vv_X + \beta^2 v_{XXXX} = 0,$$

where $y(x', t') = u(X, T)$ and $v = u_X$.

(iii) By rescalings, reduce the equation in (ii) to the KdV equation of Eq. (10.3).

Hint: see Zabusky and Kruskal [19].

10.2. Show that $I_4 = 5\theta^4 + 10\theta\theta_x + \theta_{xx}$ is a conserved density of the KdV equation of Eq. (10.3) with $\alpha = 6$.

10.3. Show that the large time limits of the 2-soliton of Eq. (10.12) is given by

$$\lim_{t \to \mp\infty} \theta(x,t) = 4 \tan^{-1}\{ \pm \exp[\pm(x - ct \mp \alpha)/(1 - c^2)^{1/2}$$
$$\mp \exp[\mp(x - ct \pm \alpha)/(1 - c^2)^{1/2}]\},$$

where $\alpha \equiv (1 - c^2)^{1/2} \ln(1/c)$. Sketch the behavior of Eq. (10.12) for $t < 0$ and $t > 0$, with t very large, and for $t = 0$.

References

[1] M.V. Nezlin, Sov. Phys. Usp. **29**, 807 (1986).

[2] A.R. Osborn and T.L. Burch, Science **208**, 451 (1980).

[3] E. Infeld and G. Rowlands, *Nonlinear Waves, Solitons and Chaos* (Cambridge University, Cambridge, 1990).

[4] A.S. Davydov, *Solitons in Molecular Systems* (Reidel, Boston, 1991).

[5] A. Hasegawa, *Optical Solitons in Fibers* (Springer-Verlag, New York, 1990).

[6] K. Maki, in *Quantum Fluids and Solids*, edited by S.B. Trickey, E.D. Adams, and J.W. Dufty (Plenum, New York, 1977).

[7] R.D. Parmentier, in [31].

[8] H.-J. Mikeska and M. Steiner, Adv. Phys. **40**, 191 (1991).

[9] J.A. Krumhansl and J.R. Schrieffer, Phys. Rev. B **11**, 3535 (1975).

[10] *Solitons in Liquid Crystals*, edited by L. Lam and J. Prost (Springer-Verlag, New York, 1992).

[11] A.J. Heeger, B. Kivelson, J.R. Schrieffer, and W.-P. Su, Rev. Mod. Phys. **60**, 781 (1988).

[12] L. Lam, Mol. Cryst. Liq. Cryst. **155**, 531 (1988).

[13] G.B. Whitham, *Linear and Nonlinear Waves* (Wiley, New York, 1974).

[14] *Solitons and Particles*, edited by C. Rebbi and G. Solianai (World Scientific, Singapore, 1985).

[15] J.R. Schrieffer, X.-G. Wen, and S.-C. Zhang, Phys. Rev. B **39**, 11663 (1989); P.W. Anderson, in *Frontiers and Borderlines in Many-Particle Physics*, edited by R.A. Broglia and J.R. Schrieffer (North-Holland, New York, 1987); T.D. Lee, Nature **330**, 460 (1987).

[16] *What Is Integrability?*, edited by V.E. Zakharov (Springer-Verlag, New York, 1991).

[17] D. Campbell, in *Lectures in the Sciences of Complexity*, edited by D.L. Stein (Addison-Wesley, Menlo Park, 1989).

[18] L. Lam, *Nonlinear Physics for Beginners* (World Scientific, River Edge, 1996).

[19] N.J. Zabusky and M.D. Kruskal, Phys. Rev. Lett. **15**, 240 (1965).

[20] M. Figenbaum, J. Stat. Phys. **19**, 25 (1978); ibid. **21**, 69 (1979).

[21] A.C. Scott, in *Nonlinear Electromagnetics*, edited by P.L.E. Uslenghi (Academic, New York, 1980).

[22] R.K. Bullough, in *Interaction of Radiation with Condensed Matter*, Vol. 1. (International Atomic Energy Agency, Vienna, 1977).

[23] A.R. Bishop, J.A. Krumhansl, and S.E. Trullinger, Physica D **1**, 1 (1980).

[24] P.G. Drazin and R.S. Johnson, *Solitons: An Introduction* (Cambridge University, Cambridge, 1989).

[25] M. Remoissenet, *Waves Called Solitons: Concepts and Experiments* (Springer-Verlag, New York, 1992).

[26] R.K. Dodd, J.C. Eilbeck, J.D. Gibbon, and H.C. Morris, *Solitons and Nonlinear Wave Equations* (Academic, New York, 1982).

[27] G. Eilenberger, *Solitons: Mathematical Methods for Physicists* (Springer-Verlag, New York, 1981).

[28] G.L. Lamb, Jr., *Elements of Soliton Theory* (Wiley, New York, 1980).

[29] A.C. Newell, *Solitons in Mathematics and Physics* (Society for Industrial and Applied Mathematics, Philadelphia, 1985).

[30] *Solitons*, edited by R.K. Bullough and P. J. Caudrey (Springer-Verlag, New York, 1980).

[31] *Solitons in Action*, edited by K. Lonngren and A.C. Scott (Academic, New York, 1978).

[32] G.S. Emmerson, *John Scott Russell* (John Murray, London, 1977).

[33] K.K. Perring and T.H.R. Skyrme, Nucl. Phys. **31**, 550 (1962).

[34] E.N. Lorenz, J. Atmos. Sci. **20**, 130 (1963).

[35] M. Tabor, *Chaos and Integrability in Nonlinear Dynamics* (Wiley, New York, 1989).

[36] J.M. Ottino, *The Kinematics of Mixing: Stretching, Chaos and Transport* (Cambrige University, Cambridge, 1989).

[37] S.I. Ben-Abraham, Phys. Lett. **55A**, 383 (1976). There are a number of misprints in this paper; for example, the exponents $1/2$ in Eq. (10) should be replaced by $-1/2$.

[38] N.C. Freeman, Adv. Appl. Mech. **20**, 1 (1980).

[39] B.B. Kadomtsev and V.I. Petviashvili, Sov. Phys. Dokl. **15**, 539 (1970).

[40] J.A. Krumhansl, Phys. Today **44**(3), 33 (1991).

[41] C.S. Gardner, J.M. Greene, M.D. Kruskal, and R.M. Miura, Phys. Rev. Lett. **19**, 1095 (1967).

[42] P.D. Lax, Commun. Pure Appl. Math. **21**, 467 (1968).

[43] V.E. Zakharov and A.B. Shabat, Zh. Eksp. Teor. Fiz. **61**, 118 (1971) [Sov. Phys. JETP **34**, 62 (1972)].

[44] M.J. Ablowitz, D.J. Kaup, A.C. Newell, and H. Segur, Phys. Rev. Lett. **31**, 125 (1973).

[45] V.G. Makhankov, Phys. Rep. **35**, 1 (1978).

11

Nonintegrable Systems

Lui Lam

11.1   Introduction

Almost all the nonlinear differential equations describing real physical systems are nonintegrable. There are at least two origins. First, quite often the integrable nonlinear partial differential equations (PDEs) representing some physical situations, such as those mentioned in chapter 10, are derived under certain approximations. When higher-order terms are included, the integrability is destroyed (e.g., the optical fibers in section 11.8.3). Second, there exist physical systems that can only be described by intrinsically nonintegrable equations, such as the strongly dissipative liquid crystals (section 11.8.1), and the many reacting and diffusing systems in biology and chemistry [1–3].

In nonintegrable systems, the solitons (quite often under the names of fronts and pulses, with kink and bell shapes, respectively) usually do not possess the same pairwise elastic-collision property as those in integrable systems (see Fig. 10.1), and are sometimes referred to as "nonrigorous" solitons [4]. However there are exceptional cases (see sections 11.2.2 and 11.9).

To see how a propagating front can arise, let us consider the diffusion equation,

$$\theta_t = \theta_{xx} - F(\theta), \tag{11.1}$$

as an example. In the linear case, $F = a\theta$ with $a = $ const, the solution is given by

$$\theta = \theta_0 \exp[ikx - (k^2 + a)t], \tag{11.2}$$

which decays with time and cannot be a permanent wave. In the nonlinear case, Eq. (11.1) is called the nonlinear diffusion equation, the evolution equation, or the reaction-diffusion equation. For example, when

$$F(\theta) = \theta(\theta - a)(\theta - 1), \quad 0 < a < 1, \tag{11.3}$$

there exists an analytic solution (Fig. 11.1a)

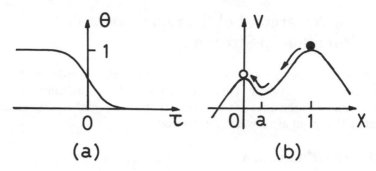

FIGURE 11.1. (a) A soliton solution of the nonlinear evolution equation given by Eqs. (11.1) and (11.3). (b) Physical picture of the soliton solution.

$$\theta = \{1 + \exp[(x - ct)/\sqrt{2}]\}^{-1} \tag{11.4}$$

with

$$c = (1 - 2a)/\sqrt{2}. \tag{11.5}$$

The existence of this solution may be understood by first turning the PDE of Eq. (11.1) into an ordinary differential equation (ODE) with the assumption of a traveling wave solution, and then using a mechanical analogue (section 10.4.3). Specifically, let $\theta(x, t) = X(\tau), \tau \equiv x - ct, F(\theta) - \partial V(\theta)/\partial\theta$. Equation (11.1) becomes

$$X_{\tau\tau} = -cX_\tau - \partial V/\partial X, \tag{11.6}$$

which represents the motion of a particle of unit mass moving in a potential V with damping coefficient c. When c is suitably chosen, the particle may roll down from the high hilltop (at $X = 1$) with zero velocity ($X_\tau = 0$), pass through the valley (at $X = a$), and then stop exactly at the lower hilltop (at $X = 0$). (see Fig. 11.1b). This solution corresponds to a soliton which, in this case, is a front connecting two stable uniform steady states of Eq. (11.6). Note that because the two hilltops in V have different heights, the damping coefficient c must be nonzero; the front must be propagating. More general solutions of Eq. (11.1) are given in section 11.3, including fronts connecting stable to unstable states.

In this chapter, examples of nonintegrable soliton equations are given in sections 11.2 and 11.3. A general method of constructing solitonic nonlinear evolution equations is presented (section 11.4). For the observation of solitons in physical systems, the problem of generation or excitation of the solitons (section 11.5), the effect of perturbations (section 11.6), and the statistical mechanics of a many-soliton system (section 11.7), are three important issues to be reckoned with. Four special examples of solitons in condensed matter are presented in section 11.8, which is followed by a discussion of the coexistence of solitons and chaos, and other related topics (section 11.9).

11.2   Nonintegrable Soliton Equations with Hamiltonian Structures

Even though the integrable soliton equations presented in chapter 10 can be derived from a Hamiltonian, the mere existence of a Hamiltonian structure for a solitonic equation does not guarantee that it is integrable. Two examples illustrating this point are given in this section.

11.2.1   The θ^4 Equation

The nonlinear Klein–Gordon PDE,

$$\theta_{xx} - \theta_{tt} = \theta(\theta + 1)(\theta - 1), \tag{11.7}$$

can be derived obviously from the continuum Hamiltonian

$$H = \int dx \left[\tfrac{1}{2}\theta_t^2 + \tfrac{1}{2}\theta_x^2 + \tfrac{1}{4}(\theta^2 - 1)^2 \right], \tag{11.8}$$

where $\theta = \theta(x, t)$. Equation (11.7) has the kink and antikink solutions

$$\theta = \pm \tanh\left[a(x - ct)/\sqrt{2} \right], \tag{11.9}$$

with $a = (1 - c^2)^{-1/2}$. The kink (antikink) solution assumes the "+" ("−") sign in Eq. (11.9), and has asymptotic states ∓ 1 (± 1) as $x \to \mp\infty$. Similar to those in the sine–Gordon equation (section 10.4.3), both kinks and antikinks here can move in either a positive or negative direction (i.e., c can assume any value between -1 and $+1$). However, when a kink–antikink pair (Fig. 11.2a) collide head-on, they do not penetrate [as is the case in the sG equation described by Eq. (10.12)] but rebound from each other (Fig. 11.2b). Furthermore, contrary to the numerical results, the existence of an oscillating breather state resulting from such a head-on collision is ruled out by rigorous mathematical analysis, using a multiscale asymptotic expansion. See [5] for a detailed discussion.

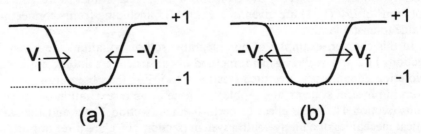

(a)   (b)

FIGURE 11.2. Head-on collision of a kink–antikink pair from the θ^4 equation. (a) Before collison. (b) Rebound of the pair after collison. Note that $v_i \neq v_f$; the collision is inelastic and there is no penetration.

Note that the θ^4 equation and the sG equation differ only on the form of the nonlinear function on the right-hand side of Eq. (11.7). The nonexistence of penetration and a breather upon the collision of a kink–antikink pair from the θ^4 equation, in contradistinction to the sG equation, highlights the subtleties that exist between nonintegrable and integrable solitonic systems.

11.2.2 Double Sine–Gordon Equation

The double sine–Gordon equation is given by

$$\theta_{xx} - \theta_{tt} = \pm [\sin \theta + \tfrac{1}{2} \sin(\theta/2)] \tag{11.10}$$

For the positive sign case, there exists a soliton solution of double-kink form,

$$\theta = 4 \tan^{-1}[\exp(u + a)] + 4 \tan^{-1}[\exp(u - a)], \tag{11.11}$$

where $u = (\sqrt{5}/2)(x - ct)(1 - c^2)^{-1/2}$ and $a = \ln(\sqrt{5} + 2)$; the spatial derivative of θ has a twin peak structure. Numerical studies [6, 7] suggest that the collision of two kinks can penetrate each other without deformation, like rigorous solitons, whereas the collision of a kink and an antikink results in loss of energy through radiation. See [6] for further interesting results, such as wobbling when a in Eq. (11.11) deviates slightly from the value given above.

For the negative sign case of Eq. (11.10), two kinds of kink solitons exist [8]; namely,

$$\theta = 2\pi + 4 \tan^{-1}[\sqrt{3/5} \tanh(v/2)] \tag{11.12}$$

and

$$\theta = 4 \tan^{-1}[\sqrt{5/3} \tanh(v/2)], \tag{11.13}$$

where $v = \sqrt{15/16}(x - ct)(1 - c^2)^{-1/2}$. The kinks in Eqs. (11.12) and (11.13) have a jump of $4\pi - 2b$ and $2b$, respectively, with $b = 2 \cos^{-1}(-1/4)$. Numerical studies show that both kink–kink and kink–antikink collisions lose energy due to radiation.

The Hamiltonian corresponding to Eq. (11.10) is given by

$$H = \int dx \{\tfrac{1}{2}\theta_t^2 + \tfrac{1}{2}\theta_x^2 \mp [\sin \theta + \cos(\theta/2)]\}. \tag{11.14}$$

The double sine–Gordon equation with the positive sign has applications in nonlinear optics; with negative sign it has applications in nonlinear optics and liquid ^{3}He.

11.3 Nonlinear Evolution Equations

Here are three examples of the nonlinear evolution equation in the form of Eq. (11.1).

11.3.1 Fisher Equation

The Fisher equation,

$$\theta_t = \theta_{xx} - \theta(\theta - 1) \qquad (11.15)$$

has an explicit soliton solution [9],

$$\theta = \{1 + \exp[\varepsilon/\sqrt{6})(x - \varepsilon c t - x_0)]\}^{-2}, \qquad (11.16)$$

where $c = 5/\sqrt{6}$, $\varepsilon = \pm 1$, and x_0 is an arbitrary constant. Equation (11.15) may be rewritten as

$$\theta_t = \theta_{xx} + \partial V/\partial \theta, \qquad (11.17)$$

with

$$V = \tfrac{1}{2}\theta^2 - \tfrac{1}{3}\theta^3. \qquad (11.18)$$

The soliton with $\varepsilon = +1$ has $\theta = 1$ at $\tau = -\infty$ and $\theta = 0$ at $\tau = +\infty$. Here $\tau \equiv x - \varepsilon c t - x_0$. The one with $\varepsilon = -1$ connects $\theta(-\infty) = 0$ to $\theta(+\infty) = 1$. Note that $\theta = 1$ is a stable state and $\theta = 0$ is an unstable state of Eq. (11.15), opposite to the impression conveyed by the diagram in Fig. 11.3. [A uniform steady state of Eq. (11.1), $\theta = \theta_0$, is stable (unstable) if a deviation from this state decays (grows) with time. θ_0 is stable (unstable) if $(\partial F/\partial \theta)_{\theta=\theta_0} > 0$ (< 0).]

Note that when the "particle" rolls down from the hilltop ($\theta = 1$) to the valley ($\theta = 0$), the critical damped case corresponds to $c = 2$. The soliton shown in Eq. (11.16) has $c = 5/\sqrt{6} \approx 2.04$, and hence is overdamped. It is just one of an infinite number of overdamped solutions [see Eq. (11.39)].

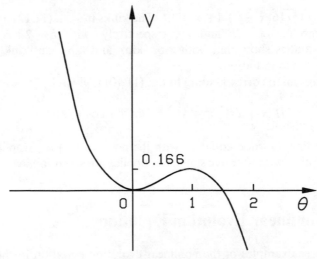

FIGURE 11.3. The potential, $V = \tfrac{1}{2}\theta^2 - \tfrac{1}{3}\theta^3$, corresponding to the Fisher equation.

The Fisher equation occurs in flame propagation, biological growth problems [1], nuclear reactor theory, and so forth. It is probably the simplest nonlinear diffusion equation with an explicit soliton solution. More importantly, because the soliton depends essentially on the shape of V in the region, $0 \leq \theta \leq 1$, the V in Fig. 11.3 or the Fisher equation itself may be used to approximate more complicated Vs for which an analytic (though approximate) soliton solution connecting a hilltop in V to an adjacent valley is desired.

11.3.2   The Damped θ^4 Equation

In Eq. (11.1), when $F(\theta) = a\theta^3 + b\theta^2 + d\theta + e$ with $a \neq 0$, solitons exist only when F has two or three distinct real roots. The corresponding V (defined by $F \equiv -\partial V/\partial \theta$) has one maximum and one inflection point, and two maxima and one minimum, respectively. Because the former may be considered as a special case of the latter, we need only consider the equation (with rescaling to remove the factor a),

$$\theta_t = \theta_{xx} - (\theta - \theta_1)(\theta - \theta_2)(\theta - \theta_3), \tag{11.19}$$

where θ_i $(i = 1, 2, 3)$ is real, and at least two θ_is are distinct. Equation (11.19) is called the damped θ^4 equation because V is quartic in θ (Fig. 11.4). For a traveling wave solution with $\tau \equiv x - ct - x_0$, Eq. (11.19) becomes

$$\theta_{\tau\tau} + c\theta_\tau - (\theta - \theta_1)(\theta - \theta_2)(\theta - \theta_3) = 0. \tag{11.20}$$

Equation (11.19) or (11.20) was found by Lam [10] to have general soliton solutions given by

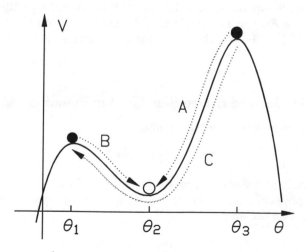

FIGURE 11.4. The θ^4 potential with $\theta_2 < \frac{1}{2}(\theta_1 + \theta_3)$. The origins of three types of solitons (A, B, and C) are shown.

$$\theta = (\theta_j - \theta_i)/\{1 + \exp[a(x - ct - x_0)]\} + \theta_i, \tag{11.21}$$

where $a = (\varepsilon/\sqrt{2})(\theta_j - \theta_i)$, $\varepsilon = \pm 1$, x_0 is an arbitrary constant, and

$$c = (\varepsilon/\sqrt{2})(\theta_j + \theta_i - 2\theta_k), \tag{11.22}$$

where $i, j, k = 1, 2, 3$, $\theta_i \neq \theta_j$, and θ_k is the third one in the set $\{\theta_n\}$, which may or may not be equal to θ_i or θ_j. Note that Eq. (11.21) is equivalent to

$$\theta = \tfrac{1}{2}(\theta_j - \theta_i)\{1 - \tanh[(a/2)(x - ct - x_0)]\} + \theta_i, \tag{11.23}$$

which is static ($c = 0$) when $\theta_k = \tfrac{1}{2}(\theta_i + \theta_j)$.

For the related hyperbolic equation

$$\theta_{xx} - \theta_{tt} - \lambda\theta_t - (\theta - \theta_1)(\theta - \theta_2)(\theta - \theta_3) = 0, \quad \lambda \neq 0, \tag{11.24}$$

the solitons are still given by Eq. (11.21) or (11.23), but now

$$a = (\varepsilon/\sqrt{2})(\theta_j - \theta_i)[1 + (\theta_j + \theta_i - 2\theta_k)^2/(2\lambda^2)]^{1/2} \tag{11.25}$$

and

$$c = (\varepsilon/\sqrt{2})(\theta_j + \theta_i - 2\theta_k)[\lambda^2 + \tfrac{1}{2}(\theta_j + \theta_i - 2\theta_k)]^{-1/2}. \tag{11.26}$$

From Fig. 11.4 one can expect to have four types of single solitons for Eq. (11.19) (see Fig. 11.5). These are listed in Table 11.1 for $c > 0$. Types A_0 and B_0 correspond to the critical damped or overdamped motion of the "particle" moving in V; types A_1 and B_1 correspond to the underdamped cases. For $\theta_1 < \theta_2 < \theta_3$, $\theta = \theta_1$ or θ_3 is a stable state and $\theta = \theta_2$ is an unstable state. Again, as in the Fisher equation, the solitons given explicitly in Eqs. (11.21) and (11.26) are overdamped.

The damped θ^4 equation, Eq. (11.19) or (11.24), is useful partly because in many cases the $F(\theta)$ in Eq. (11.1) may be expanded into the form of the θ^3 function of Eq. (11.19) when θ is small (see section 2.7.2 of [13] for an example).

11.3.3   The Damped Driven Sine-Gordon Equation

The damped driven sine–Gordon equation,

$$\theta_t = \theta_{xx} + \gamma + \cos 2\theta, \tag{11.27}$$

occurs naturally in a shearing nematic liquid crystal (see section 11.8.1) and many other physical systems. For a traveling wave solution $\theta = \theta(x - ct) \equiv \theta(\tau)$, the equation becomes

$$\theta_{\tau\tau} = -c\theta_\tau - \partial V/\partial\theta, \tag{11.28}$$

where $V = \gamma\theta + \tfrac{1}{2}\sin 2\theta$ (Fig. 11.6). Similar to the case of the damped θ^4

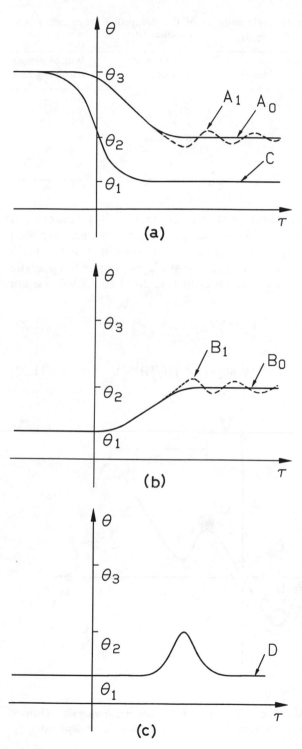

FIGURE 11.5. Four types of solitons of the damped θ^4 equation. Types A, B, and C are propagating; type D is static.

TABLE 11.1. Classification of single solitons of the damped θ^4 equation. $c > 0$, $\theta_1 < \theta_2 < \theta_3$, $c_0 = 2(\theta_2 - \theta_1)(\theta_3 - \theta_2)$, $c_1 = \frac{1}{\sqrt{2}}(\theta_3 + \theta_1 - 2\theta_2)$.

Type		Particle Starts at	Ends at	Condition of existence	With or without Oscillating Tails
A	A_0	θ_3	θ_2	$c_0 \leq c$	No
	A_1			$0 < c < c_0$	Yes
B	B_0	θ_1	θ_2	$c_0 \leq c$	No
	B_1			$0 < c < c_0$	Yes
C		θ_3	θ_1	$c = c_1$	No
D		θ_1	θ_1	$c = 0$	No

equation, here there are also four types of single solitons [11]. However, in spite of the simplicity of Eq. (11.28), no analytic soliton solutions have been found. For the physically important case of $c \gg 1$ (in shearing nematic liquid crystals), Eq. (11.28) can be expanded in $1/c$ and approximate but analytic soliton solutions do exist. To second order in $1/c$, the A and B solitons are given, respectively, by

$$\theta = \tan^{-1}\{W \tanh[(\gamma - 1)W\tau/c]\} \tag{11.29}$$

and

$$\theta = \cot^{-1}\{W^{-1} \tanh[-(\gamma + 1)\tau/(Wc)]\} \tag{11.30}$$

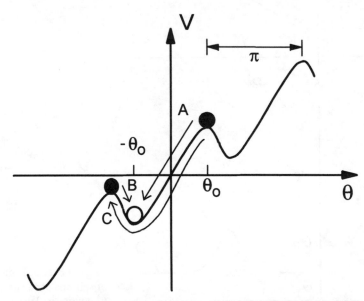

FIGURE 11.6. The potential $V(\theta)$ corresponding to the damped driving sine–Gordon equation. Origins of the three types of single solitons (A, B, and C) are depicted.

where $W \equiv \tan \theta_0 = [(1 + \gamma)/(1 - \gamma)]^{1/2}$, and $\pi/4 \leq \theta_0 \equiv \frac{1}{2} \cos^{-1}(-\gamma) \leq \pi/2$, assuming $\gamma \leq 1$. The shape of the A (B) soliton is like the antikink (kink) depicted in Fig. 10.6. See [11] for further results on Eq. (11.28).

Note that the hyperbolic counterpart of Eq. (11.27), with θ_{xx} replaced by $\theta_{xx} - \theta_{tt}$, is the equation describing the dynamics in a long Josephson junction [12].

11.4   A Method of Constructing Soliton Equations

There is no general method of constructing soliton equations with rigorous solitons. But for the nonlinear diffusion type of Eq. (11.1) there does exist a general class of soliton equations with explicit soliton solutions that can be easily constructed [4].

It is easy to see that the equation $\theta_t = \theta_{xx} - F(\theta)$ with

$$F(\theta) = -f(\theta)[1 - c^{-2}f'(\theta)] \tag{11.31}$$

does possess a traveling-wave solution, $\theta = \theta(x - ct - x_0) \equiv \theta(\tau)$, if

$$f(\theta) = -c\theta_\tau \tag{11.32}$$

where $f'(\theta) \equiv df/d\theta$. The recipe to turn Eq. (11.31) into a soliton equation is (i) to start with any soliton-like function $\theta = \theta(\tau)$ (e.g., kink or bell-shaped); (ii) use Eq. (11.32) to find $f(\theta)$; and (iii) obtain the soliton equation, Eq. (11.31). Of course, there is no guarantee that the solitons so obtained are rigorous solitons. In fact, they usually are not.

Example 11.1 (i): Let $\theta(\tau) = (1 + e^{a\tau})^{-2}$. (ii): By Eq. (11.32) we then have $f(\theta) = 2ac(\theta - \theta^{3/2})$. (iii): One has $F(\theta) = 2a[(c - 2a)\theta + (5a - c)\theta^{3/2} - 3a\theta^2]$ where a is arbitrary, by Eq. (11.31). For the special case of $a = c/5$, $F(\theta) = (6c^2/25)\theta(1 - \theta)$; the Fisher equation is recovered if $c = 25/6$.

For the chosen

$$\theta = 2 \tan^{-1}\{\exp[a(x - ct - x_0)]\}, \tag{11.33}$$

the recipe gives the soliton equation

$$\theta_t = \theta_{xx} - ac \sin \theta - \tfrac{1}{2}a^2 \sin 2\theta, \tag{11.34}$$

where a, c and x_0 are arbitrary constants; a and c may assume the same or opposite signs. Equation (11.34) describes the switching process in a ferroelectric C^* liquid crystal display [13].

11.5   Formation of Solitons

In a soliton the height, width, and velocity are intimately related to each other. Obviously, one cannot easily obtain such a wave by adjusting all three quantities simultaneously from scratch. Fortunately, there is a basic property of any soliton-bearing medium that an initial profile will, under suitable conditions, evolve automatically into one or more soltions. One can thus take advantage of this property in the experimental generation of solitons; one can simply prepare a suitable initial profile and let it evolve into a soliton. In this regard, one needs to know what the suitable initial profile should be, or, how an initial profile will evolve in time.

For an integrable system, such as the KdV case for which the ISM is applicable, analytic results governing the emergence of solitons from arbitrary initial profiles are available [14]. For the nonlinear diffusion equations some rigorous but limited mathematical results exist. Specifically, given

$$\theta_t = f(\theta_{xx}, \theta_x, \theta), \quad -\infty < x < \infty, \ t > 0, \qquad (11.35)$$

$$\lim_{x \to \pm\infty} \theta(x, t) = \theta_{\pm},$$

$$\theta(x, 0) = \theta_0(x).$$

Equation (11.35) is parabolic with the assumption that for all real a, b and c, $\partial f(a, b, c)/\partial a \geq 1$. The constant states at $\pm\infty$ further satisfy

$$f(0, 0, \theta_{\pm}) = 0, \qquad f_\theta(0, 0, \theta_{\pm}) = 0, \qquad (11.36)$$

and the initial profile satisfies $\lim_{x \to \pm\infty} \theta_0(x) = \theta_{\pm}$. Hagen [15] generalized the results of Aronson and Weinberger [16] and showed that, for example,

 (i) In general, nonmonotonic traveling waves are unstable.
 (ii) If traveling waves exist, a large class of initial profiles $\theta_0(x)$ will evolve to the same traveling wave.
 (iii) Infinitely many wave speeds c of traveling waves may be allowed.
 (iv) The traveling wave to which the initial profile evolves depends on the asymptotic behavior of the initial profile at infinity.

Note that (i) implies that the solitons A_1 and B_1 with oscillating tails in Fig. 11.5 are unstable. In the special case of the Fisher equation, Eq. (11.15), $\theta_- = 1$ and $\theta_+ = 0$; one has the following specific results:

 (i) The existing traveling-wave solutions can assume all wave speeds $c \geq 2$.
 (ii) A positive initial profile $\theta_0(x)$, decaying at least exponentially as $x \to \infty$, evolves into a unique traveling wave.
 (iii) If $\theta_0(x) \sim \exp(-\beta)$ as $x \to \infty$, then the initial profile evolves to a traveling wave of speed $c(\beta)$ given by

$$c(\beta) = \begin{cases} (1 + \beta^2)/\beta, & \beta \leq 1 \\ 2, & \beta \geq 1. \end{cases} \qquad (11.37)$$

The speeds in Eq. (11.37) correspond to the overdapmed and critical damped cases of a viscous particle rolling down a hill discussed in section 11.3.1. These results are confirmed by numerical calculations [17] in which the boundary conditions at the ends of the finite grid are handled with extreme care. It seems that the fact that different initial profiles (depending on their decay rate at infinity) may evolve into traveling waves of different speeds has been overlooked by the proponents [18] of the marginal stability hypothesis in pattern formation, in which the lowest velocity in Eq. (11.37) is always taken to be the only one possible.

The conditions for these results to hold are sometimes very stringent. This is demonstrated in Figs. 11.7 and 11.8 for the damped driven sine–Gordon equation of Eq. (11.27). In both figures, $\gamma = 0.96$, $\theta_\mp = \pm \theta_0$ with $\theta_0 = 81.87°$. In Fig. 11.7, the initial profile is a step function bounded between θ_+ and θ_- everywhere; it evolves into a traveling wave as expected. In contrast, the initial profile in Fig. 11.8 has a small portion below θ_+; it does not evolve to any solitary wave but to the constant state θ_-. In any real experimental situation, it is difficult to control the initial profile precisely and the mathematical results quoted above have to be applied with utmost care [11, 19].

11.6 Perturbations

In a real system, due to the existence of boundaries, defects, impurities, dissipation, external fields, and so forth, the system usually experiences some kind of perturbation. Or, when the soliton behavior in the system is being observed or measured, the measurement itself is bound to alter the state of the system to some extent. In fact, the appearance of the soliton always implies that some kind of external excitation has been applied to the system (such as that due to the pushing plate in Fig. 10.4). Furthermore, the thermal igitation due to temperature is a perturbation that no system can avoid.

When all these perturbations are relatively weak, they may be handled by perturbation methods. The main spirit of soliton purturbation is illustrated with an example below. Let us assume that the equation consists of two parts. The first part has soliton solution, and the second part is small. For example,

$$\theta_{xx} - \theta_{tt} - \sin\theta = \varepsilon R(\theta, \theta_t), \tag{11.38}$$

where ε is a small parameter. The solution of Eq. (11.38) is assumed to be $\theta = \theta_s + \varepsilon\theta_1$, with $\theta_s = 4\tan^{-1}\{\exp[\pm(x - X)/(1 - V^2)^{1/2}]\}$, $X = X(t)$, and $V = V(t)$. Here θ_s represents a kink or antikink with time-dependent velocity and width, the form of which is inspired by Eqs. (10.9) and (10.10). Obviously, when $\varepsilon = 0$ one should have $X = ct$ and $V = c = $ const. When this assumed form of θ is substituted into Eq. (11.38), by the use of the ISM [20], variational method [21], or Green function [22], one may obtain $X(t)$, $V(t)$, and even θ_1. In general, such a solution represents a shape-changing, accelerating soliton with a tail (Fig. 11.9). However, there does exist a perturbed soliton with constant shape and without tail [23].

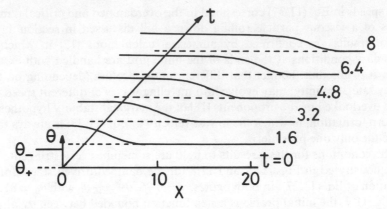

FIGURE 11.7. Numerical solution of Eq. (11.27) with a step-down function from θ_- to θ_+ as the initial profile, which evolves to a soliton at large time t.

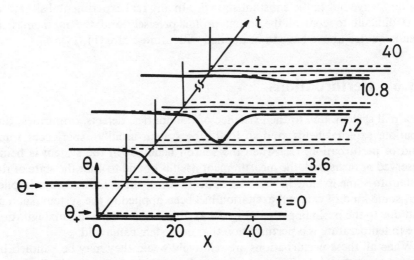

FIGURE 11.8. Numerical solution of Eq. (11.27) with an initial profile that slightly exceeds (θ_+, θ_-). In contrast to Fig. 11.7, the profile evolves to a horizontal line.

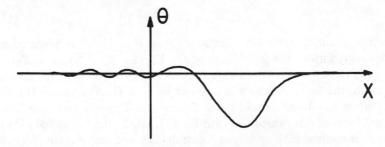

FIGURE 11.9. Soliton solution of the perturbed KdV equation. The unperturbed soliton is shown in Fig. 10.3.

The singular perturbation method is also used in obtaining perturbed solitons. For example, the perturbed KdV equation was treated by Ko and Kuehl [24]; the damped driven sG equation by Xu, Shu, and Lin [23].

For the Fisher equation, Eq. (11.15), corresponding to *every* velocity $c \geq 2$ there is an antikink-like soliton connecting the state $\theta(-\infty) = 1$ to $\theta(+\infty) = 0$. The soliton given analytically in Eq. (11.16) with $c = 5/\sqrt{6}$ is only one of many. But there is no exact solution for any other c. However, a good approximate form derived from the perturbation method is available [25]; namely,

$$\theta(x, t) = \theta(\tau; c)$$

$$= \frac{1}{1 + \exp(\tau/c)} - \frac{1}{c^2} \frac{\exp(\tau/c)}{1 + \exp(\tau/c)^2} \left[1 - \ln \frac{4 \exp(\tau/c)}{1 + \exp(\tau/c)^2} \right]$$

$$+ O\left(\frac{1}{c^4}\right). \tag{11.39}$$

Here, the small parameter of expansion is $1/c^2$, and $\tau \equiv x - ct - x_0$.

Qualitative and sometimes even quantitative results of perturbed solitons may be obtained through simple consideration of energy balance [22]. For a general discussion of soliton perturbations, see [26].

11.7 Soliton Statistical Mechanics

In a soliton-bearing medium at finite temperature, the heat bath coupling to the medium acts as a perturbation and can excite solitons in the medium. The physical properties of the medium are then modified by the excited solitons.

For the sake of discussion, let us consider a perfect crystal. The molecules vibrate around their equilibrium sites in the lattice and give rise to so-called elementary excitations, called phonons [27], which are spatially-extensive linear waves. The solitons, or localized nonlinear waves, cannot be obtained by a superposition of these phonons because the superposition principle does not work in a nonlinear system. Consequently, the solitons have to be treated separately from the phonons. The question is, can the solitons be treated as a new kind of elementary excitation? The motivations behind this desired approach are twofold: (i) Elementary excitations are particle-like entities that provide a very appealing and intuitive physical picture in predicting the physical properties of the medium. (ii) The direct calculation of thermal properties through the use of statistical mechanics [28] is notoriously difficult. Yet a whole machinary has been devised in treating a many-particle system [29], especially in dealing with elementary particles.

Through the work of Krumhansl and Schrieffer [30] and many others [31–33], it is now clear that solitons (in condensed matter) can indeed be viewed as elementary excitations; there is evidence to support this claim. However, there are still some unresolved theoretical and experimental issues; the study of soliton statistical mechanics remains an active field of research [33]. The basic ideas of soliton statistical mechanics will be illustrated in the following subsection using the θ^4 system as an example. Theoretical results for the sG system (related to magnetic systems) are briefly summarized in section 11.7.2, and some examples in condensed matter are presented in section 11.8.

11.7.1   The θ^4 System

Historically, the θ^4 equation is the system considered by Krumhansl and Schrieffer [30] in relation to the central peak problem observed in structural phase transitions. A simplified Hamiltonian representing a (displacive) structural transition is given by

$$H_0 = \sum_i \left[\left(-\frac{1}{2} A u_i^2 + \frac{1}{4} B u_i^4 \right) + \sum_j \frac{1}{2} C_{ij} (u_i - u_j)^2 \right] + \sum_i \frac{1}{2} m_i \dot{u}_i^2, \quad (11.40)$$

where A, B, C_{ij} and m_i are positive constants. In the continuum limit H_0 is transformed to

$$H = \int \frac{dx}{l} \left[\frac{p(x)^2}{2m} + V(u) + \frac{1}{2} m c_0^2 \left(\frac{du}{dx} \right)^2 \right], \quad (11.41)$$

where $V[u(x)] \equiv -\frac{1}{2} A u(x)^2 + \frac{1}{4} B u(x)^4$, l is the lattice spacing and $x \equiv x_j = jl$. The equation of motion for $u(x,t)$ following from Eq. (11.41) is

$$m c_0^2 \frac{\partial^2 u}{\partial x^2} - m \frac{\partial^2 u}{\partial t^2} = -Au + Bu^3, \quad (11.42)$$

which is the same as the θ^4 equation, Eq. (11.7), after proper redefinition of the symbols. There are two types of *physically* important solutions of Eq. (11.42). The first type is the phonons derived from the linearized version of Eq. (11.42) by dropping the u^3 term,

$$u = \pm u_0 + \alpha u_0 \sin(qx - \omega_q t + \phi), \quad (11.43)$$

with the dispersion relation

$$\omega_q^2 = c_0^2 q^2 + 2A/m, \quad (11.44)$$

where $u_0 = (A/B)^{1/2}$, and α and ϕ are constants. The second type is the solitons

$$u = u_0 \tanh[(x - vt)/\sqrt{2}\xi] \quad (11.45)$$

where $\xi^2 = m(c_0^2 - v^2)/A$.

In the following, the basic arguments and results of [30] are presented, followed by the improvements due to Currie et al. [34]. At this point it helps to understand a little bit of the physics behind the Hamiltonian Eq. (11.40) or

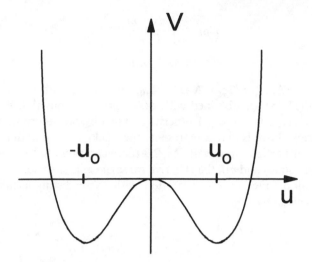

FIGURE 11.10. A double-well potential.

(11.41). In Eq. (11.40) u_i and $\dot{u}_i$ are the displacement and velocity of the ith ion with respect to some heavy ion or reference lattice. Each displacing ion moves in a double-well potential $V(u)$ with minima at $u = \pm u_0$ (Fig. 11.10). The coupling between neighboring ions gives rise to the $(du/dx)^2$ term in Eq. (11.41); the $p(x)^2$ term comes from the kinetic energy. The phonons of Eq. (11.43) correspond to the ions oscillating around either one of the two minima in V. The soliton of Eq. (11.45) corresponds to a domain wall with ions on the two sides of the wall occupying a different minimum of V.

In the ideal gas phenomenological model valid at very low temperatures, the phonons and solitons are treated as noninteracting entities, that is, any interaction among the phonons, the solitons, and between the phonons and solitons are ignored. The free energy of the system is then the sum of the free energies of its components; namely,

$$F = F_{ph} + F_D, \tag{11.46}$$

where F_{ph} is due to the phonons and F_D is due to the domain walls, the solitons. To evaluate F_D, let us first calculate the potential energy E_{DP} and the kinetic energy E_{DK} of a single domain wall. The results are given by

$$E_{DP} = \int \frac{dx}{l} \left[-\frac{A}{2}(u^2 - u_0^2) + \frac{B}{4}(u^4 - u_0^4) + \frac{mc_0^2}{2}\left(\frac{d\mu}{dx}\right)^2 \right]$$

$$\approx (\Delta/l)(A^2/2B)(1 - 7/60) \tag{11.47}$$

and

$$E_{DK} = \int \frac{dx}{l} \left(\frac{m}{2} \dot{u}^2 \right)$$

$$\approx (m_D^*/2)v^2, \tag{11.48}$$

where $m_D^* \equiv m(\Delta/l)(u_0^2/2\xi_0^2)$, $\Delta \equiv 2\sqrt{2}\xi_0$, and ξ_0 is ξ with $v = 0$. Equations (11.47) and (11.48) are obtained with the approximations that $v^2 \ll c^2$, and $\tanh y \approx y$ if $|y| < 1$ and $\approx \pm 1$ otherwise. The obvious interpretation of the different terms above is that Δ represents the thickness of a domain wall, Δ/l the number of particles in a wall, $A^2/2B$ the mean potential energy (relative to the ground state) and m_D^* the effective mass of each particle.

The partition function Z_D associated with these distinguishable "quasi-particles" is given by

$$Z_D = \sum_{n_w} \left(\int \frac{dv}{b} \, e^{-\beta m_D^* v^2/2} \right)^{n_w} \frac{n_s!}{n_w!(n_s - n_w)!} \, e^{-\beta n_w E_{DP}}$$

$$\approx (2\pi kT/b^2 m_D^*)^{\bar{n}_w/2} \exp(\bar{n}_w) \tag{11.49}$$

where $n_s \equiv L/\Delta$, L the length of the lattice, $\bar{n}_w \equiv n_s \exp(-\beta E_{DP})$ the most probable n_w, b, an appropriate space normalization, and $\beta \equiv 1/(kT)$ is the inverse temperature (k the Boltzmann constant). From $F_D = -kT \ln Z_D$, one obtains

$$F_D = -NkT \left(\frac{l}{\Delta} \right) e^{-E_{DP}/kT} \left(1 + \frac{1}{2} \ln \frac{2\pi kT}{b^2 m_D^*} \right). \tag{11.50}$$

From the expression of F, one can calculate other thermodynamic quantities (such as the internal energy, entropy, and specific heat) and correlation functions [30, 34].

It turns out that the free energy of the system described by Eq. (11.41) can be evaluated by the transfer integral operator approach. Within this approach F can be expressed *exactly* in terms of ε_0, the lowest eigenvalue of a pseudo-Schrödinger equation. However, ε_0 can only be calculated approximately. An approximate WKB calculation by Krumhanhl and Schrieffer [30] gives

$$F = F_{osc} + F_{tunn} \tag{11.51}$$

where

$$F_{osc} = F_{ph}$$

and

$$F_{tunn} = -NkT(l/\Delta) \exp(-E_{DP}/kT). \tag{11.52}$$

The fact that Eq. (11.52) agrees with Eq. (11.50) up to a factor of order unity gives support to the validity of the ideal gas phenomenological model. Here, F_{osc} is identified with the harmonic oscillator states around the minima of the double-well potential V; F_{tunn} is associated with the tunneling splitting of the lowest oscillator level.

Subsequently, a more consistent calculation of ε_0 by Currie et al. [34] gives a revised form of F_{tunn}; namely,

$$F_{tunn} \sim (E_{DP}/kT)^{-1/2} \exp(-E_{DP}/kT). \tag{11.53}$$

When the ideal gas model is modified to take into account the interaction between the phonons and the domain walls, exact agreement between Eq. (11.53) and the phenomenological model is achieved. The phenomenological approach treating solitons as elementary excitations is thus justified.

11.7.2 The Sine–Gordon System

For the sG chain in the continuum limit, the Hamiltonian is given by

$$H = E_0 a \int_{-\infty}^{\infty} dz \left[\frac{1}{2} \left(\frac{\partial \Phi}{\partial z} \right)^2 + \frac{1}{2c^2} \left(\frac{\partial \Phi}{\partial t} \right)^2 + m^2 (1 - \cos \Phi) \right]. \tag{11.54}$$

The transfer integral operator approach is still applicable. The free energy is found to be (see, e.g., [35])

$$F/N = F_1 + F_2 + F_3 + \cdots, \tag{11.55}$$

where

$$F_1 = kT \ln\left(\frac{\beta \hbar}{A}\right) + kTma\left(\frac{1}{2} - \frac{1}{4}t - \frac{1}{8}t^2 - \cdots\right), \tag{11.56}$$

$$F_2 = kTma\left[-\left(\frac{8}{\pi t}\right)^{1/2} \exp\left(-\frac{1}{t}\right)\left(1 - \frac{7}{8}t - \cdots\right)\right], \quad \text{and} \tag{11.57}$$

$$F_3 = kTma\left[\frac{8}{\pi t} \exp\left(-\frac{2}{t}\right)\left\{\ln\left(\frac{4\gamma}{t}\right) - \frac{5}{4}t\left[\ln\left(\frac{4\gamma}{t}\right) + 1 - \cdots\right\}\right], \tag{11.58}$$

where $t \equiv kT/(8maE_0)$ and $A \equiv E_0 a/c$.

Similar to the case of the θ^4 case, the phenomenological low-temperature approach of treating the phonons and solitons as independent elementary excitations gives

$$F = NkT\left[\ln\left(\frac{\hbar c}{akT}\right) + \frac{1}{2}ma - 2ma\left(\frac{2}{\pi t}\right)^{1/2} \exp\left(-\frac{1}{t}\right)\right], \tag{11.59}$$

which agrees with Eq. (11.55) to lowest order. On the right-hand side of Eq. (11.59), the first two terms come from the phonons and the third term from the solitons and antisolitons. It thus provides a physical interpretation of the different terms in Eq. (11.58); namely, F_1 gives both harmonic and anharmonic phonon contributions. F_2 results from one-soliton contributions, with the first term from the noninteracting solitons, and higher terms from magnon–soliton interferences. F_3 is due to soliton–soliton overlap effects.

As shown in Eqs. (11.50) and (11.59), the soliton contribution to thermodynamic properties is characterized by its activated temperature dependence. In the case of the sG system, the breathers do not contribute to the free energy; the phonons and breathers are found to be two aspects of the same phenomenon, resulting from the fact that the breathers do not have an energy gap. This conclusion is reached by different authors after a decade of work using the methods of Bethe ansatz approach, functional integration, and so forth.

11.8   Solitons in Condensed Matter

The manifestation of solitons in four important examples of condensed matter systems are presented here. Other examples can be found in the many conference proceedings on this subject [36–41].

11.8.1   Liquid Crystals

Liquid crystal is a state of matter intermediate between liquid and crystal. In the liquid crystal phase, the material is optically anisotropic and can flow in at least one spatial dimension. The molecules of the organic compound showing liquid crystal phases can be either rod-like, disc-like or bowl-like in shape [42–45].

Take the rod-like molecules as an example. At low temperature, both the orientations and positions of the molecules are in (long range) order and we have the crystal phase. At high temperature, both types of degree of freedom are in disorder and the material is in isotropic liquid phase. However, within a certain temperature range there *may* exist an intermediate state—the nematic phase—in which the orientations are in order (corresponding to molecules more or less parallel to each other) but the positions are in disorder. There exist in fact a large number of other liquid crystal phases, such as the cholesterics and the smectics (Fig. 11.11). The rich variety of phases and phase transitions between them make liquid crystals the testing ground for many of the interesting ideas proposed in the theory of critical phenomena, including the bond-orientational order [46]. Moreover, liquid crystals provide interesting results in the study of pattern formation [47], the analog of superconducting states [48], and even as a model for the early development of the universe [49].

In liquid crystals, because the molecules have both orientational and translational degrees of freedom, the hydrodynamic equations of motion are coupled nonlinear equations of **n** and **v**. Here, **n** is the director, a unit vector representing the local average orientation of the molecules, **v** is the velocity of the center of mass of the molecule, and both **n** and **v** are functions of space and time. Because the orientation of molecules can be detected optically, ori-

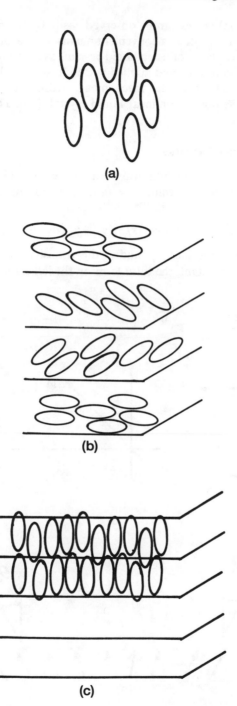

FIGURE 11.11. Some liquid crystal phases: (a) nematic; (b) cholesteric; and (c) smec-tic A.

entational (soliton) waves can be observed easily by the naked eye and provide, in addition to the flow, a convenient means of measuring the waves. (See
[50] for other advantages of using liquid crystals as a nonlinear medium.) Not
surprisingly, in both basic and applied research, solitons have been found to
have important effects in the mechanical, hydrodynamical, and thermal
properties in these highly nonlinear liquid crystals [13]. Three such examples
are presented below.

Solitons in Shearing Nematics

In a one-dimensional (1D) uniform shear of nematic (Fig. 11.12), the orientation angle of the molecule, $\theta = \theta(x, t)$, obeys the driven damped sG
equation [11]

$$K\theta_{xx} - \gamma_1\theta_t + (s/2)(\gamma_1 - \gamma_2 \cos 2\theta) = 0, \tag{11.60}$$

where $s \equiv dv/dy$ is the shear, γ_1 and γ_2 the viscosity ($\gamma_1 > 0, \gamma_2 < 0$), and K
the Frank elastic constant; the third term on the left-hand side of Eq. (11.60)

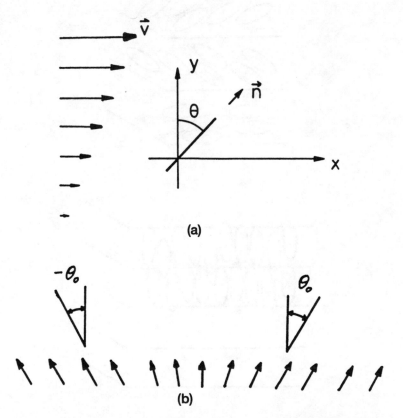

(a)

(b)

FIGURE 11.12. (a) Nematic with 1D uniform shear. (b) Molecular orientations in a
type A soliton.

shows the coupling between the flow and the molecular orientation. In dimensionless form, Eq. (11.60) reduces to Eq. (11.27) with $\gamma = \gamma_1/|\gamma_2|$.

Experimentally, the A solitons can be generated by the pressure gradient method. For example, in a linear homeotropic nematic cell (i.e., a thin layer of nematic sandwiched between two parallel glass plates, with the molecules at the two inner surfaces arranged perpendicular to the surfaces), the pressure at one end of the cell is maintained constant while that at the other end is varied in time with a rectangular wave form [51]. Shearing states are generated in different stages and the molecular orientations change accordingly (Fig. 11.13). The region of vertical molecules, the center of the A soliton, will appear as a dark line under white light as observed experimentally (Fig. 11.14). Propagating dark or white rings representing 2D solitons, a rare occurrence in any system, are also observed in circular cells under pressure gradients [52]. See [11] for further discussion of solitons in shearing nematics.

Discommensurations

In the Frenkel–Kontorova model of crystal dislocations as shown in Fig. 10.12, the atoms in the upper chain are under the influence of two kinds of force: one due to the interatomic springs and the other due to the fixed potential provided by the lower chain, the substrate. Let us assume that the springs are weak and prefer a lattice constant b for the atoms, while the substrate favors a separation of a, which is slightly greater than b. At low pressure, that is, when the springs are slightly compressed from both ends, the atoms would sit above the potential minimima with the springs slightly stretched forming a so-called commensurate phase. This commensurate phase is stable as the pressure is further increased until at a certain point, the atoms rearrange themselves so that most of them are at the potential minima while others are squeezed into localized regions, which are called discommensurations or solitons. This configuration of atoms forms the discommensurate phase, which has a lower energy than the configuration with the atoms uniformly spaced across the substrate. Incommensurability and the commensurate–incommensurate phase transition have been observed in various condensed matter systems such as modulated crystals, superlattices, and quasicrystals [53].

Like many other phenomena studied elsewhere, it turns out that the idea of discommensurations can also be illustrated nicely in liquid crystals. Specifically, a cholesteric may be considered as a helical structure with planes of parallel molecules perpendicular to the helical axis; the common molecular axis in each plane is represented by the director $\mathbf{n}$ and varies continuously from plane to plane. When a uniform magnetic field $\mathbf{H}$ is applied perpendicular to the helical axis, and assuming that the molecules prefer to align themselves with the magnetic field, then for a large enough magnetic field strength the helix will unwind itself completely and the cholesteric transforms into the nematic phase with uniform molecular orientations. There are two interesting

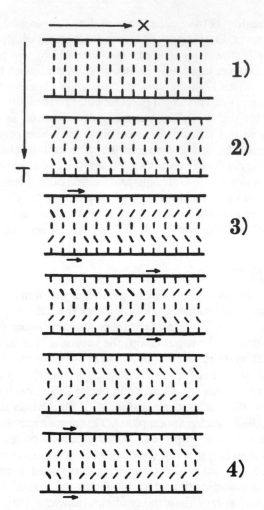

FIGURE 11.13. Sketch of molecular orientations during the four stages [denoted by 1) to 4)] in a pressure-gradient experiment of a linear homeotropic nematic cell [51]. The pressure at the right end of the cell p_R is maintained constant; the pressure at the left end p_L is varied in time such that in stage 1), $p = 0$; stage 2), $p < 0$; stage 3) $p > 0$; stage 4), $p < 0$. Here $p \equiv p_L - p_R$. The arrows indicate the locations and moving directions of the region of vertical molecules, the center of the A soliton.

points about this cholesteric nematic transition. First, as H is increased gradually the helix unwinds partially but not uniformly. A periodic structure is formed in which sections of completely unwound planes (with **n** parallel or antiparallel to **H**) are separated by sections of helical structure. In other words, an incommensurate phase is formed. Second, there is a threshold value of H at which the cholesteric completely unwinds into a nematic; the transition is like a second order phase transition.

The mathematical description of this phenomenon is particularly simple.

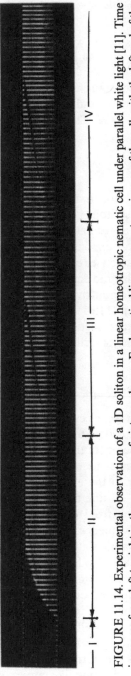

FIGURE 11.14. Experimental observation of a 1D soliton in a linear homeotropic nematic cell under parallel white light [11]. Time increases from left to right in the sequence of pictures shown. Each vertical line represents an image of the cell, with the left end of the cell at the bottom of the line. The pressure p_R is maintained at one atmospheric pressure. The four stages I to IV correspond to 1) to 4) in Fig. 11.13, respectively. In Stage I, $p = 0$; Stage II, $p = -6.0\,\mathrm{cm\,Hg}$; Stage III, $p = 6.0\,\mathrm{cm\,Hg}$; Stage IV, $p = -5.8\,\mathrm{cm\,Hg}$. Note that the soliton appears as a propagating dark line and appears only in Stages III and IV as expected.

Let the helical axis of the cholesteric be in the z direction, and the magnetic field be in the y direction. Assume $\mathbf{n} = (\cos\phi, \sin\phi, 0)$, with $\phi = \phi(z)$. The free energy of the system is given by

$$F = \frac{K_2}{2} \int \left[\left(\frac{d\phi}{dz} - q_0 \right)^2 - \xi^{-2} \sin^2\phi \right] dz, \tag{11.61}$$

where K_2 is the twist elastic constant, $q_0 = 2\pi/p_0$ with p_0 representing the pitch of the undistorted cholesteric, and ξ is the magnetic coherence length, which is inversely proportional to H. The Lagrange equation corresponding to Eq. (11.61),

$$\frac{d^2(2\phi)}{dz^2} + \xi^{-2} \sin 2\phi = 0 \tag{11.62}$$

is the static sG equation. The incommensurate phase described above is nothing but the periodic multisoliton solution shown in Fig. 10.8 (with $\theta = 2\phi$). The threshold value of H arises naturally when the period of the multi-soliton is set to infinity. Critical exponents similar to those in equilibrium phase transitions can be calculated (see chapter 10 of [13]).

In smectics consisting of polar molecules, the mass density wave and the dipolar density wave couples in a nontrivial way. Under suitable conditions, a free energy similar to that in Eq. (11.61) can be derived, leading to the possible existence of incommensurate smectic phases (see chapter 6 of ref. [13]). However, these incommensurate smectics have not been found experimentally [54].

Solitons in Electroconvection of Nematics

Consider a nematic cell with an AC electric field (of voltage V and frequency f) applied perpendicular to the cell plates. The molecules are aligned parallel to each other and to the cell plates (in the x direction, say) forming a planar cell. The type of nematic molecules used has the property that they prefer to align themselves perpendicular to the electric field. As the voltage and frequency are changed, a wealth of different dynamical patterns are obtained [55]. For example, with a suitable fixed f, as V is increased very slowly from zero the nematic undergoes a sequence of changes, from the rest state to a series of convective rolls to a chaotic state. The series of convective rolls starts from normal rolls (NR), undulatary rolls (UR), oblique rolls (OR), varicose, and then bimodal, similar to the case of Rayleigh–Bénard convections in a simple liqiud under a temperature gradient. Part of these bifurcation transitions in the (f, V^2) plane is shown in Fig. 11.15 [56].

When viewed from the top of the cell, the NR consists of a series of dark and bright straight lines along the y axis, that is, perpendicular to both the x and z axes, where the z axis is the normal of the cell. For $f > f_M$, the frequency at the point M in Fig. 11.15, sine-like curves of the UR first appear at the threshold V_{uz}, the upper solid line in Fig. 11.15. The UR then become

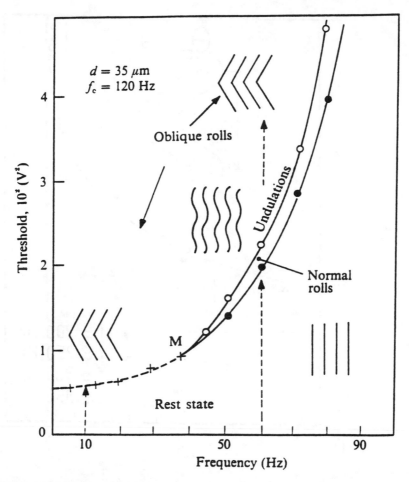

FIGURE 11.15. Bifurcation diagram in the (f, V^2) plane due to the electroconvection of a nematic [56]. The material used is a Merck Phase V compound. The broken line represents a first-order–like, subcritical transition; the two solid lines second-order–like, supercritical transitions.

more angular in shape with increasing wavelength Λ until finally a zigzag structure composed of rectilinear oblique rolls (the OR) is reached. For fixed f, as V is increased, the measured Λ of the UR as a function of θ_m^2/θ_l^2 is plotted in Fig. 11.16 [57]. Here θ_m is the maximum tilt angle of the UR with respect to the y axis, which is found to increase with V. At high voltage, θ_m tends to θ_l, the tilt angle of the OR, and Λ seems to diverge.

An interesting observation [57] is that the gradual transformation of the UR to the OR, as V is increased with f fixed above f_M, can be described phenomenologically by the evolution of a θ^4 nonlinear oscillator, and the OR comes out naturally as a soliton represented by the hypertangent function.

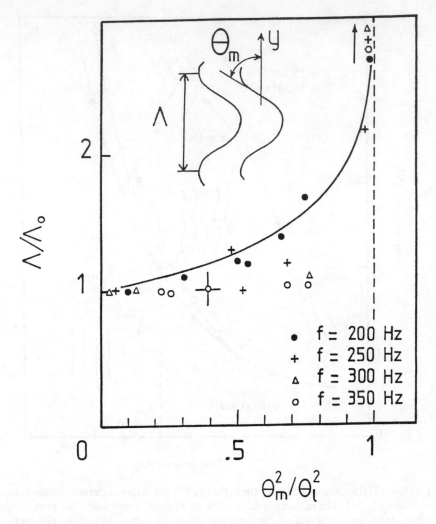

FIGURE 11.16. Reduced spatial wavelength Λ/Λ_0 of the undulatory rolls as a function of the reduced maximum tilt angle squared θ_m^2/θ_l^2 [57]. The solid line represents the theoretical curve.

Let $\theta(y)$ denote the tilt angle of the UR along the y axis; $\theta(y) \leq \theta_m$. The most simple functional consistent with the physical symmetry $\theta \rightarrow -\theta$ of the system is

$$F_0 = \int dy[-\tfrac{1}{2}A\theta^2 + \tfrac{1}{4}\theta^4 + \tfrac{1}{2}L(d\theta/dy)^2],   (11.63)$$

where $A, L > 0$. The corresponding Lagrange equation is given by

$$Ld^2\theta/dy^2 = -A\theta + \theta^3,   (11.64)$$

which is a static version of Eq. (11.19). The periodic solutions are the Jacobian elliptic functions [58] $\theta = \theta_m \mathrm{sn}(u|k^2)$, with $u = yL^{-1/2}\theta_l(1+k^2)^{-1/2}$, $\theta_l = A^{1/2}$ and $k = [2(\theta_l/\theta_m)^2 - 1]^{-1/2}$. The period of $\theta(u)$ is $4\mathrm{K}(k^2)$, where $\mathrm{K}(k^2)$ is the complete elliptic integral of the first kind. For $0 \leq \theta_m \leq \theta_l$ we have $0 \leq k \leq 1$, and $-\theta_m \leq \theta \leq \theta_m$ for each k. We note that for $k \to 0$ ($\theta_m \to 0$), one has $\mathrm{sn}(u|k^2) \to \sin u$; for $k \to 1$ ($\theta_m \to \theta_l$), $\mathrm{sn}(u|k^2) \to \tanh u$. The latter is a soliton, with infinite wavelength corresponding to the infinitely large domains of rectilinear rolls tilted by $\pm\theta_l$. These shapes are exactly those observed for the UR and the OR if we identify θ_m and θ_l here with that defined in the observed experimental patterns. More quantitatively, we may transform θ from the u space into the physical y space and obtain the wavelength of the UR,

$$\Lambda/\Lambda_0 = (1+k^2)^{1/2}\mathrm{K}(k^2)/\mathrm{K}_0, \qquad (11.65)$$

where $\mathrm{K}_0 \equiv \mathrm{K}(0)$ and $\Lambda_0 \equiv \Lambda(0)$. By Eq. (11.65), for the OR (where $\theta_m \to \theta_l$), Λ diverges logarithmically. The theoretical universal curve, Λ/Λ_0 versus θ_m^2/θ_l^2, is plotted as the solid line in Fig. 11.16 and is in good agreement with the experiments. The curve should be valid for different samples and different materials, and is independent of f.

As noted by Joets, Ribotta, and Lam [57], the bifurcation diagram in Fig. 11.15 can be reproduced like a phase diagram of an equilibrium system. The two order parameters are θ_m and ϕ, where ϕ is the amplitude of the deviation of the director from the x axis. The rest state corresponds to $\phi = 0 = \theta_m$; the NR to $\phi \neq 0$, $\theta_m = 0$; the UR and OR to $\phi \neq 0 \neq \theta_m$. The Landau-type free energy required is given by

$$F = a\phi^2 + \tfrac{1}{2}b\phi^4 + \tfrac{1}{3}c\phi^6 + A\theta_m^2 + \tfrac{1}{2}B\theta_m^4 - \gamma\phi^2\theta_m^2. \qquad (11.66)$$

Analytic results are obtained. For the particular choice of parameters, $b = 94.77(ec)^{1/2}$, $\gamma^2 B = 2.31(ec)^{1/2}(80.026 - f)$ and $a/e = -1 + 10.71[-V^2 + 86.81 + 0.124(f - 29.826)^2]$, a bifurcation diagram in excellent agreement with that in Fig. 11.15 is obtained. Here $e \equiv \gamma A/B$; both e and c are arbitrary positive constants; V is in volt and f in hertz.

11.8.2 Polyacetylene

Polyacytylene is a crystalline polymer with metallic conductivity comparable to that of copper, whereas most other polymers are insulators. Polyacetylene is of much importance in its potential applications; for example, as a substitute for metals as light-weighted, anisotropic electric conductors and as a new type of high-energy–density battery. Physically, the magnetic, electrical, and optical properties of polyacetylene are believed to be dominated by solitons [59–61]. Polyacetylene is probably the most-studied solid state system, experimentally and theoretically, by the soliton community.

Polyactylene has the chemical formula $(CH)_x$. It is quasi-one–dimensional and has one unpaired electron per carbon atom. For such a system, the lattice will spontaneously distort (as indicated by the arrows in Fig. 11.17) and

262   L. Lam

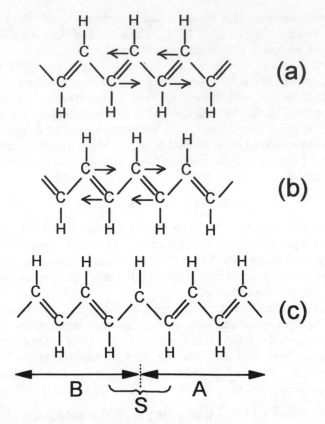

FIGURE 11.17. Schematic formation of the soliton state in an "idealized" *trans*-polyacetylene. (a) The perfectly dimerized *A* phase. (b) The perfectly dimerized *B* phase. Four representative distortions of the carbon atoms are shown as arrows in (a) and (b); alternative distortions actually occur for all the carbon atoms. The double bond is formed when the two adjacent distortions are toward each other. (c) The single soliton state. The central region of the soliton, formed from the *A* and *B* phases, is indicated by *S*. Energetically, a pair of soliton and antisoliton comprised of *ABA* or *BAB* phases will be favorably generated over the single soliton state shown in (c).

results in a chain of alternative double and single bonds between the carbon atoms. (This spontaneous distortion of the lattice is called the Peierls instability or the dimerization process.) For an infinite chain, there are obviously two degenerate states, the *A* phase and the *B* phase, that can result from this process. It is then not difficult to imagine the existence of a soliton state that can be formed by linking up the *A* and *B* phases (Fig. 11.17).

The situation described above is in fact an idealized picture. In reality, in the soliton state, every carbon–carbon bond is neither fully single nor double, but is something in between, as indicated by the broken lines in Fig. 11.18a.

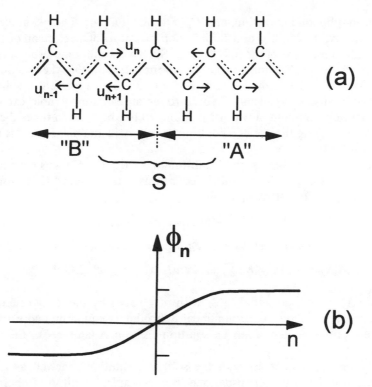

FIGURE 11.18. (a) Schematic formation of the soliton state in a real *trans*-poly-acetylene. (b) The corresponding staggered displacement ϕ_n.

The corresponding staggered displacement $\phi_n \equiv (-1)^n u_n$ is a kink, as shown in Fig. 11.18b, where n is the discrete site index. It turns out that such a soliton has a small effective mass, about six times the electron mass, indicating that quantum effects are important.

Theoretically, a Hamiltonian to describe the *trans*-$(CH)_x$ system has been constructed by Su et al. [62] and is given by

$$H = H_\pi + H_{\pi-ph} + H_{ph} \tag{11.67}$$

where

$$H_\pi = -t_0 \sum_{n,s} (c_{n+1,s}^+ c_{n,s} + c_{n,s}^+ c_{n+1,s}) \tag{11.68}$$

$$H_{\pi-ph} = \alpha \sum_{n,s} (u_{n+1} - u_n)(c_{n+1,s}^+ c_{n,s} + c_{n,s}^+ c_{n+1,s}) \tag{11.69}$$

$$H_{ph} = \sum_n p_n^2/2M + K/2 \sum_n (u_{n+1} - u_n)^2. \tag{11.70}$$

Here H_π describes the hopping of the electrons along the chain, $H_{\pi-ph}$ the electron–phonon interaction, and H_{ph} the phonon part. The $c_{n,s}^+$ and $c_{n,s}$ are, respectively, the creation and annihilation operators of an electron of spin s at site n; M is the mass of the (CH) group. Such a Hamiltonian problem cannot be solved exactly, but numerical calculations [62] show that the function $\phi_n \sim \tanh(na/\xi)$, in the form shown in Fig. 11.18b, does represent a soliton excitation that is energetically favored (over an electron or hole excitation). Furthermore, the soliton width ξ is found to be about seven times the lattice constant, making the use of a continuum theory justifiable. Here a is the lattice constant.

The continuum limit of the Hamiltonian of Eq. (11.67) was developed by Takayama et al. [63]. The resulting equations of motion lead to the following set of coupled consistent equations

$$-iv_F \partial u_{ms}(x)/\partial x + \Delta(x)v_{ms}(x) = E_m u_{ms}(x)$$

$$iv_F \partial v_{ms}(x)/\partial x + \Delta(x)u_{ms}(x) = E_m v_{ms}(x) \tag{11.71}$$

$$\Delta(x) = -(4\alpha^2 a/K) \sum_{m,s} [u_{ms}(x)v_{ms}^*(x) + v_{ms}(x)u_{ms}^*(x)],$$

where $\Delta(x)$ is proportional to ϕ_n with n replaced by the x coordinate. The problem is analogous to the nonlinear Schrödinger equation, and Eq. (11.71) indeed possesses the analytic solution $\Delta(x) = \Delta \tanh(x/\xi)$, the soliton expected.

The charge–spin relation of the soliton excitation—viewed as a quasi-particle—differs from the usual one. For example, a soliton of charge zero has spin $1/2$; solitons with charge $\pm e$ have spin 0 (in contrast to an electron of charge -e and spin $1/2$). This problem is related to the general phenomenon of fractional charge and fractional statistics [59].

Solitons in polyacetylene, a solid, do not appear individually and cannot be observed directly as a wave as in liquid crystals. The existence and effects of solitons in polyacetene can only be inferred indirectly from experimental data. Calculations involving the statistical mechanics of solitons (see section 11.7) are sometimes involved in the theoretical studies [59–61]. In the next subsection, we turn to an opposite example, an optical fiber, in which individual soliton waves can be generated and measured with very high precision.

11.8.3   Optical Fibers

Total internal reflection, a simple optical phenomenon taught to high school students, is at the foundation of a multibillion-dollar industry; namely, the industry of optical communications. Consider a plane interface separating two media of different reflective indices. A light beam on the side with the smaller reflective index will be reflected totally (without transmission) from the interface if the incident angle of the beam is larger than a critical angle.

Thus, light can be trapped within and propagate along a glass fiber through total internal reflection when the reflective index in the core part of the fiber is made higher than that in a surrounding outer layer. When light pulses are sent along such a fiber, a string of digital signals and information can be transmitted, with a light pulse representing 1 and a null 0.

Optical fiber communication became practical only after 1979 when silica fibers of very low energy loss at the wavelength near 1.55 μm were manufactured. Still, the small but finite loss plus the dispersion effect of any linear wave gives rise to distortion of the light pulses; reshaping of the pulses is needed as they propagate along the fiber. An effective way of doing this appeared in 1987, with the advent of erbium-doped fiber amplifiers [64]. Since then, all-optical transocean or transcontinent communications have become feasible. The advantage of using optical fibers in communications is showcased by these figures: a conventional cable, three inches thick and containing 1200 pairs of copper wires, can carry 14,400 telephone conversations; whereas a single pair of hair-thin optical fibers can transmit three times that number of conversations.

The performance of optical communications can be improved further when the optical pulse is transmitted in the form of a soliton [65]. Indeed, the transmission of soliton pulses at 10 Gbits/s over 10^6 km with zero error has been demonstrated. In principle, being a soliton, the dispersion effect is balanced by the nonlinearity of the fiber, leaving energy loss to be the only factor in pulse distortion, which can be ignored in first approximation. In reality, for long-distance communications, higher-order effects need to be taken into account.

Wave optics is required to describe the propagation of an optical pulse in a single-mode fiber with a radius of a few microns. For a fiber with an inhomogeneous refractive index, the condition $\nabla \cdot \mathbf{E} = 0$ of Eq. (10.38) is not satisfied. The Maxewell equations cannot be reduced to a scalar equation and the calculation is more complicated than that shown in section 10.6.3 for a homogeneous medium. A perturbation calculation gives the following equation [65]

$$i\,\frac{\partial q}{\partial Z} + \frac{1}{2}\,\frac{\partial^2 q}{\partial T^2} + |q|^2 q = i\varepsilon\left[\beta_1\,\frac{\partial^3 q}{\partial T^3} + \beta_2\,\frac{\partial}{\partial T}\,(|q|^2 q) + i\sigma_R q\,\frac{\partial}{\partial T}\,|q|^2\right], \quad (11.72)$$

where q is proportional to the amplitude of the electric field; the expansion coefficient ε is proportional to the inverse of the product of the pulse width and the carrier frequency of the light wave. The β_1 and β_2 terms are due to dispersion effects, and the σ_R term represents the self-induced Raman effect. For $\varepsilon = 0$, Eq. (11.72) reduces to the standard nonlinear Schrödinger equation of Eq. (10.6). It turns out that without the σ_R term, Eq. (11.72) is still integrable and a soliton solution similar to that in Eq. (10.7) is available. It is the nonintegrable σ_R term that gives rise to completely new phenomena, such as the decrease of the central frequency of a soliton in proportion to the distance of propagation [65].

The theory of optical solitons in fibers was proposed by Hasegawa and Tappert [66] in 1973. The first experimental confirmation by Mollenauer et al. [67] was published in 1980. Since then there have been a series of successful intertwining theoretical and experimental developments in this field, including the invention of soliton lasers and the creation of femtosecond solitons [65]. Optical solitons in fibers is probably the most directly and thoroughly studied system among all soliton systems. And, with the anticipated use of these solitons in optical communications in the near future, optical solitons will represent the most important application of any kind of solitons.

11.8.4   Magnetic Systems

The simplest model of soliton in a magnetic system is that of a domain wall in a linear chain of spins. The spins are aligned in opposite directions on the two sides of the wall; the wall represents a smooth transition region between these two alignments. This simple picture applies to a classical Ising (easy-axis) chain, and the corresponding equation is the sine–Gordon equation.

Strong interaction between theory and experiment of solitons in magnetic systems became possible since the availability of real magnetic compounds approximating one-dimensional systems, in 1970. In the late 1970s, the statistical mechanics of solitons as elementary excitations in a magnetic chain were considered, suggesting the existence of a central peak observable in neutron scattering experiments. Quantum effects have since been studied, which is necessitated by the fact that although classical spins correspond to the spin $S \to \infty$, the actual spins are often quite small; namely, $S = 1/2$ or 1. These and other important issues regarding magnetic solitons are summarized by Mikeska and Steiner [35].

11.9   Conclusions

Research in recent years has seen the integration of solitons to other areas of nonlinear physics such as soliton chaos [68], solitons in cellular automata [69, 70], and solitons in pattern formation [71]. Soliton chaos is particularly interesting.

From the second law of thermodynamics it is well known that the entropy (or disorder) of a *closed* system always increases. This, of course, does not preclude the formation of ordered structures in an *open* system, which may be considered as part of a larger closed system. Furthermore, in an open system with different degrees of freedom, it is possible for ordered structures to be formed spontaneously in some physical quantities while letting other physical quantities take care of the increase of entropy [72]. In other words, coherent soliton structures may exist in the presence of chaos. A possible example is the coexistence of the giant Red Spot and the turbulent atmosphere in Jupiter. Theoretically, in soliton systems chaos can exist in the form of chaotic behavior of soliton parameters, such as amplitude or phase, so the shape of the soliton can be well preserved [72]. A typical equation used for this type of investigation may look like this:

$$\theta_{tt} - \theta_{xx} + \sin\theta = \Gamma\sin(\omega t) - \varepsilon\theta_t, \tag{11.73}$$

where the left-hand side of the equation corresponds to the sine–Gordon equation, giving rise to the formation of solitons, while the dissipation θ_t term is the source of chaos.

Although solitons in nonintegrable systems generally do not possess the elastic-collision property, there are exceptions. Apart from the case of the double sine–Gordon equation (see section 11.2.2), there are more known exceptions: the two-dimensional "modons" of the Hasegawa–Mima equation [73], the localized solutions of the Davey–Stewartson equations [74] and some envelope equations [75]. Analytical studies of the stability characters of these exceptional solitons are still lacking.

Localized solutions of the complex Ginzburg–Landau equation encountered in pattern formation problems—in the form of fronts and pulses—have been much studied in recent years [76].

Although nonintegrable systems have been discussed more extensively here because of their practical applications, it does not mean that the study of solitons in integrable systems has ceased to be exciting. On the contrary, for integrable systems, new mathematical structures are found, and intriguing techniques used in the study of soliton equations are related to those in conformal field theory and string theory [77].

Research of solitons in both integrable and nonintegrable systems, for example, solitons relating to ocean surface waves [78], shallow water [79], free electron laser [80], optical systems [81, 82], film draining experiments [83], and traffic jams and granular flow [84], are still going on strongly. In addition, new soliton systems have been proposed [85] and old problems are being further developed [86]. A collection of recent reviews can be found in [87].

Problems

11.1. Show that θ of Eq. (11.21) indeed is a solution of Eq. (11.19).

11.2. (i) Given: the soliton function $\theta(\tau) = [\tanh(a\tau)]^k$, $\tau = x - ct$, where a, c are arbitrary constants. Use Eq. (11.31) to show that $F(\theta) = -ka[ak(k+1)\theta^{2k+1} - 2ak^2\theta^{2k-1} + ak(k-1)\theta^{2k-3} - c\theta^{k+1} + c\theta^{k-1}]$. Here k is any number such that $F(\theta)$ is finite. (ii) Show that if k is a positive integer, then θ is the kink type when k is odd, and is bell-shaped when k is even.

References

[1] P.C. Fife, *Mathematical Aspects of Reacting and Diffusing Systems* (Springer-Verlag, New York, 1979).

[2] *Oscillations and Traveling Waves in Chemical Systems*, edited by R.J. Field and M. Burger (Wiley, New York, 1985).

[3] D. Walgraef, *Spatial-temporal Pattern Formation, with Examples from Physics, Chemistry and Materials Science* (Springer-Verlag, New York, 1997).

[4] L. Lam, in [13].

[5] D.K. Campbell and M. Peyrard, in *Chaos*, edited by D.K. Campbell (American Institute of Physics, New York, 1990).

[6] R.K. Bullough, P.J. Caudrey, and H.M. Gibbs, in *Solitons*, edited by R. K. Bullough and P.J. Caudrey (Springer-Verlag, New York, 1980).

[7] M.J. Ablowitz, M.D. Kruskal, and J.F. Ladik, SIAM J. Appl. Math. **36**, 428 (1979).

[8] P.W. Kitchenside, P.J. Caudrey, and R.K. Bullough, Phys. Scr. **20**, 673 (1979).

[9] M.J. Ablowitz and A. Zeppetella, Bull. Math. Biol. **41**, 835 (1979).

[10] L. Lam (unpublished, 1981).

[11] L. Lam and C.Q. Shu, in [13].

[12] A. Barone and G. Peternó, *Physics and Applications of the Josephson Effects* (Wiley, New York, 1982).

[13] *Solitons in Liquid Crystals*, edited by L. Lam and J. Prost (Springer-Verlag, New York, 1992).

[14] R.M. Miura, SIAM Rev. **18**, 412 (1976); W. Eckhaus and P. Schuur, Math. Meth. Appl. Sci. **5**, 97 (1983).

[15] P. Hagen, Stud. Appl. Math. **64**, 57 (1981); SIAM J. Math. Anal. **13**, 717 (1982).

[16] D.G. Aronson and H.F. Weinberger, Adv. Math. **30**, 33 (1978); in *Partial Differential Equations and Related Topics*, edited by J.A. Goldstein (Springer-Verlag, New York, 1975).

[17] T. Hagstrom and H.B. Keller, SIAM J. Sci. Stat. Comput. **7**, 978 (1986).

[18] G. Dee and J.S. Langer, Phys. Rev. Lett. **50**, 383 (1983); W. van Saarloos, Phys. Rev. A **37**, 211 (1988).

[19] L. Lin (L. Lam), C.Q. Shu, and G. Xu, Phys. Lett. **109A**, 277 (1985).

[20] V.I. Karpman, Phys. Scr. **20**, 462 (1979).

[21] A. Bondeson et al., in *Solitons in Physics*, edited by H. Wilhelmsson, Phys. Scr. **20**, 289 (1979).

[22] D.W. McLaughlin and A.C. Scott, Phys. Rev. A **18**, 1652 (1978).

[23] G. Xu, C.Q. Shu, and L. Lin, Phys. Rev. A **36**, 277 (1987).

[24] K. Ko and H.H. Kuehl, Phys. Rev. Lett. **40**, 233 (1978); Phys. Fluids **25**, 1688 (1982).

[25] P.L. Sachdev, *Nonlinear Diffusive Waves* (Cambridge University, Cambridge, 1987).

[26] Y.S. Kivshar and B.A. Malomed, Rev. Mod. Phys. **61**, 763 (1989); G.L. Lamb, Jr., *Elements of Soliton Theory* (Wiley, New York, 1980); A.R. Bishop, in *Solitons in Action*, edited by K. Lonngren and A.C. Scott (Academic, New York, 1978).

[27] C. Kittel, *Introduction to Solid State Physics* (Wiley, New York, 1996).

[28] K. Huang, *Statistical Mechanics* (Wiley, New York, 1987).

[29] G. Mahan, *Many-Particle Systems* (Plenum, New York, 1991).

[30] J.A. Krumhansl and J.R. Schrieffer, Phys. Rev. B **11**, 3535 (1975).

[31] C.M. Varma, Phys. Rev. B **14**, 244 (1976).

[32] A.R. Bishop, in [37]; N. Theodorakopoulos, in [39].

[33] R.K. Bullough, Y.Z. Chen, and J. Timonen, in *Nonlinear and Turbulent Processes in Physics*, edited by V.E. Zakharov, A.G. Sitenko, N.S. Erokhin, and V.M. Chernousenko (World Scientific, Singapore, 1990); S.G. Chung, Int. J. Mod. Phys. B **8**, 2447 (1994).

[34] J.F. Currie, J.A. Krumhansl, A.R. Bishop, and S.E. Trullinger, Phys. Rev. B **22**, 477 (1980).

[35] H.-J. Mikeska and M. Steiner, Adv. Phys. **40**, 191 (1991).

[36] *Solitons and Condensed Matter Physics*, edited by A.R. Bishop and T. Schneider (Springer-Verlag, New York, 1978).

[37] *Physics in One Dimension*, edited by J. Bernasconi and T. Schneider (Springer-Verlag, New York, 1981).

[38] *Nonlinearity in Condensed Matter*, edited by A.R. Bishop, D.K. Campbell, P. Kumer, and S.E. Trullinger (Springer-Verlag, New York, 1987).

[39] *Dynamical Problems in Soliton Systems*, edited by S. Takeno (Springer-Verlag, New York, 1985).

[40] *Nonlinear Science: The Next Decade*, edited by D. Campbell, R. Ecke, and J.M. Hyman (North-Holland, Amsterdam, 1991) [Physica D **51**, Nos. 1–3 (1991)].

[41] *Nonlinearity in Materials Science*, edited by A. Bishop, R. Ecke, and J. Gubernatis (North-Holland, Amsterdam, 1993) [Physica D **66**, Nos. 1 2 (1993)].

[42] P.G. de Gennes and J. Prost, *The Physics of Liquid Crystals* (Clarendon, Oxford, 1993).

[43] S. Chandrasehar, *Liquid Crystals* (Cambridge University, Cambridge, 1992).

[44] *Liquid Crystalline and Mesomorphic Polymers*, edited by V.P. Shibaev and L. Lam (Springer-Verlag, New York, 1994).

[45] For the bowlics, see: L. Lam, Wuli (Beijing) **11**, 171 (1982); Mol. Cryst. Liq. Cryst. **146**, 41 (1987); and the review in [44].

[46] *Bond-Orientational Order in Condensed Matter Systems*, edited by K.J. Strandburg (Springer-Verlag, New York, 1992).

[47] *Pattern Foundation in Liquid Crystals*, edited by A. Buka and L. Kramar (Springer-Verlag, New York, 1996).

[48] K.J. Ihn, J.A.N. Zasadzinski, R. Pindak, A.J. Slaney, and J. Goodby, Science **258**, 275 (1992); P.G. de Gennes, Solid State Commun. **10**, 753 (1972).

[49] I. Chuang, N. Turok, and B. Yurke, Phys. Rev. Lett. **66**, 2472 (1991); F. Flam, Science **252**, 649 (1991).

[50] L. Lam, in *Wave Phenomena*, edited by L. Lam and H.C. Morris (Springer-Verlag, New York, 1989).

[51] C.Q. Shu and L. Lin, Mol. Cryst. Liq. Cryst. **131**, 47 (1985).

[52] C.Q. Shu, R.F. Shao, S. Zheng, Z.C. Liang, G. He, G. Xu, and L. Lam, *Liq. Cryst.* **2**, 717 (1987); R.F. Shao, S. Zheng, Z.C. Liang, C.Q. Shu, and L. Lin, Mol. Cryst. Liq. Cryst. **144**, 345 (1987); Z.C. Liang, R.F. Shao, S.L. Yang, and L. Lam, in *3rd Asia Pacific Physics Conference*, edited by Y.W. Chan, A.F. Leung, C.N. Yang, and K. Young (World Scientific, Singapore, 1988).

[53] P. Bak, Rep. Prog. Phys. **45**, 587 (1982); T. Janssen and A. Janner, Adv. Phys. **36**, 519 (1987).

[54] S. Kumar, L. Chan, and V. Surendranath, Phys. Rev. Lett. **67**, 322 (1991); P. Patel, S. Kumar, and P. Ukleja, Liq. Cryst. **16**, 351 (1994). See also L. Lam, Chaos Solitons Fractals **5**, 2463 (1995).

[55] A. Joets and R. Ribotta, J. Stat. Phys. **64**, 981 (1991); R. Ribotta, in chapter 9 of [13].

[56] R. Ribotta, A. Joets, and Lin Lei (L. Lam), Phys. Rev. Lett. **56**, 1595 (1986); **56**, 2335 (E) (1986).

[57] A. Joets, R. Ribotta, and L. Lam (unpublished, 1988). This article is reprinted in *Nonlinear Physics for Beginners*, edited by L. Lam (World Scientific, River Edge, 1996).

[58] *Handbook of Mathematical Functions*, edited by M. Abramowitz and I.A. Stegun (Dover, New York, 1972).

[59] A.J. Heeger, S. Kivelson, J.R. Schrieffer, and W.-P. Su, Rev. Mod. Phys. **60**, 781 (1988).

[60] S. Roth and H. Bleier, Adv. Phys. **36**, 385 (1987).

[61] L. Yu, *Solitons and Polarons in Conducting Polymers* (World Scientific, Singapore, 1988).

[62] W.P. Su, J.R. Schrieffer, and A.J. Heeger, Phys. Rev. B **22**, 2099 (1980).

[63] H. Takayama, Y.R. Lin-Liu, and K. Maki, Phys. Rev. B **21**, 2388 (1980).

[64] E. Desurvire, Phys. Today **47**(1), 20 (1994).

[65] A. Hasegawa and Y. Kodama, *Solitons in Optical Communications* (Oxford University, New York, 1995). See also H.A. Haus and W.S. Wong, Rev. Mod. Phys. **68**, 423 (1996).

[66] A. Hasegawa and F.D. Tappert, Appl. Phys. Lett. **23**, 142 (1973).

[67] L.F. Mollenauer, R.H. Stolen, and J.P. Gorden, Phys. Rev. Lett. **45**, 1095 (1980).

[68] F.Kh. Abdullaev, Rev. Mod. Phys. **179**, 1 (1989).

[69] J.K. Park, K. Steiglitz, and W.P. Thurston, Physica D **19**, 423 (1986).

[70] Y. Aizawa, I. Nishikawa, and K. Kaneko, Physica D **45**, 307 (1990).

[71] A.J. Adams, J. Bechhoefer, and A. Libchaber, Phys. Rev. Lett. **61**, 2574 (1988).

[72] A. Hasegawa, Adv. Phys. **34**, 1 (1985).

[73] M. Makino, T. Kamimura, and T. Taniuti, J. Phys. Soc. Jpn. **50**, 980 (1981).

[74] A.S. Fokas and P.M. Santini, Phys. Rev. Lett. **63**, 1329 (1989); Physica D **44**, 99 (1990).

[75] H.R. Brand and R.J. Deissler, Phys. Rev. Lett. **63**, 2801 (1989); R.J. Deissler and H.R. Brand, Phys. Rev. A **44**, 3411 (1991).

[76] M.C. Cross and P.C. Hohenberg, Rev. Mod. Phys. **65**, 851 (1993); W. van Saarloos, in *Spatio-Temporal Patterns in Nonequilibrium Complex Systems*, edited by P.E. Cladis and P. Palffy-Muhoray (Addison-Wesley, Menlo Park, 1995).

[77] H. Segur, Physica D **51**, 343 (1991); A.S. Fokas, Nonlinear Sci. Today **1**(3), 6 (1991).

[78] A.R. Osborne, E. Segre, G. Boffetta, and L. Cavaleri, Phys. Rev. Lett. **67**, 592 (1991).

[79] R. Camassa and D.D. Holm, Phys. Rev. Lett. **71**, 1661 (1993).

[80] A. Sen and G.L. Johnston, Phys. Rev. Lett. **70**, 786 (1993).

[81] M. Segev, B. Crosignani, A. Yariv, and B. Fischer, Phys. Rev. Lett. **68**, 923 (1992); M. Segev, G. Salamo, B. Crosignani, G. Duree, P. Di-Porto, and A. Yariv, in *Novel Laser Structures and Applications*, edited by J.F. Becker, A.C. Tam, J.B. Gruber, and L. Lam (SPIE, Bellingham, WA, 1994).

[82] K. Hayata and M. Koshiba, Phys. Rev. Lett. **71**, 3275 (1993);

[83] F. Melo and S. Douady, Phys. Rev. Lett. **71**, 3283 (1993).

[84] D.A. Kurtze and D.C. Hong, Phys. Rev. E **52**, 218 (1995).

[85] P. Rosenau and J.M. Hyman, Phys. Rev. Lett. **70**, 564 (1993); P. Rosenau, Phys. Rev. Lett. **73**, 1737 (1994).

[86] A.R. Osborne, Phys. Rev. Lett. **71**, 3115 (1993); F.J. Alexander and S. Habib, Phys. Rev. Lett. **71**, 955 (1993).

[87] *Solitons in Science and Engineering: Theory and Applications*, edited by M. Lakshmanan, a special issue in Chaos Solitons Fractals **5**, No. 12 (1995).

V
Special Topics

12

Cellular Automata and Discrete Physics

David E. Hiebeler and Robert Tatar

12.1 Introduction

An *automaton* is a simple computer that has a finite number of internal states (*discrete values*). Through a specific function, or "rule," the internal state of the automaton depends on one or more inputs. In the simplest case, the "output" of the automaton consists of an integer representing the internal state. For the present discussion, we will call such a simple automaton a "cell."

A *cellular automaton* consists of a regular lattice (array) of identical cells (*discrete space*). As an imaginary clock ticks (*discrete time*), every cell is simultaneously updated according to a rule that receives as input the state of the cell, as well as the states of some of its closest neighbors (see Fig. 12.1).

Recently, there has been much interest in using cellular automata to model physical systems. Part of the reason for this is that cellular automata provide representations of physical systems that are different from traditional ones. This new perspective often provides additional insights into the behavior of the system under study. CA can also provide a more direct representation that explicitly embodies many of the physical properties of space, such as locality of interactions, finite speed of information propagation (finite speed of light), and relativity. Many practitioners also point out that the CA's restriction to a finite amount of information per cell (i.e., information per unit volume) is also a reasonable assumption for space [1].

As abstract ideas, these properties may not be very exciting. It is very exciting, however, that CA models have been used to simulate physical processes, such as diffusion, diffusion-reactions, wave propagation, fluid flow, growth mechanisms, phase transitions on spin lattices, the evolution of non-equilibrium and thermodynamic systems, and other phenomena. Effective modeling of these processes creates the possibility of using CA simulations for complex engineering design, such as the blades of a high-temperature turbine, a complex chemical reactor bed, or exotic lithographic processes for electronic device fabrication [2]. Some advocates claim that cellular automata models offer the promise of performing such simulations much faster and more accurately than conventional models.

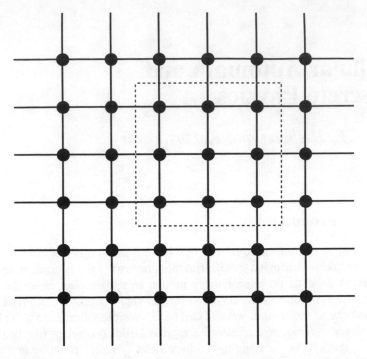

FIGURE 12.1. A two dimensional (2D), rectangular array of cells. The dashed line identifies the set of cells that define the "MOORE" neighborhood for the center cell. This neighborhood is also called a "3 × 3" neighborhood.

Although many CA practitioners use high-end workstations or super-computers such as Crays and Connection Machines to perform simulations, the intrinsic regularity of cellular automata permits the construction of high-speed, low-cost hardware coprocessors called "cellular automata machines." CAMs have excellent cost/performance characteristics for CA simulations. Examples of such hardware are RAP-1, LGM-1, CAM-6, CAM-PC, CAM-8, and others [3–8].

Cellular automata machines can also be used for animated, real-time visu-alizations of the process under study. Traditionally, many researchers observe their simulations with a short sequence of still images. It is much easier, how-ever, to comprehend the temporal evolution of a simulation by using high-speed computation and display, which is provided by appropriately equipped supercomputers or a low-cost CAM. Cellular automata machines also allow the user to interactively experiment with different rules and boundary con-ditions.

Much larger CA machines are already being designed and built [9]. While CA models are intellectually stimulating and even entertaining, it is the potential engineering applications that provide the primary justification for constructing large machines [10].

Before using CA models to simulate physical systems that have not yet been built, it is necessary to compare several critical CA simulations with laboratory experiments to gauge how well they approximate the systems under study. One of the most successful CA flow simulations to date was reported recently by the Lattice Gas group at Los Alamos. A flow simulation was performed using the digitized geometry of a thin slab of rock. The measured and simulated two dimensional (2D) flow permeability were found to agree within 5%, which is remarkable for *any* model. The Los Alamos group has challenged practitioners of more conventional methods to obtain equal or better results [11].

It is important to realize that CA models in general, and for physical systems in particular, are really still in their infancy. It is possible, but not likely, that the empirical results described above are fortuitous. The work must be repeated several times elsewhere before it is accepted as solid scientific evidence. In addition, most large corporations already use conventional methods for process and structural engineering design simulations; however, no one is yet using CA models for this purpose. Furthermore, the tools for using CA methods are rather primitive, whereas engineering tools that incorporate traditional methods have been evolving for decades. The many barriers to overcome before CA models effectively compete with traditional methods in applications are providing worthy challenges to the present generation of scientists.

The above discussion was intended to show the reason for interest in CA and CAMs by physicists, materials scientists, and chemists, among others. In the remainder of the chapter, we describe in more detail how CA and CAMs can be used.

12.1.1   A Well-Known Example: Life

To understand CA for physical simulations, it is helpful to have a concrete example in mind, which illustrates the relationship between the underlying lattice and the rule. The example we have chosen is John Conway's "Life," because of its simplicity and the fact that it is widely known.

Life is an example of a two-state cellular automaton on a 2D rectangular lattice. Each cell becomes active or inactive ("alive" or "dead," or takes values 1 or 0) based upon its current state and the number of active cells nearby. More specifically, each cell counts how many of its eight neighbors are alive. The new value of the cell depends on its current value and on the sum of active neighbors as follows:

- 0 or 1 neighbors alive: die from isolation;
- 2 neighbors alive: stay as you currently are;

- 3 neighbors alive: become alive, or stay alive (trisexual mating?); and
- 4 or more neighbors alive: die from overpopulation.

Because it is two dimensional, we can think of the state of this CA as a binary image. Perhaps suprisingly, this simple rule can lead to very complex evolving patterns when the sequence of states are observed on a computer or CAM display. Because the rule is fixed, the behavior depends entirely on the initial pattern of living and dead cells in the array. Starting with a random initial configuration, the system usually appears "active" for some time and then eventually stabilizes into either a static or periodic pattern.

A great deal of effort went into studying all aspects of the Life rule, particularly during the 1970s. See [12–14] for more in-depth discussions about Life.

The emergence of complex behavior from massive quantities of very simple components, as in Life, encouraged thinking about the possibility of modeling complex physical systems with more general cellular automata.

12.1.2   Cellular Automata

A cellular automaton as defined in the introduction is an abstract, mathematical object. There are precise, formal ways to define a cellular automaton, but for the present purpose, it is convenient to think of an array of "cells" with the following properties:

- discrete space: N-dimensional array of cells;
- discrete states: finite number of bits at every cell;
- discrete time: clock ticks away, every cell updates on each tick;
- synchronous updating;
- homogeneous rule: update rule is the same everywhere;
- deterministic rule;
- spatially local rule: only look at nearby cells; and
- temporally local rule: only current state (and maybe previous state) affects new state.

Occasionally, a few of these properties are relaxed for particular simulations. For example, one might use a stochastic update rule or allow each cell to have a continuous value rather than a discrete one. Such extensions to the basic notion of cellular automata are important but will not be discussed in this chapter.

At first, cellular automata may seem very restricted because every parameter is discretized, and one might wonder how it is possible to have much variety. In fact, tremendous variation is possible, as the following argument suggests. Consider an n-state cell, which is updated according to the states of k neighbors (possibly including itself). There are n^k possible state combinations that a cell's neighborhood may assume. For each possible neighborhood configuration, the cell may take on one of the n states. Thus, there are a total of n^{n^k} rules. ,For the 3×3 neighborhood used in Life ($n = 2$, $k = 9$), this number is 2^{2^9}, or approximately 10^{154}. Running out of rules to try doesn't seem to be an immediate concern. On the other hand, randomly trying different rules isn't very constructive. Some way is needed to generate rules that do something interesting or useful.

The first comprehensive study to classify cellular automata was conducted by Stephen Wolfram [15]. In particular, Wolfram examined one-dimensional (1D), two-state CA with a "radius" of one, meaning the value of the current cell plus both of its immediate neighbors determine the new value for the cell (1D, $n = 2$, $k = 3$). (With a radius of two, the new value would also depend on the next-nearest neighbors, and so on.) Because this class of CA has only $2^{2^3} = 256$ rules, Wolfram classified the typical behavior of all of them, starting with various initial configurations. He showed that there are several distinct types of cellular automata which all have physical analogs. He also illustrated comprehensive approaches to analyzing the algebraic and statistical mechanical properties of cellular automata (mostly based on studies of 1D systems).

Others have attempted to classify higher-dimensional automata. One approach is Chris Langton's *lambda* parameter, a simple analogy of temperature, to describe the space of CA rules. Results indicate that there is some kind of "phase transition" in the region of the rule space between rules that settle down to periodic behavior, and rules that exhibit chaotic behavior. At the transition between the two, the system will follow extremely long transients, which may be a sign of computational capacity [16–18].

Such studies provided more specific evidence for the belief that cellular automata could model various types of complex physical systems.

12.1.3   The Information Mechanics Group

Although Wolfram and others emphasized the ensemble properties of cellular automata, several individuals focused on the "microscopic" properties. In particular, the Information Mechanics Group at MIT contributed a series of papers that discussed the characteristics that CA have in common with physical space-time:

- spatially local rule—there is no action at a distance;
- there is a finite speed of information propagation—"speed of light";
- homogeneous rule—the laws of physics are the same, no matter where you are; and
- deterministic rule—rules are generally forward-deterministic; by making them backward-deterministic as well, it is possible to create microscopically reversible rules.

Toffoli and Margolus of the Information Mechanics Group made the bold claim that these factors could allow construction of much more efficient computers for physical simulations. They then designed and built a series of cellular automata machines (CAM-1 through CAM-8) to demonstrate their points!

The Information Mechanics Group has studied a wide variety of applications of CA with their various machines. Many of these applications are described in their book [19], which is an excellent introductory book on the subject of physical modeling with cellular automata. Although it heavily uses CAM-6 notation to describe the rules, it is still a valuable introduction to the subject in general.

12.2 Physical Modeling

12.2.1 CA Quasiparticles

Until now, most CA applications to physics generally use collections of CA "quasiparticles" for simulation of both statistical ensembles and "continuum" fields.

The simplest example of a CA simulation for a statistical ensemble is a spin lattice, where each site has a spin variable that can take on values "up" or "down." A variety of multistate spin lattices and Potts models have been studied with CA. Several references can be found in [20].

For simulating "continuum" fields, it is convenient to think of each quasiparticle as representing an imaginary particle whose collective properties are similar to those of the field. Feynman had a very nice way of expressing this:

We have noticed in nature that the behavior of a fluid depends very little on the nature of the individual particles in that fluid. For example, the flow of sand is very similar to the flow of water or the flow of a pile of ball bearings. We have therefore taken advantage of this fact to invent a type of imaginary particle that is especially simple for us to simulate. This particle is a perfect ball bearing that can move at a single speed in one of six directions. The flow of these particles on a large enough scale is very similar to the flow of natural fluids. [21, p. 88.]

The model described by Feynman is known as a *lattice gas*. Note that *any* integrable vector field, such as an electrostatic field, or a classical gravity field, can be described in terms of the flow of imaginary particles.

A cellular automaton provides a direct way to implement such a set of simple particles. Figure 12.1 shows a cellular automaton that can be used to implement fluid-like particles that can move in one of four directions, instead of six. In fact, by playing tricks with the rules, a square-lattice automaton can also be used to implement the hexagonal lattice gas described by Feynman.

By using a particle representation of a "continuous" field, the techniques of statistical physics are applicable, just as particles are used in quantum field representations of electromagnetic fields. For fluid simulations, however, CA quasiparticles are much simpler than the photons that make up an electromagnetic field, which provides an enormous computational simplification. See [22] for an introduction to the statistical properties of 1D CA.

12.2.2   Physical Properties from CA Simulations

In simulation of statistical ensembles, properties over the whole system are important, such as the number of particles in a particular state, or conservation of a particular type of particle.

A necessary component of a model is that it has certain properties that can be measured and studied, such as mass, velocity, momentum, and so forth. There are both intrinsic and extrinsic properties of a cellular automaton system. The latter are very straightforward. Consider mass as an example. In a two-state CA, the total mass of the system can simply be defined as the number of cells in State 1. (Vacuum is then defined as State 0.) In a system with more than two values per cell, it may be desirable to determine the mass from particular states. For example, cells with an even value might count toward the total mass, whereas odd cells might not. In this case, odd values would have another interpretation, such as a quantum of energy.

A next step might be the definition of momentum. Suppose the current rule is one in which particles (nonzero values) propagate in various directions. The momentum of a cell could simply be its mass multiplied by its direction of motion (a unit vector). These vector quantities could be summed over the entire array, thus giving a value for the total momentum of the system. By summing only over particular regions or clusters of active cells, it would be possible to measure the momentum of a particular region or structure in space.

Intrinsic properties such as pressure may also be defined. To use the propagating particle example again, suppose that the model also has fixed walls that deflect the particles. The force at any given instant might be defined as the total number of particles currently adjacent to the wall—either about to collide or just after colliding. The pressure would then be defined as the average number of such particles per unit area.

This leads to an interesting observation about the measurement of certain properties of a cellular automaton; namely, that an *averaging* technique is often very useful. This technique is used regularly in studies of lattice gas models, which will be described shortly. In some cases, it is only meaningful to speak in terms of local averages. For example, to measure the density of a particular area in a particle model, it wouldn't be very useful to speak of the density of a single cell, which is either zero or one. Instead, such quantities are measured in a local area, for example an $N \times N$ block of cells (where finding the optimum value for N is a problem in itself). Similar techniques are used in kinetic theory, where it is desirable to use a volume that is large enough to contain many particles, but small enough to treat as infinitesimal compared to the size of the system. It is partly for this reason that simulations with enormous numbers of cells are desirable.

12.2.3 Diffusion

A class of CA applications to ensembles is the study of the behavior of a large number of diffusing particles. This is important for simulation of fluid flow, electron scattering in solids, or chemical reactions. In this case, the CA quasiparticle might be thought of as representing an individual molecule of the ensemble, even though this is not achievable in practice. Although the model described below is very simple, it demonstrates a few of the techniques and limitations that often arise in particle-based CA models.

Note that in traditional models of particle systems, a list of structures is maintained, with each structure describing the attribute of a particle, such as position, velocity, and so forth. In a CA model, a very different approach is used. Instead of simulating many particles that move around in space, we simulate many regions of space through which particles move. Literally, a particle or a "bit of matter" moves through space by passing a value from one cell to the next. In this sense, each cell is "a piece of space."

Consider a lattice in which particles move horizontally and vertically, as in Fig. 12.1. All particles move at the same speed. Each cell will be able to hold up to four particles at once. No two particles, however, may be in the same location moving in the same direction (like the Fermi exclusion restriction). Thus, a "full" cell would have one particle moving in each of the four orthogonal directions. We must also be sure that under the update rule, it will not be possible for more than four particles to converge upon one cell.

One way to mathematically describe this implementation is to use an array of 4-vectors, where each 4-vector represents a cell. The components of the vector can take on values 0 or 1.

Thus, for example, a state like $(0, 1, 0, 0)_{ij}$ might represent the presence of a particle moving "South" in $cell_{ij}$, while for the same cell, $(1, 0, 0, 0)_{ij}$ might represent a particle moving "North." The existence of particles moving "East" and "West" can be represented using the other vector components. Note that state $(0, 0, 0, 0)_{kl}$ represents the absence of any particles, or a vacuum, at $cell_{kl}$. With the help of this notation, it should be easy to see how this simple lattice gas can be implemented with a 16-state CA on a rectangular lattice.

For behavior that resembles diffusion to occur, particles must interact with each other in some way, or else the particles would move independently and the model would not be very interesting. Some sort of *collision rule* is needed to describe what happens when particles meet. It we require that momentum is conserved in all collisions, there is only one possible collision that can alter the paths of particles. Namely, when two particles hit head-on, they can both be deflected by 90°; the net momentum before and after is zero.[1] With any

---

[1] Note that it is irrelevant whether particles are deflected clockwise or counter-clockwise, because all particles are identical.

other combination of particles, deflecting any of them would require a change in total momentum.

Next, a method of applying or organizing the rule is needed. The *propagation-collision* cycle is a very common and simple technique to accomplish this. Two different rules are applied on alternating time-steps. First, a rule is used in which every particle moves one cell in whichever direction it is going; then, the collision-rule is applied. This changes the direction of motion of some of the particles. By alternating between the two, we get motion and interaction. In fact, these two steps can be combined into a single rule, but it is often useful to think of the two phases separately, which also makes code development easier. Figure 12.2 illustrates a complete update-cycle.

This model is called the HPP lattice gas, from the initials of those who first extensively studied its properties [23, 24]. Besides its pedagogical utility, the HPP model is of historical importance because it introduced the concept of the propagation-collision update cycle, which is a key step in implementing lattice-gases.

Let's now run an experiment using this rule. We will start with a pattern of

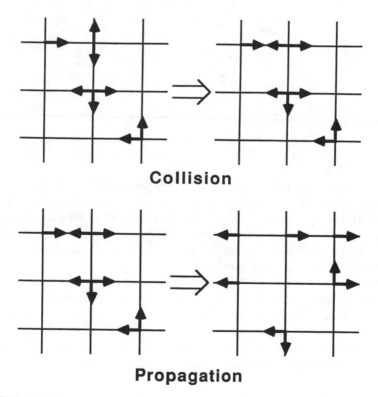

FIGURE 12.2. The complete HPP update cycle, shown in two parts. First, the collision transformation is applied, and then the propagation rule.

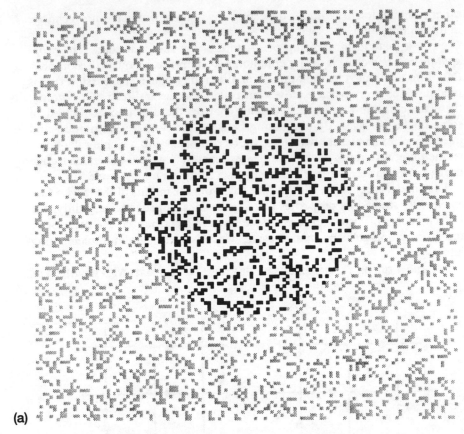

(a)

FIGURE 12.3. (a) Two types of particles in the array.

uniform density, but color a portion of the particles black, and the rest gray[2] (see Fig. 12.3a). After about 50 time-steps, the two types of particles are somewhat mixed, as shown in Fig. 12.3b. The number of time steps required to thoroughly mix the particles depends on the size of the system and on how the black particles are distributed. This behavior is very similar to the mixing of two gases with identical dynamics.

Extensions to the basic HPP model address some of its deficiencies. One important extension was discovered by Frish, Hasslacher, and Pomeau (the FHP model) [25]. They found that by allowing six velocities on a hexagonal lattice, the resulting lattice gas accurately approximated the behavior of the Navier–Stokes equation. This derived from the improved set of symmetry properties of rules on this lattice. Later additions of a small number of addi-

---

[2] To trace individual particles through collisions, we can simply assume all head-on collisions rotate particles clockwise, or counter-clockwise, or flip a coin in each case.

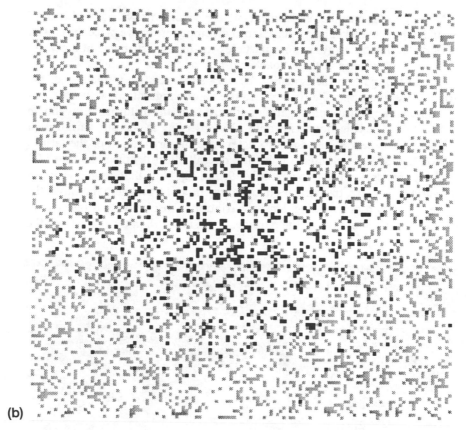

(b)

Figure 12.3 *(continued)*. (b) The particles have begun to mix.

tional speeds, including particles "at rest" were shown to further improve the behavior for many types of fluid simulations [26, 27].

12.2.4   Sound Waves

The HPP model has been modified to model many things besides molecular gas diffusion. One such application is the motion of sound waves.

In Section 12.2.3, we discussed systems at equilibrium (the two colors of particles obeyed the same rules, and so the system was always at dynamical equilibrium). Now, we investigate the results of disturbing the equilibrium in one region of the array.

Consider an experiment where we place a solid block of particles in an otherwise empty array, and apply the HPP rule. For a brief time, there will be an interaction between the particles. Shortly thereafter, however, we will see four groups of particles rush out in the four directions of propagation, never to interact again (assuming an infinite lattice). Obviously, the behavior of a dense cluster of particles in a vacuum is very anisotropic.

Now consider a similar experiment, where we again have a solid block of particles, but instead of being placed into an empty array, we place the cluster of particles into an array that has particles randomly distributed at a density of 40%. The presence of these "background" or "noise" particles will cause the mean free path in the system to be very small relative to the size of the array. If we now apply the HPP rule to the new configuration, we get a *circular* compressional wave propagating outward from the disturbance, as shown in Fig. 12.4. The behavior is isotropic—by observing the density variation of the wave, it is nearly impossible to determine the orientation of the lattice. Toffoli and Margolus observe that "the overall effect is much like that of a bucket of water dumped into a pond" [19, p. 172]. Note that, as in a wave on the surface of water, the particles in the array do not travel with the wave; the wave is simply a propagating density variation whose velocity is faster than the random-walk motion of the particles, but slower than the velocity of the

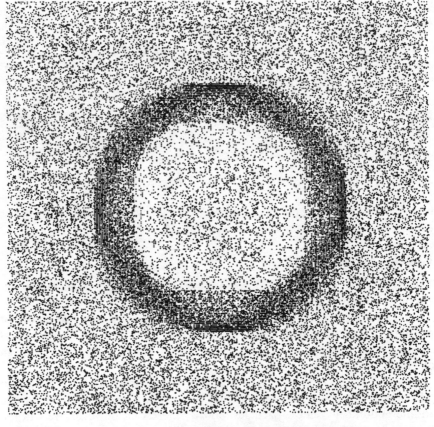

FIGURE 12.4. A circular wave emerges from the dynamics of particles on a square lattice.

individual particles in the system. The random background therefore serves as a medium in which the sound waves may travel.

Other properties of fluids have been studied with lattice gas models as well. In fact, lattice gas methods of hydrodynamics is currently one of the most actively studied applications of cellular automata, around the world. As mentioned in the introduction, a group at Los Alamos has been very successful with flow simulations through porous media. Many other experimental results and mathematical analyses of CA hydrodynamics have appeared [10, 25–28].

Other approaches to acoustic wave propagation are described in [29, 30].

12.2.5 Optics

The same HPP model, with minor modifications, has also been used to perform simulations of optical phenomena, such as reflection and refraction. In fact, in this section the wave phenomena is more similar to sonics than optics because a background gas or "ether" is used, but the description is easier to follow in the language of optics. A different CA approach to optics with a demonstration of interference in a "double-slit experiment," is described in [31].

Reflection is fairly simple to accomplish by introducing a new type of quasiparticle that doesn't move, but with which the normal particles interact. Either specular collisions or a reversal of momentum ("no-slip") can be used. These "wall" particles are considered to have infinite mass. Thus they don't move during collisions. Reflective surfaces of arbitrary shape (such as a parabolic mirror) made of these special particles can be drawn in the array.

Note that these extra particles can be mathematically described by extending the vector representation discussed previously. Thus, a 5-vector can be used to describe the lattice gas particles and the immobile barrier ($2^5 = 32$ states/cell).

Implementing refraction might seem like a more difficult task, but actually only requires another very simple modification. A way to give different regions different *indices of refraction* is needed. This means that the "speed of light" should be different in these different regions (i.e., slower in a lens than in a vacuum). This can be accomplished by marking certain sites of the array with a flag. These sites will have the special property that they will only be updated *half as often as normal cells*. Although at first this may seem like a very nonstandard extension, it is possible to implement this as part of the local rule, by sufficiently increasing the number of states per cell. By updating the cells in the lens region half as often, the index of refraction, which is defined as $n_2 = c/v_2$, should have a value of 2, while the vacuum's index of refraction n_1 will simply be 1. Note that other update-frequency ratios can be used, giving other relative indices of refraction.

With the addition of these "slow cells," it is possible to draw lenses of any

FIGURE 12.5. A planar wave, originally vertically oriented at the left edge of the array, is reflected and refracted as it moves toward the right. The white area represents a vacuum, the grey area a lens. To improve the clarity of the image, several copies of the system were run. Each copy began with an identical initial wave, but a different random distribution of the background noise. By correlating the different copies of the system, some of the background noise was filtered out in this image, and the clarity of the waves was improved.

shape, such as circular or planar lenses. By sending planar "light" waves through these lenses, their properties can be observed and measured. It is possible to construct a planar "light" wave by activating an entire row or column with particles ("photons"). Several adjacent columns can be activated in this way for a stronger signal. As usual, the random background noise will be present in the array.

The result of a planar wave passing from a vacuum into a lens that has an index of refraction of 2 is shown in Fig. 12.5. The angles θ_i, are the angles between the propagation vectors and the normal to the interface. When these angles are measured from the figure, the empirical index of refraction can be calculated from $n_1 \sin(\theta_1) = n_2 \sin(\theta_2)$. The experiment shown yields $n_2 = 1.99 \pm 0.07$, which is very close to the expected value. Note also that there is a faint *reflection* from the surface of the lens; this too matches the behavior in a physical system. Although at first it seems that there is no reason for a reflection, there is a simple explanation. The HPP rule is reversible; when a wave passes from vacuum into lens, it slows down, thus compressing as it moves into the lens. However, the waves we have been using are so dense that to compress, they would have to exceed the maximum possible particle density. Because this cannot happen, and because in a reversible system no information can be lost, there is necessarily a reflection.

The empirical index of refraction corresponds to the predicted value over a wide range of angles, whether the planar wave is moving from vacuum into lens or lens into vacuum. However, the measured values begin to deviate as the critical angle[3] is approached. Beyond the critical angle, no wave exits from the lens, which is the correct behavior.

A shortcoming in this model is the following. When a wave passes from lens into vacuum, there is no internal reflection from the surface. The above argument for the necessity of a reflection when moving from vacuum to lens does not apply here, because the wave *expands* upon exiting the lens, and will not exceed the maximum density. However, by adjusting the boundary conditions between the two media, and making a small extension to the rule, this problem can be overcome as well. If we allow a *third* medium that has a higher index of refraction than either of the others, and then place an "infinitesimal" (single-cell) layer of this new medium at the interface between lens and vacuum, then our information-loss argument will again apply (because the wave has to compress to pass through the new layer). A lens that has had its surface treated in this way seems to produce results that are correct within the measured precision. Further analysis will determine how successful it really is under different conditions.

12.2.6 Chemical Reactions

In addition to the physical systems discussed in earlier sections, there is also interest in using cellular automata to model the evolution of chemical reactions [20, 32–36]. We will present one example in some detail below. This is a model of a surface-reaction that exhibits a kinetic phase transition. This system is of interest for its collective (statistical) properties rather than its detailed spatial behavior. Thus, this example illustrates the use of CA for simulation in an abstract parameter space.

CO–O$_2$ Surface Reactions

A system that has been studied a great deal is the oxidation of carbon monoxide. This system consists of the following components: (i) adsorption of the reactants on the surface, (ii) reaction of the species, and (iii) desorption of the product. More specifically, O_2 and CO are adsorbed onto the surface, where the reaction $CO + O \rightarrow CO_2$ can take place, and the resulting CO_2 is desorbed. In our cellular automaton model, each cell in the two-dimensional

---

[3] When the wave is moving inside the lens toward the vacuum, this is the angle at which the outgoing ray is refracted to 90° from the normal to the surface. Thus, the wave travels along the lens surface and the situation is referred to as *total internal reflection*.

array represents a site on the surface of the catalyst. It will be a 3-state CA, the states being (A) *empty*, (B) O, and (C) CO. The model will be probabilistic. The gas above the surface consists of a CO concentration X_A, and O_2 concentration $X_B = 1 - X_A$. We will ignore the effects of the surface reactions on the composition of the gas, and thus assume that the concentration X_A (and therefore X_B) is constant.

CA models of this surface-reaction system have been studied by Ziff et al. [32], Dickman [33], and Droz and Chopard [34]. We will discuss the version described in [34]. A simplified version of their rule is as follows:

- If there are two adjacent empty sites, they will become state **B** (containing an O atom) with probability $1 - X_A$. Upon adsorption, O_2 dissociates into two O atoms residing on adjacent sites.
- An empty site (regardless of its neighbors) will assume state **C** (a CO molecule) with probability X_A.
- A site in state **B** will become empty if it has a neighbor in state **C**, and a cell in state **C** will become empty if a neighbor is in state **B**. This represents the reaction of CO with O on adjacent sites, and desorption of the product.

Analysis and experiments were done by Droz and Chopard using a CAM, to study the properties of this model. The results of experimentation indicate that two second-order phase transitions exist near two critical concentrations X_{A1} and X_{A2}. When the concentration of CO exceeds the critical value X_{A1}, the surface will become permanently "poisoned" with CO; a similar poisoning with O occurs when the concentration X_A of CO falls below the threshold X_{A2}. The behavior of the system near these two values exhibits second-order transitions between zero production-rate of CO_2 and nonzero rates. Further analysis indicates that the model corresponds well with mean-field theory analyses. The mean-field-like behavior emerges as a result of the "randomness"[4] of the gases above the surface. This prevents correlations on the surface that might emerge in a model that explicitly simulated the gas above the surface.

12.3   Hardware

A wide variety of machines have been built that can perform CA computations at high speeds. Some of the more widely known past and future machines are described below. A good bibliography of machines built by various research groups is included in [37].

---

[4]The probability concentration X_A can be implemented either with external hardware or as a diffusing gas on a CAM.

- CAM-PC, based on the CAM-6 specifications but with several enhancements, plugs into an IBM PC or compatible and updates a 256×256 array of 16-state cells, 60 times per second [38].
- The CAM-8 [7, 8] is currently available from the Information Mechanics Group at MIT. This is an expandable modular achitecture also capable of simulating CA in more than two dimensions.
- Connection Machines or other massively parallel computers are well suited for cellular automata simulation; indeed CA are often referred to as "embarrasingly parallel" because it is so easy to simulate them efficiently on parallel machines.
- A lattice gas machine has been proposed [9], which will run at approximately 10^8 times the speed of a Cray Y-MP, at about two-thirds the cost.

12.4   Current Sources of Literature

Major research institutions around the world support research in cellular automata: For example, the Los Alamos Alamos National Laboratory, IBM Research, the French National Center for Scientific Research, and the Trieste International School for Advanced Studies, in addition to science and engineering departments in major universities around the world.

CA research is published in diverse places. Several regular sources of information are:

- *Complex Systems*—This journal is edited by S. Wolfram, who has made many contributions to CA science. CA articles appear here regularly.
- *Physica D*—CA articles are frequently published in this journal. Special issues of this journal contained the proceedings of the 1986 and 1989 CA conferences as well.
- *Scientific American*—The various incarnations of "Computer Recreations," "Mathematical Games," and similar columns by A.K. Dewdney, Martin Gardner, and others have often featured CA as the special topic (see [14, 39] for example).
- In general, CA-related articles have begin to appear with increasing frequency in various specialized journals [40, 41].

12.5   An Outstanding Problem in CA Simulations

Many fields of research have difficult problems whose solution opens or closes many avenues of investigation and application. The field of cellular automata is still relatively young and there are many such problems. Although most of the CA applications research focuses on gases and incompressible fluids, one of the outstanding problems in cellular automata research is the representation of solid elastic bodies. This requires that many "particles" move in a correlated way. A solution to this problem has been described for 1D objects [42] and it has been implemented on a CAM [43]. A general rule that allows a representation of arbitrarily shaped 2D and 3D bodies is desired. This would enable complex simulations that combine objects having different length scales. For example, it may be desirable to simulate the deformation of an elastic tube in a fluid stream, or the buckling of a support under a high pressure gas. By treating both solid bodies and particles or fluids on the same footing, some very complex transformations within the simulation are possible, such as a volcanic explosion that scatters a large fractal structure into small solids that fly and spin through space.

The solid body problem is a correlation problem. A solution to this problem may also have an impact on the simulation of other systems which can be described in terms of correlations of massive numbers of particles, such as superfluids and the electronic structures in metals and semiconductors [44]. There are other representation problems that need to be addressed, but in the authors' view, a solution to the "solid body" problem would provide tremendous rewards.

Problems

Note: Many of the following exercises are for cellular automata on a square lattice. To facilitate a discussion of the rules, the following nomenclature, which is identical to that of [19], is introduced. For two-state automata, the value of the state of a cell is referred to as CENTER. The values of the states of the nearest neighboring cells are called NORTH, SOUTH, EAST, and WEST. The values of the states of the diagonally nearest neighboring cells are called NORTHEAST, SOUTHEAST, SOUTHWEST, and NORTHWEST.

Most of the rules here make use of the "MOORE" neighborhood, which uses the values CENTER, NORTH, SOUTH, EAST, WEST, NORTH-EAST, SOUTHEAST, NORTHWEST, and SOUTHWEST.

12.1. How would you extend the representation discussed in section 12.2.3 to describe the lattice gas on an hexagonal array?

12.2. Write a rule to simulate a self-diffusing gas on a hexagonal lattice. Assume that momentum is conserved. Is more than one rule possible?

12.3. Write a rule to simulate three diffusing gases on a square lattice. Assume that momentum is conserved and that all particles have the same mass. Is more than one rule possible?

12.4. Consider the following rule on a square lattice of two-state automata: For a given cell, compute the sum of the values in its MOORE neighborhood. If this number is greater than 4, set the cell state to 1, otherwise set the cell state to 0. This is known as a voting rule. Given a random initial condition, what happens to the pattern over many time steps?

12.5. Describe what happens when two sound waves collide in the HPP model.

12.6. How would you extend the mathematical representation of the HPP model to handle both reflecting and refracting materials?

12.7. Consider the following rule on a square lattice of two-state automata. For a given cell, examine NORTH and SOUTH. If SOUTH is 0, set the state of the cell to NORTH. If SOUTH is 1, don't change the state of the cell. What does this rule do to an initial condition consisting of a row of six consecutive cells in State 1 if all the other cells are in State 0?

12.8. Describe how to set up a cellular automata simulation of an equilibrium reaction between a diatomic molecule and its atomic constituents. Assume that the atomic species freely diffuse and that two atoms form a diatomic molecule whenever they are in adjacent cells. Also assume that there is a catalyst at a finite number of locations, which decomposes the molecule into atoms. Also conserve the total number of atoms.

12.9. Compute the mean free path as a function of density for the self-diffusion model on both square and hexagonal lattices.

12.10. Write a rule on a square lattice that computes the instantaneous force on a movable wall.

12.11. Describe how to implement the correlated motion of three particles arranged in a row on a 2D square lattice (hint: use a short period sequence of rules).

12.12. Describe how to implement the correlated motion of an elastic body of arbitrary shape on a 2D lattice. How would you handle a collision between two such bodies?

12.13. In a 2D system in which each cell has an integer value between 0 and 255 (8-bit cells), consider the following rule: add the values of the center cell and its eight neighbors, and divide by nine. In other words, each cell's new value will be the average value in its current neighborhood. This has been used to model *heat flow* (see [45], for example). Why do you think this model works? (i) What will happen if we start with random initial configurations, and let the system run for a very long period of time? Will the system ever become stable or periodic? If so, what is the physical analogy of the stable configuration? (ii) When finding the average value of the local neighborhood, some method of truncation or rounding must be used. What effect will different methods have on the dynamics of the system? (iii) How can the system be generalized (i.e., what simple change could we make to introduce one or more parameters with physical meaning)?

References

[1] R. Feynman, Int. J. Theor. Phys. **21**, 219 (1982).

[2] R. Guerrieri and A. Neureuther, IEEE Trans. on Computer-Aided Design **7**, 755 (1988).

[3] T. Toffoli, Physica D **10**, 195 (1984).

[4] A. Clouquer and D. d'Humieres, Helvetica Physica Acta **62**, 525 (1989).

[5] S.D. Kugelmass and K. Steiglitz, Proceedings of the 22nd Annual Conference on Information Sciences and Systems, Princeton University, March 16–18 (1988).

[6] H.J. Herrman, Physica **140A**, 421 (1986).

[7] N. Margolus and T. Toffoli, in Lattice Gas Methods for Partial Differential Equations, edited by G.D. Doolen et al. (Addison-Wesley, Boston, 1990).

[8] N. Margolus, in *Pattern Formation and Lattice-Gas Automata*, edited by A. Lawniczak and R. Kapral (American Mathematical Society, Providence, RI, 1996).

[9] A. Despain, C. Max, G.D. Doolen, and B. Hasslacher, in *Lattice Gas Methods for Partial Differential Equations*, SFI SISOC, edited by G.D. Doolen et al. (Addison-Wesley, Boston, 1990).

[10] N. Margolus, T. Toffoli, and G. Vichniac, Phys. Rev. Lett. **56**, 1964 (1986).

[11] G.D. Doolen, "What Can We Hope for Cellular Automata?" To be published.

[12] E.R. Berlekamp, J.H. Conway, and R.K. Guy, *Winning Ways for Your Mathematical Plays* Vol. 2, (Academic, New York, 1982).

[13] W. Poundstone, *The Recursive Universe*, (Morrow, New York, 1985).

[14] M. Gardner, *Sci. Am.*, Oct. **223**, 120 (1970); Feb. **224**, 112 (1971); March **224**, 106 (1971); April **224**, 114 (1971); Jan. **226**, 104 (1972).

[15] *Theory and Applications of Cellular Automata*, edited by S. Wolfram (World Scientific, River Edge, 1986).

[16] C. Langton, Physica D **22**, 120 (1986).

[17] C. Langton, Physica D **42**, 12 (1990).

[18] C. Langton and W. Wootters, Physica D **45**, 95 (1990).

[19] T. Toffoli and N. Margolus, *Cellular Automata Machines: A New Environment for Modeling* (MIT, Cambridge, 1987).

[20] M. Droz and B. Chopard, Physica D **10**, 195 (1984).

[21] W.D. Hillis, Physics Today, Feb. 78 (1989).

[22] S. Wolfram, Rev. Mod. Phys. **55**, 601 (1983).

[23] J. Hardy, Y. Pomeau, and O. de Pazzis, J. Math. Phys. **14**, 1746 (1973).

[24] J. Hardy, O. de Pazzis, and Y. Pomeau, Phys. Rev. A **13**, 1949 (1976).

[25] U. Frisch, B. Hasslacher, and Y. Pomeau, Phys. Rev. Lett. **56**, 1505 (1986).

[26] K. Diemer, K. Hunt, S. Chen, T. Shimomura, and G. Doolen, in *Lattice Gas Methods for Partial Differential Equations*, edited by G.D. Doolen et al. (Addison-Wesley, Boston, 1990).

[27] Shiyi Chen, Hudong Chen, and Gary Doolen, Complex Syst. **3**, 243 (1989).

[28] S. Wolfram, J. Stat. Phys. **45**, 471 (1986).

[29] J.-I. Huang, Y.-H. Chu, and C.-S. Yin, Geoph. Res. Lett. **15**, 1239 (1988).

[30] H. Chen, S. Chen, and G.D. Doolen, Phys. Lett. **140**, 161 (1989).

[31] H. Chen, S. Chen, G. Doolen, and Y.C. Lee, Comp. Syst. **2**, 259 (1988).

[32] R. Ziff, E. Gulari, and Y. Barshad, Phys. Rev. Lett. **56**, 2553 (1986).

[33] R. Dickman, Phys. Rev. A. **34**, 4246 (1986).

[34] B. Chopard and M. Droz, J. Phys. A: Math. Gen. **21**, 205 (1988).

[35] B. Madore and W. Freedman, Science **222**, 615 (1983).

[36] M. Droz and B. Chopard, Helvetica Physica Acta **61**, 801 (1988).

[37] D. d'Humieres, in *Cellular Automata and Modeling of Complex Physics Systems*, edited by P. Manneville, N. Boccara, G.Y. Vichniac, and R. Bidaux (Springer-Verlag, New York, 1989).

[38] CAM-PC is available from Automatrix, Inc., P.O. Box 196, Rexford, NY 12148-0196, USA.

[39] A.K. Dewdney, Sci. Am., Aug. **259**, 104 (1988); Aug. **261**, 102 (1989); Sept. **261**, 180 (1989).

[40] K. Perry, Byte, Dec., **11**, 181 (1986).

[41] P. Wayner, Byte, May, **13**, 253 (1988).

[42] B. Chopard, J. Phys. **23**, 1671 (1990).

[43] N. Margolus has programmed this for CAM-6 in a software file called "glider1.exp." This file is available from N. Margolus and the authors.

[44] R.P. Messmer, R.C. Tatar, and C.L. Briant, in *Alloying*, edited by J. Walters (ASM International, Ohio, 1988).

[45] N. Packard, in *Theory and Applications of Cellular Automata*, edited by S. Wolfram (World Scientific, River Edge, 1986).

13

Visualization Techniques for Cellular Dynamata

Ralph H. Abraham

13.1   Historical Introduction

Complex dynamical systems (CDSs) [1] and cellular dynamata (CDs) [2] are classes of dynamical models that have evolved in the context of applied work over a period of years. The literature on one-dimensional cellular dynamata is extensive [3]. Recently, two-dimensional lattices have become the focus of several exploratory studies [4]. In this chapter, we describe some methods for creating visual displays of important qualitative features of cellular dynamata in one and two spatial dimensions.

Reaction/diffusion equations constitute a special class of partial differential equation (PDE) systems of evolution type. They were introduced and used by the pioneers of biological morphogenesis: Fisher (1930), Kolmogorov-Petrovsky-Piscounov (1937), Rashevsky (1940), and Turing (1952). Fisher introduced the logistic/diffusion equation, particularly important for CD research, as a model for the diffusion of mutants in a population of flys [5]. Rashevsky introduced spatial discretization corresponding to biological cells in his work on embryogenesis [6]. Discretized reaction/diffusion systems provide examples of cellular dynamata, probably the first in the literature. Reaction/wave systems are another source of important CDs. Further developments were made by Southwell (1940–1945), Turing (1952), Thom (1966–1972), and Zeeman (1972–1977). The latter includes a heart model [7], and a simple brain model exhibiting short- and long-term memory [8]. The arrival of scientific computation in Los Alamos stimulated Ulam and Von Neumann to develop cellular automata (CAs) in connection with their efforts to solve the heat equation. During the 1980s, under the influence of the growing availability of computer graphic workstations in the scientific community, experimental work on 1D/CDs (one spatial dimension) began. Later, as computational power became adequate, 2D/CD studies began to appear in the literature. The ideas of cellular dynamaton theory, inspired by these pioneers, are summarized in the next section.

13.2   Cellular Dynamata

Cellular dynamata are complex dynamical systems characterized by two special features: the nodes (component schemes) are all identical copies of one scheme, and they are arranged in a regular spatial lattice. We now recall the basic definitions.

13.2.1   Dynamical Schemes

By *dynamical system* we mean an autonomous system of coupled ordinary differential equations of the first order. More generally, we include vector-fields on manifolds, both finite and infinite dimensional, which we call *state spaces*. The *phase portrait* is a visualization of the dynamical system within its state space. Thus, systems of coupled partial differential equations of evolution type are included, along with integro-differential-delay equations, and so on. By *dynamical scheme* we mean a dynamical system depending upon parameters in a supplementary manifold, the *control space*. The visualization of a dynamical scheme (in low dimensions) is provided by its *response diagram*. The familiar pictures from elementary catastrophe theory (ECT) [9], dynamical bifurcation theory (DBT) [10], and geometric function theory (GFT) [11], are exemplary response diagrams. The chief features are the *attractrix* (or locus of attractors) and the *separatrix* (or locus of separators). The most widely known of these diagrams are the *cusp* (shown in Fig. 13.1), and the *logistic period doubling sequence* (shown in Fig. 13.2). We will make use of these to illustrate visualization techniques for cellular dynamata.

13.2.2   Complex Dynamical Systems

Dynamical schemes may be serially coupled in various ways. The simplest, which suffices for most applications, is called a *static coupling*. This is a function from the state space of one dynamical scheme to the control space of another. The canonical example is the driven pendulum. In this way, a finite set of dynamical schemes (nodes) may be serially coupled by an appropriate set of static couplings (directed edges) in a network (directed graph). This is the definition of a *complex dynamical system*, the primary object of complex dynamical systems theory. Exemplary models for several physiological systems have been developed and run, producing convincing simulated data [12].

13.2.3   CD Definitions

By a *cellular dynamical system, cellular dynamaton,* or *CD*, we mean a complex dynamical system in which the nodes are all identical copies of a single dynamical scheme, the *standard cell*, and are associated with specific loca-

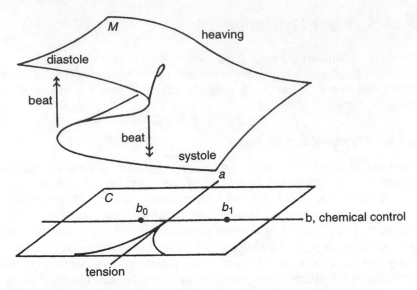

FIGURE 13.1. The cusp.

tions in a supplementary space, the *physical space*. Exemplary systems have been developed from reaction/diffusion systems by numerical methods, such as Southwell's relaxation method, which proceed by discretization of the spatial variables. Other important examples of this construction are the heart and brain models of Zeeman. CDs have something in common with the *cellular automata*, or *CAs*, of Ulam and Von Neumann, but possess more structure, and are in some ways more general.

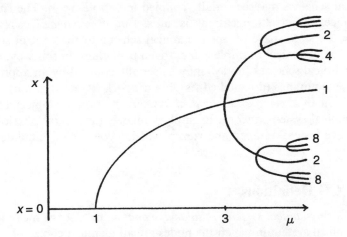

FIGURE 13.2. The logistic period doubling sequence.

13.2.4   CD States

An *intantaneous state* of a CD consists of a map of the lattice in its physical space, L, into the state space of its standard cell, S, assigning an local state to each node. From this data, one may compute the control parameters at each node via the coupling functions, and thus obtain a derived map from L into the space, CxS, of the response diagram of the standard cell. We regard this map, or perhaps its image (a finite point set within CxS), as the *global state* of the CD.

13.2.5   CD Simulation

Simulations of a CD are done as follows. In case the local dynamics of the standard cell are specified by differential equations, a time interval and fixed time-step integration algorithm are chosen. Thus, the continuous dynamics are replaced by a discrete time scheme. Alternatively, a variable time-step method may be used, but forced to report the trajectory at fixed (larger) time steps. This is the method of *periodic reportage*, commonly used to provide the Poincaré section of a periodic attractor.

Our recipe applies to the discrete-time case only. We begin with an initial, global CD state, and fixed values of any free control parameters. Then, parallel computations are performed at each node, with the local values of initial state and control parameters, to obtain the next local state at that node. This determines the next global state. Iteration of this global step produces a sequence of global states, the global trajectory. This approaches a global attractor, in the space of all global states.

13.2.6   CD Visualization

The behavior of a cellular dynamical system under simulation may be visualized by various methods, here we mention three. In *Zeeman's method*, an image of the lattice in physical space is projected into the response diagram of the standard cell, where it moves about, close (more or less) to the locus of attraction. (It is closer if the coupling is weak.) Alternatively, the behavior may be visualized by the *graph method*, in which we attach a separate copy of the standard response diagram to each cell of the location space. Within this product space, the instantaneous state of the model may be represented by a graph, showing the local state occupied by each cell, within its own response diagram. A related method of particular importance in our current work with two-dimensional CDs is the *isochron method*, in which isochrons (phase basins) are used to color the physical space. In any case, the behavior of the complete CD system may be tracked, as the controls of each cell are separately manipulated, through an understanding of the response diagram of the standard cell provided by dynamical systems theory, in terms of attractors, basins, separatrices, and their bifurcations [10]. We now give some examples of these methods.

13.3 An Example of Zeeman's Method

This method appeared first in a model by Zeeman for the human heart. Organs typically contain many different types of cells. In the unusual case that there is only one type of cell, one could imagine a model for the organ consisting of a single CD system. This is the case with Zeeman's heart model. An explicit CD model for the organ would require an explicit model for the standard cell, which might be found in the specialized literature devoted to that cell. In this case, Zeeman uses a well-known qualitative model, with unspecified coupling.

13.3.1 Zeeman's Heart Model: Standard Cell

In Zeeman's heart model, the standard cell is the cusp catastrophe. Each node corresponds to a muscle fiber, and the standard cell is a model for the muscle fiber dynamics. The two control parameters are muscle tension and the concentration of some transmitter chemical. The state variable is the length of the fiber. The upper sheet of the locus of attraction corresponds to an elongated state, the lower sheet to a contracted state.

13.3.2 Zeeman's Heart Model: Physical Space

Let us imagine a heart-shaped region in the plane as the physical space of the heart model. The spatial lattice of our CD model resides in this heart-shaped region, and its standard cell is the cusp. Rather than map this lattice into the space of the cusp, we will map the whole region. Assuming adiabatic conditions (weak coupling), the image moves close to the locus of attraction.

13.3.3 Zeeman's Heart Model: Beating

In Zeeman's model, the beating of the heart consists of a sliding motion of the image of the physical space over the fold catastrophe of the cusp, falling from elongated to contracted states, as shown in Fig. 13.3. Presumably, this would be the result of a simulation of his model. In any case, this figure illustrates his novel method of visualization, which may be applied successfully in many other cases.

13.4 The Graph Method

We proceed with our examples of visualization techniques, using the one-dimensional logistic lattice, the object of much recent research. First, we must present this object as a CD.

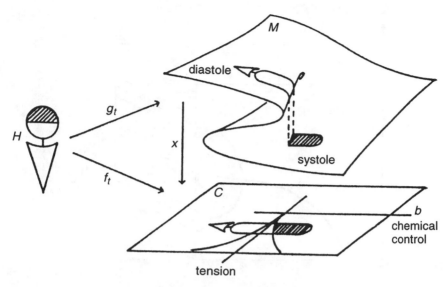

FIGURE 13.3. Zeeman's heart.

13.4.1   The Biased Logistic Scheme

The logistic map is the real-valued function of a single real variable,

$$f(x) = rx(1 - x). \tag{13.1}$$

Here, the control parameter, *gain*, r, should be in the range $[1, 4]$, so f may be regarded as a map of the closed unit interval to itself, $f : I \mapsto I$. It serves our purpose to add to this function a constant called the *bias*, c, which is regarded as a second control parameter. Thus,

$$f(x) = rx(1 - x) + c \tag{13.2}$$

If this is to be regarded as a function from I to itself also, we must add clipping:

$$\text{if } f(x) > 1, \text{ replace it by one,}$$

$$\text{if } f(x) < 0, \text{ replace it by zero.}$$

Thus, we have an iterated function scheme with two control parameters. The response diagram of this scheme is a simple extension of the period doubling sequence of Fig. 13.2, as shown in Fig. 13.4, at least if the bias is small. (In our example and discussion below, following section 3 of Crutchfield and Kaneko [3], the range of the bias will be within ± 0.001.)

13.4.2   The Logistic/Diffusion Lattice

In our application, the simple logistic/diffusion lattice, we will build a CD by coupling the bias parameter of each node to the states of the closest neigh-

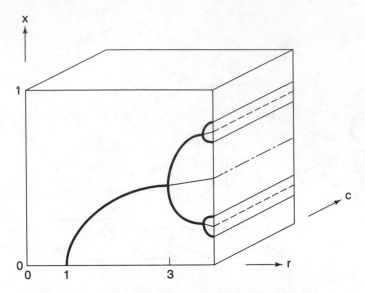

FIGURE 13.4. Response diagram of biased logistic scheme.

boring nodes by Laplacian coupling, leaving the gain free. The physical space will be the closed real interval, $I = [0, N - 1]$, where N is the number of nodes, equally spaced in the physical space. We will set N and r (at a common value for each node) at the start of any simulation. Below, we will use $N = 128$ and $r = 3.5$.

At the ith node, $i = 1 \cdots N - 2$, the next state, x_i^+, is given by

$$x_i^+ = rx_i^0(1 - x_i^0) + c(i), \qquad (13.3)$$

where x_i^0 denotes the current local state, and the the local bias is determined by

$$c(i) = \frac{\gamma}{2N^2}(x_{i+1}^0 + x_{i-1}^0 - 2x_i^0), \qquad (13.4)$$

and γ is set to a constant value at the start of a simulation. In our example, $\gamma = 4.8\ k$ or about 5000.

These formulas are applied at the endpoints, $i = 0, N - 1$, by special rules, called the *boundary conditions*. In our example, we will use the *toral* boundary conditions, in which the 0th and $N - 1$st nodes are identified. Thus $c(-1)$ is replaced by $c(N - 2)$ in Eq. (13.4) when $i = 0$, and so on.

13.4.3 The Global State Graph

Choosing the exemplary values mentioned above, we may now visualize the global states of this 1D/CD, the logistic/diffusion lattice, by the graph method.

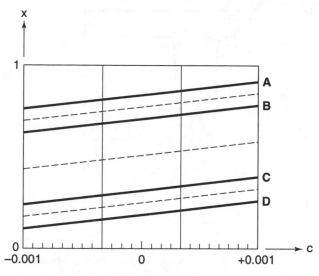

FIGURE 13.5. Response diagram, gain fixed at 3.5.

Fixing $r = 3.5$ at each node, we may reduce the response diagram shown in Fig. 13.4 to two dimensions, as shown in Fig. 13.5. The locus of attraction comprises four nearly straight lines, slightly rising to the right, corresponding to the four points of the periodic attractor. These are visited in the order: A, C, B, D. In between are the two curves of the Period two repellor, and the curve of the fixed repellor. All three curves are shown dashed. Attaching this 2D response diagram (visualized as a vertical plane) to each node in the 1D physical space (shown as a horizontal line) creates the 3D space in which the graph of a global state, $i \mapsto (x(i), c(i))$, may be plotted.

In Fig. 13.6 we show the graph of a state of the exemplary CD, taken from Crutchfield and Kaneko, [3, Fig. 2 (top)]. The projection into the plane of (i, x) is shown in Fig. 13.6a. The initial state leading to this global attractor is the sinewave (shown dashed). The global attractor, in this case, is a periodic trajectory of Period four. The periodic trajectory (superimposition of four successive global states) is shown by the lighter solid curves. One of the four instantaneous global states is shown as a heavier, solid curve. Note that although the coupling is weak, nodes in large regions of physical space are pulled far from their attractors by the combined influence of their nearest neighbors. In Fig. 13.6b is shown the projection of the heavy graph onto the (i, c) plane. In Fig. 13.6c is shown the projection onto the (c, x) plane. Note that this plane contains the reduced response diagram of the standard cell, as seen previously in Fig. 13.5. The graph of the instantaneous state is a curve, in the full 3D space of (i, c, x).

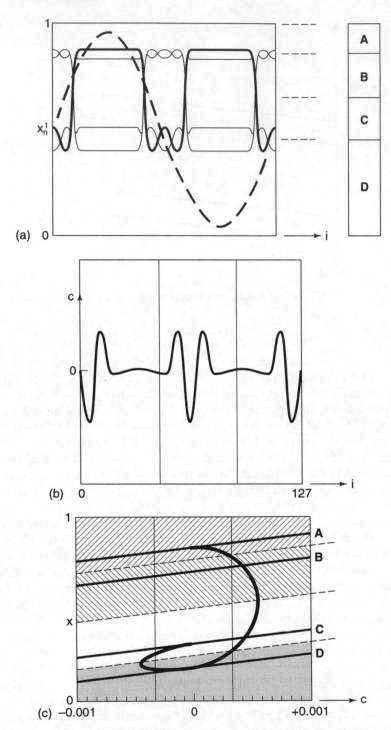

FIGURE 13.6. (a) Graph projected on (i, x) plane. (b) Graph projected on (i, c) plane. (c) Graph projected on (c, x) plane.

13.5   The Isochron Coloring Method

The graph method, for even slightly more complicated CDs, will require too many dimensions for direct visualization. Thus, we seek algorithms for pulling back the essential information from the graph to some kind of a map in the physical space. In our current example taken from Crutchfield and Kaneko [3], this will be a division of the physical space into interval regions. These could be colored for graphical presentation. The algorithm we are using currently in our research with 2D CDs is based on the projection of global states into the response diagram of the standard cell, as shown in Fig. 13.6c. It may be explained as follows.

13.5.1   Isochrons of a Periodic Attractor

Note that in the response plane, Fig. 13.5, the dashed curves (loci of the repellors, which are virtual separators) separate the plane into four strips. Each contains one of the four solid curves of the locus of attraction. These represent its phases: A, B, C, and D. Given an initial state of the biased logistic scheme in one of these phases, say A, the eventual state after many iterations (a multiple of four) will be phase A. These strips, which are basins of attraction of the four-fold iteration of the logistic map, are called *isochrons* of the response diagram. Only periodic attractors have isochrons, which constitute a decomposition of the basin of attraction of the periodic attractor into disjoint pieces. Periodic trajectories of continuous dynamical systems have isochrons also, but we will not discuss these here.

13.5.2   Coloring Strategies

A simple coloring strategy is used in Zeeman's method, Fig. 13.3. There, the lower sheet of the locus of attraction is colored black, the upper sheet, white. Note that nodes of the physical space, which are far from their attractors, are not colored by this rule. Thus, we would like to extend the rule outward from the loci of attraction, into the entire 3D response diagram of the standard cell. For control points inside the bifurcation set, the cusp curve in the horizontal control plane, this is easy, We use the locus of separation (the intermediate sheet of the cusp surface) to divide the wedge-shaped column into white (upper) and black (lower) regions. For points outside the bifurcation set, however, there is only one attractor, or sheet of the locus of attraction. We may use gray for this entire region, which is neither black nor white.

In the case of the logistic/diffusion lattice, we may color the four isochrons in different colors. Additional information may be encoded by varying the hue or saturation of a color according to the distance of the point from the attractor. This method is easily extended to the case of a CD with 2D physical space. We color the isochrons of the response diagram of the standard cell. Then, we map the physical space into the standard response diagram by the graph method, projected, and imagine the physical space moving around in the colored diagram as the simulation progresses. At each step, the colors encountered by the moving image are pulled back to the physical space, creating an animated color movie within it, *the isochron movie*. This is a substantially different movie than the one usually seen, in which the color indicates the values of the state variable, x, which we call *the state movie*. Another useful visualization is the *control movie*, in which the control space is colored, and these colors are pulled back to the physical space. A useful still picture view (for periodic global attractors) is the *period view*, in which each point of the physical lattice is colored according to the period of that node. This may be obtained from the isochron movie, and is analogous to the Fourier transform of a real function.

13.6   Conclusions

Here we have described three methods for obtaining animated color movies from a cellular dynamaton simulation; the familiar state movie method, and two new ones: the isochron and control movie methods. Suitable for periodic states primarily, the extension to some chaotic attractors having approximate isochrons, such as the Rössler attractor, may prove useful. Other color movie methods, based on entropy, Lyapunov exponents, spectral shape, symbolic dynamics, and so on, may be more appropriate for chaotic CDs [13]. The research results on experimental dynamics of physically 2D CDs will require video publications, such as the new aperiodical from Aerial Press, *Dynamics Showcase: a Hypermedia Journal*.

References

[1] R.H. Abraham, in *Mathematical Modelling in Science and Technology*, edited by X.J.R. Avula, R.E. Kalman, A.I. Leapis, and E.Y. Rodin, (Pergamon, Oxford, 1984).

[2] R.H. Abraham, in *Mathematics and Computers in Biomedical Applications, Proceedings IMACS World Congress, Oslo, 1985*, edited by J. Eisenfeld and C. DeLisi (North-Holland, Amsterdam, 1986).

[3] J.P. Crutchfield and K. Kaneko, in *Directions in Chaos*, edited by Hao Bai-Lin, (World Scientific, Singapore, 1987).

[4] R.H. Abraham, J.B. Corliss, and J.E. Dorband, Int. J. Chaos and Bifurcations **1**, 227 (1991).

[5] R.A. Fisher, *The Genetical Theory of Natural Selection*, (Dover, New York, 1930/1958).

[6] N. Rashevsky, *Mathematical Biophysics* (University of Chicago, Chicago, 1938).

[7] E.C. Zeeman, in *Catastrophe Theory*, edited by E.C. Zeeman (Addison-Wesley, Reading, 1977).

[8] E.C. Zeeman, in *Catastrophe Theory*, edited by E.C. Zeeman (Addison-Wesley, Reading, 1977).

[9] E.C. Zeeman, *Catastrophe Theory* (Addison-Wesley, Reading, 1977).

[10] R.H. Abraham and C.D. Shaw, *Dynamics, the Geometry of Behavior* four vols. (Aerial, Santa Cruz, 1982–1988).

[11] R.L. Devaney, *An Introduction to Chaotic Dynamic Systems*, 2nd ed. (Addison-Wesley, Reading, 1989).

[12] R.H. Abraham and A. Garfinkel, in *Ultradian Rhythms in Life Processes*, edited by D. Lloyd and E.L. Rossé (Springer-Verlag, New York, 1992).

[13] K. Kaneko, Physica D **37**, 60 (1989).

14

From Laminar Flow to Turbulence

Geoffrey K. Vallis

14.1 Preamble and Basic Ideas

Fluid dynamics in general, and turbulence in particular, is a most nonlinear field. The subjects are enormous; one need only browse through Monin and Yaglom [1] to see this. This chapter is not a review of turbulence, for in a few pages any attempt at an overview would certainly be ambitious and perhaps even foolish; rather, it is didactic introduction into just a couple of aspects, one of fairly recent origin and one with more traditional roots. We will first describe various routes involved in the *transition* to turbulence. This is a subject that has undergone a revolution in the past 20 or 30 years. It has greatly affected how we think about, although not how we calculate the properties of, our second topic of discussion, namely *fully developed*, or (especially in the Russian literature) *strong* turbulence. In fact, statistical theories of strong turbulence seem to be oblivious (perhaps with good reason) to theories and routes regarding transition. In both areas our discussion is rather selective, concentrating on the utility of scaling arguments and self-similarity. These properties allow the application of renormalization arguments in simple maps (e.g., the logistic map), enabling quantitative predictions to be made regarding the onset of chaos; remarkably, some of these have been verified in some real fluid dynamical experiments. Scaling arguments also lie at the heart of phenomenological theories of strong turbulence of the "Kolmogorov type," for which there is also observational support. We give rather short shrift to descriptions of experimental evidence and observations, simply referring the reader to the original literature, not because the area is not important but because it deserves much more space than we can give it here. In other ways, too, our discussion is incomplete, for nowhere do we discuss certain well-known prototypical problems such as the Lorenz equations, partly because such discussions abound elsewhere.

In turbulence, nonlinear interactions connect motions on different scales, leading to unpredictability on all scales of flow even when the error in the initial conditions is confined to small scales. This provides motivation (if any were needed) for our travails by way of a very practical question, for how long can we forecast the weather? The rapid loss of information in the smaller scales of motion suggests that one should attempt to treat the smaller scales *parametrically* in terms of the better-resolved and better-predicted large scales, and we will briefly discuss current approaches to this problem (the eddy diffusivity problem). Again, self-similarity and scaling allows the use of renormalization-type methods, in particular successive averaging methods, to rescale the diffusivity and obtain a parameterization of small scale motions. Space will forbid us exploring many of the side issues of these problems. Although the presentation is a little less abstract, and more overtly fluid dynamical than that often found in expositions to do with the onset of chaos, we do not approach the range of fluid dynamical problems discussed in various fluid dynamical texts (e.g. [2]). We have drawn on discussions by Hu and by Landau and Lifshitz [3] as well as original sources cited in the text. The reviews of Miles and of Eckmann, which cover similar ground in rather different styles, and of Rose and Sulem, are also recommended reading [4].

The equations describing fluid motion (14.1) are manifestly and nontrivially nonlinear, and historically much of the "early" (i.e., circa 1960–1975) development of nonlinear dynamics was in fact motivated by problems in fluid dynamics. Many of the applications of the modern theory of nonlinear dynamics (often loosely called "chaos theory") have been applied, with some success, to the theory of the it transition to turbulence (or more accurately the transition to chaos) where a few degrees of freedom might be expected to dominate the flow. A successful theory of strong turbulence has, on the other hand, proved very elusive, although the statistical approaches based on renormalized perturbation theories (beginning with the Direct Interaction Approximation [5]) have contributed a great deal. There is still much controversy about whether deterministic, "dynamical systems" approaches to turbulence are even appropriate, because of the relatively high dimensionality of fully turbulent flow. Furthermore, the parameter space over which low-dimensional, transitional behavior occurs is rather small compared to that over which fully turbulent behavior occurs. Thus, weak turbulence is the exception, rather than the rule, and the natural philosopher is led to ponder why she should study it at all. One answer, superficially trite, is that it is a field where true progress can and has been made; the investigator's hope, springing eternal, is that further investigation along the same lines may lead to substantial attacks on the theory of strong turbulence. In this chapter we first discuss the theory of transition, and then direct attention toward *statistical* attacks (based on Kolmogorovian phenomenology) into just a couple of aspects of strong turbulence: namely, the predictability problem and the eddy diffusivity problem.

14.1.1   What Is Turbulence?

The equations describing constant density flow phenomena are the Navier–Stokes equations; namely:

$$\frac{\partial \mathbf{u}}{\partial t} + (\mathbf{u} \cdot \nabla)\mathbf{u} = -\frac{1}{\rho}\nabla p + \nu\nabla^2\mathbf{u} \qquad (14.1)$$

along with the mass continuity constraint

$$\nabla \cdot \mathbf{u} = 0. \qquad (14.2)$$

Here, $\mathbf{u}(x, y, z, t)$ is the velocity field, p is the pressure, ρ the density (henceforth taken as unity), and ν the kinematic viscosity. These equations, with appropriate boundary conditions, determine the evolution of most incompressible fluid phenomena.

If U is a typical velocity magnitude, and L a typical length scale, then the Reynolds number, $Re = UL/\nu$ is a useful nondimensional measure of the ratio of the inertial terms to the viscous terms. Laminar flow may be defined as that flow for which the field variables $(\mathbf{u}, p)$ are time independent, or vary in a periodic way. In some particular flow geometry or experimental configuration, we may imagine being able to control the Reynolds number externally, perhaps by increasing the pressure difference along a pipe, or increasing the shear in a parallel flow as we discuss in section 14.2. As this is done, one typically finds that at some critical value the time-independent flow becomes unstable, and the flow bifurcates into some other configuration, perhaps a periodically varying flow. As Reynolds number increases further bifurcations occur, their precise nature and sequence depending on the flow at hand. After a small number of bifurcations the flow is frequently completely chaotic. Now, chaos is often taken as meaning the superficially random temporal behavior of a possibly small number of variables, as seen for example in the Lorenz equations or any number of well-known systems. A positive Lyapunov exponent is a common indicator of chaos. Fluid turbulence, in its usual sense, is also taken to imply *spatiotemporal* chaos, implicitly involving a larger number of degrees of freedom. That is, although strong turbulence almost certainly implies chaos, it is much more than that. Still, the distinction is sometimes *overemphasized*, because even fully developed turbulence has to all intents and purposes, a finite number of excited degrees of freedom and can be described as accurately as we wish by a system (albeit large) of ordinary differential equations. In any case, although the transition from "chaos" to "turbulence" in a fluid is not quantitatively understood, once chaos has arisen turbulence is thought to follow very shortly, or even immediately, thereafter. Because of the large number of degrees of freedom it then becomes sensible to seek a statistical approach in which a completely deterministic description of the fluid is foregone in favor of a description of averages. The reasons are twofold. First, a large number of degrees of freedom makes determinism extremely clumsy. Second, even if we could follow every eddy, the unpredictability of the fluid motion would make such knowledge useless, because the initial conditions of small scales of motion are not given. Thus, unless we were to perform Monte Carlo simulations with a range of initial conditions for the unobserved small scales, a statistical picture (or a description of mean quantities using some parameterization, or closure) is actually demanded. However, because of the nonlinearity of the equations this is very difficult; starting with any nonlinear equations of the from

$$\frac{da}{dt} = aa, \tag{14.3}$$

the equation for a mean quantity $\bar{a}$ is of the from

$$\frac{d\bar{a}}{dt} = \bar{a}\bar{a} + \overline{a'a'}, \tag{14.4}$$

where a prime denotes a deviation from the mean. The equation for $\overline{a'a'}$ involves triple correlations, those for triple correlations fourth order terms, and so on in an unclosed heirarchy. Earlier approaches to turbulence, such as the Direct Interaction Approximation [5], concentrated on "closing" this heirarchy, by the introduction of assumptions not directly deducible from the equations of motion, but without using the scaling properties of the Navier–Stokes equations described in Section 14.4. This turned out to be a very difficult road to follow, and in fact may be demanding too much. More recent approaches have *started* with the scaling properties, and built from there. Whichever, if either, approach turns out to reveal more, it is fair to say that a completely satisfactory statistical description of turbulence at this time does not exist.

14.2   From Laminar Flow to Nonlinear Equilibration

The Navier–Stokes equations are tremendously verdent, and depending on the boundary conditions or the nature of the fluid a myriad of flows are possible, each being controlled by various nondimensional parameters. For adiabatic, incompressible flow (as, for example, in water flow in a pipe) the Reynolds number is the controlling parameter in a given geometry. In problems in convection (usually fluid in a constant gravitational field heated from below, or cooled from above) a relevant control parameter is the Rayleigh number, $g\alpha\Delta T d^3/(\nu\kappa)$, where g is gravity, α measures the thermal expansion, d is a length scale, ΔT a typical temperature difference, and ν and κ are inertial and thermal diffusivities, respectively. In rotating fluids the Ekman number, $\nu/(2\Omega L^2)$, where Ω is the rate of rotation, and the Rossby number, $U/(2\Omega L)$, determine the relative importance of friction and rotation. In different geometries other parameters may play a role. Given this rich structure, it would be very surprising if there were only one "route to chaos" in fluids, and indeed there is not. However, it does seem that there may be a small number (perhaps in single digits) of recognizably different transition scenarios. By scenario we mean a qualitatively distinct path (sequence) with certain characteristic behavior; we do not mean necessarily identical behavior in each case. Now, each sequence often begins with a *linear instability*, in which a steady (i.e., time independent) flow satisfying the Navier–Stokes equations becomes unstable. Before plunging into nonlinear analysis, let us analyze a simple linear instability.

14.2.1   A Linear Analysis: The Kelvin–Helmholtz Instability

The simplest instance of a linear instability, which is also of some physical interest, is perhaps the Kelvin–Helmholtz instability. This is a shear instability, in which two fluids sliding against one another, or one fluid with a strong shear perpendicular to its mean velocity, become unstable. We will consider the simplest case, that of two fluids of equal density, with a common surface at $z = 0$, moving with velocities $-U$ and $+U$ in the x-direction, respectively (Fig. 14.1). There is no variation in the basic flow in the y-direction, and we will assume this is also true for the instability. This flow is clearly a solution of the inviscid Navier–Stokes equations (the Euler equations). The question to be asked is, what happens if the flow is perturbed slightly? We will neglect variations in the y-direction (these are unnessential to the instability) and, for the moment, also neglect friction. If the perturbation is initially small, then even if it grows we can, for small times after the onset of instability, neglect the nonlinear interactions in the governing equations because these are the squares of small quantities. The equations determining the evolution of the initial perturbation are then the Navier–Stoke equations (14.1), linearized about the steady solution. Thus, for $z > 0$

$$\frac{\partial \mathbf{u}'}{\partial t} + U \frac{\partial \mathbf{u}'}{\partial x} = -\nabla p', \tag{14.5a}$$

$$\nabla \cdot \mathbf{u}' = 0, \tag{14.5b}$$

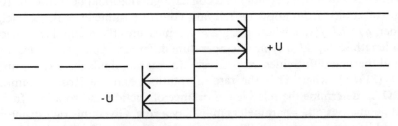

FIGURE 14.1. Basic flow giving rise to Kelvin–Helmholtz instability. For the problem as solved in the text, the boundaries are removed to infinity.

with a similar equation but with U replaced by $-U$ for $z < 0$. (The density is unity, so does not appear.)

We can represent the perturbations by a Fourier expansion of the from

$$\phi(x, z, t) = \sum_k \hat{\phi}_k(z, t) \exp[ikx], \tag{14.6}$$

where ϕ is any field variable (pressure or velocity). Because (14.5) is linear, the Fourier modes do not interact and we confine attention to just one. Taking the divergence of Eq. (14.5a), the left-hand side vanishes and the pressure satisfies Laplaces equation

$$\nabla^2 p' = 0. \tag{14.7}$$

For $z > 0$, this has solutions in the form

$$p' = \hat{p}_1 e^{ikx + \theta t} e^{-kz}, \tag{14.8}$$

where we have also introduced the explicit time dependence $e^{\theta t}$. In general θ is complex; if the soon-to-be-found dispersion relationship gives θ with a positive real component, we have an instability. Any imaginary component gives oscillatory motion. To obtain the dispersion relationship, we consider the z-component of (14.5), namely (for $z > 0$)

$$\frac{\partial w_1'}{\partial t} + U \frac{\partial w_1'}{\partial x} = -\frac{\partial p_1'}{\partial z}. \tag{14.9}$$

Substituting a solution of the form $w_1' = \hat{w}_1 \exp(ikx + \theta t)$ yields, with Eq. (14.8),

$$(\theta + ikU)\hat{w}_1 = k\hat{p}_1. \tag{14.10}$$

But the velocity normal to the discontinuity is, at the discontinuity, nothing but the rate of change of the discontinuity itself. That is, at the interface $z = +0$

$$w_1 = \frac{\partial \zeta}{\partial t} + U \frac{\partial \zeta}{\partial x}, \tag{14.11}$$

or

$$(\theta + ikU)\hat{\zeta} = \hat{w}_1. \tag{14.12}$$

Using this in Eq. (14.10) gives

$$(\theta + ikU)^2 \hat{\zeta} = k\hat{p}_1. \tag{14.13}$$

The above few equations pertain to motion on the $z > 0$ side of the interface. Similar reasoning on the other side gives (at $z = -0$)

$$(\theta - ikU)^2 \hat{\zeta} = -k\hat{p}_2. \tag{14.14}$$

But, at the interface $p_1 = p_2$ (because pressure must be continuous). The dispersion relationship then emerges from Eq. (14.13) and (14.14), giving

$$\theta^2 = k^2 U^2. \tag{14.15}$$

Thus the flow is *unstable* and the amplitude of any small perturbation will initially grow exponentially. The instability itself can be seen in the natural world when billow clouds appear wrapped up into spirals: the clouds are acting as tracers of fluid flow, indicating a shear in the atmosphere. The interested reader should peruse Ludlum's wonderful book *Clouds and Storms* [6].

In our ideal example the instability appears immediately; that is, no matter how small the shear. The reason for this is the neglect of viscosity, so that the Reynolds number is always infinite. If viscosity is retained, the analysis becomes a little more involved, in part because the basic state cannot have a velocity discontinuity. Most approaches involve allowing the velocity shear to vary smoothly. The instability now sets in only when the shear is sufficiently large; that is, when an appropriate Reynolds number reaches some critical value. Suppose that the initial value of the shear *is* slightly supercritical. Then the linear analysis tells us a perturbation will grow exponentially. But of course a real instability (if only on energetic grounds) cannot keep growing. In fact the perturbation will eventually reach finite amplitude, at which point a nonlinear analysis is necessary.

14.2.2   A Weakly Nonlinear Analysis: Landau's Equation

In this section our approach starts to get more abstract, and we deviate from a strict analysis of the governing equations. To discuss this it is helpful to know what a Hopf bifurcation is (Figs. 14.2 and 14.3). (For more detailed descriptions of bifurcations in hydrodynamics, see for example the article by Joseph in Swinney and Gollub [2].) Suppose that the linearization of a system (about a solution) has complex eigenvalues $\{\sigma_i\}$, such that the evolution of the system is proportional to $\exp[\sigma_i t]$, whose values depend on some parameter of the system, say R (e.g., the Rayleigh or Reynolds numbers in fluid dynamics). For $R < R_c$, where R_c is some critical value, suppose the eigenvalues lie in the left half of the complex plane. Then if the system is perturbed, it will damp back to equilibrium in an oscillatory fashion. If at $R = R_c$ one complex conjugate pair crosses the imaginary axis, then the system becomes unstable, and an oscillatory growth takes place on perturbation. A Hopf bifurcation has occurred. (Engineers sometimes call the ensuing kind of instability an overstability). Landau was among the first people to consider how a fluid might equilibrate after such a bifurcation; his argument may be paraphrased as follows. For values of R close to R_c and for small times after the onset of instability, the amplitude of the flow may be expected to behave something like

$$A(t) = C(t)e^{\theta t}, \tag{14.16}$$

R<R$_c$ R>R$_c$

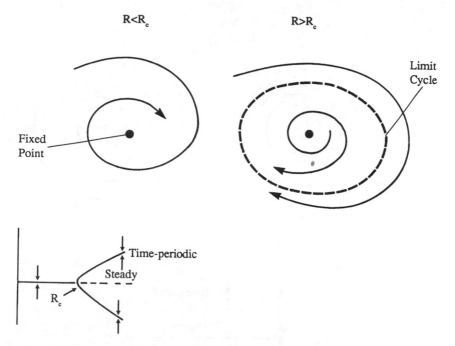

FIGURE 14.2. The supercritical Hopf bifurcation. (A positive Landau constant.) After the bifurcation at R_c the flow converges to a stable limit cycle.

where A is a measure of the amplitude of some field variable (it might, for example, be a Galerkin component). A is in general complex; implicit as usual is the addition of a complex conjugate to ultimately obtain a real field. θ is also complex and may be written $\theta = \sigma + i\omega$. We expect that $\sigma = f(R - R_c)$ and that for $R < R_c$, $\sigma < 0$. That is, the flow is stable and a small perturbation would be damped to zero. The parameter $C(t)$ will, for small times, be a constant, and the amplitude will exponentially grow. The differential equation satisfied by (14.16) is then

$$\frac{d|A|^2}{dt} = 2\sigma|A|^2. \tag{14.17}$$

For longer times this equation cannot be valid, and higher-order terms must enter. Now, if the flow is just supercritical, the growth rate, σ, can be expected to be small and in particular $\sigma \ll |\omega|$. Thus, the flow undergoes an oscillatory instability (see Fig. 14.4). We are interested in its behavior on the timescale $1/\sigma$, averaged over many oscillations. Adding in higher-order terms to Eq. (14.17)

$$\frac{d|A|^2}{dt} = 2\sigma|A|^2 + a_1|A|^2A + a_2|A|^4 \ldots$$

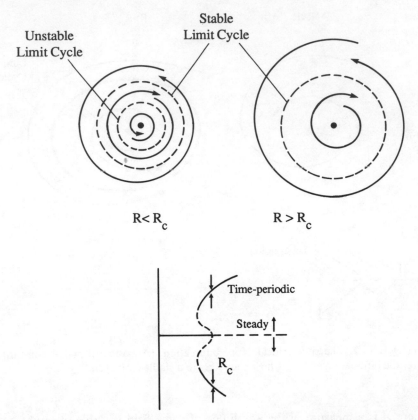

FIGURE 14.3. The subcritical Hopf bifurcation. The outer (stable) limit cycle exists only in the presence of higher-order nonlinearities, absent in our treatment of the Landau equation.

Averaging over the oscillatory time period, the third-order terms will vanish (or actually give a fourth-order contribution) because they contain an oscillatory term. We are left with the fourth-order term, and the Landau equation:

$$\frac{d|A|^2}{dt} = 2\sigma|A|^2 - \alpha|A|^4, \tag{14.18}$$

where α is an undetermined parameter (the Landau constant), which may be positive or negative.

The method of derivation of Eq. (14.18) is really just an application of the method of averaging. The observation that multiple time scales are present (i.e., the slow growth rate and the fast oscillation) suggests one might explicitly employ a multiple time-scale analysis to actually derive the Landau equation, directly from a prototype equation for hydrodynamical and other instabilities. This we shall now do, although our prototype is rather simple and not explicitly hydrodynamical.

Consider, then, the simple system:

$$\frac{dx}{dt} = \sigma x - \omega y + f_1(x, y) \tag{14.19}$$

$$\frac{dy}{dt} = +\omega x + \sigma y + f_2(x, y). \tag{14.20}$$

The function f_1 and f_2 contain nonlinear terms, and are small and analytic but otherwise arbitrary. These prototype equations therefore contain growth, an oscillation, and nonlinearity, many of the features of a real fluid insta- bility. We shall recognize that growth is slow by writing $\varepsilon = \sigma/\omega \ll 1$. Divid- ing through by ω, defining a new (nondimensional) time by ωt, and writing $z = x + iy$ we obtain:

$$\frac{dz}{dt} = \varepsilon z + iz + \varepsilon g(z), \tag{14.21}$$

where we now explicitly recognize the smallness of the nonlinear term $\varepsilon g(z)$. There is a similar equation for z^*. The analysis now proceeds by introducing

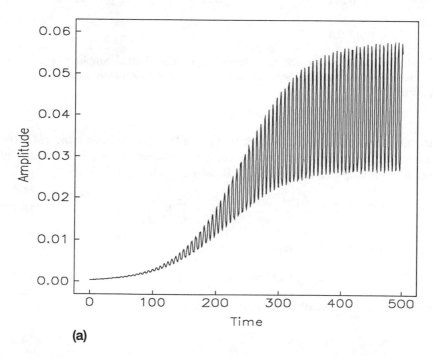

(a)

FIGURE 14.4. (a) Amplitude growth obtained by numerically integrating Eqs. (14.19) and (14.20).

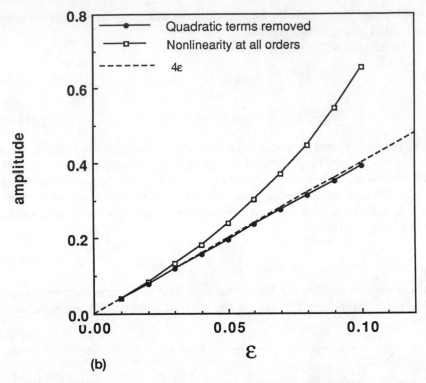

(b)

FIGURE 14.4 *(continued)*. (b) Time-averaged equilibrated amplitudes obtained numerically (dots and open squares) compared to prediction from Landau equation (dashed line).

the *slow time* $\tau = \varepsilon t$ and assuming that z is a function of both t and τ. (If the reader is not familiar with this device, it is explained further in a number of books on asymptotic methods; e.g., Bender and Orszag [7].) Thus,

$$\frac{dz}{dt} = \frac{dz}{dt} + \frac{dz}{d\tau}\frac{d\tau}{dt} = \frac{dz}{dt} + \varepsilon\frac{dz}{d\tau}. \qquad (14.22)$$

Then we expand z in the series:

$$z(t, \tau) = z_0(t, \tau) + \varepsilon z_1(t, \tau)\ldots \qquad (14.23)$$

Using Eq. (14.22) we obtain, to lowest order

$$\frac{dz_0}{dt} = iz_0, \qquad (14.24)$$

which has solutions

$$z_0 = C(\tau)e^{it}. \tag{14.25}$$

This of course is just the fast oscillation. To obtain the expression for the slow change in amplitude of z, we go to next order and obtain:

$$\frac{dz_0}{d\tau} + \frac{dz_1}{dt} = z_0 + iz_1 + g(z_0). \tag{14.26}$$

Because z is small, we assume that we may expand the function $g(z_0)$ in powers of z_0, noting that linear terms are already taken care of, whence Eq. (14.26) becomes

$$\frac{dz_1}{dt} - iz_1 = z_0 - \frac{dz_0}{d\tau} + O(z_0^2) + O(z_0^3)\ldots \tag{14.27}$$

or, explicitly in terms of x and y

$$\frac{dx_1}{dt} + y_1 = x_0 - \frac{dx_0}{d\tau} + O(x_0^2, y_0^2) + O(x_0^3, y_0^3)\ldots \tag{14.28a}$$

$$\frac{dy_1}{dt} - x_1 = y_0 - \frac{dy_0}{d\tau} + O(x_0^2, y_0^2) + O(x_0^3, y_0^3)\ldots \tag{14.28b}$$

Now, by assumption εz_1 is smaller than z_0. Thus, for our analysis to be consistent (and consistency, not uniqueness, is all that can reasonably be demanded of an asymptotic analysis) there can be no secular terms (i.e., terms that grow linearly with time) in the form of z_1. Because a solution of the homogeneous equation $dz_1/dt - iz_1 = 0$ is $z \propto e^{it}$, secular terms are eliminated if terms proportional to it vanish on the right-hand side, because secular terms arise when an oscillator is forced at its natural, unperturbed, frequency. The quadratic terms do not contribute to this, but the cubic terms do [this is evidently also seen to be true for Eqs. (14.28a) and (14.28b)], and so secular terms are avoided if

$$\frac{dz_0}{d\tau} = z_0 + a_2|z_0|^2 z_0, \tag{14.29}$$

where a_2 is determined by the projection of the cubic terms proportional to $(\exp[it])^3$ onto $\exp[it]$. Eq. (14.29) leads directly to the Landau equation, and is itself sometimes called a Landau equation.

To be more explicit, consider a particular choice of nonlinear function $f_1(x, y) = f_2(x, y) = \exp(x + y) - (1 + (x + y))$. (These arbitrarily chosen functions have nonlinearity at all orders. Any number of other functions would give similar behavior, and there is no necessity to choose $f_1 = f_2$, although it makes the algebra a little easier.) Letting $x(t) = x_0(t, \tau) + \varepsilon x_1(t, \tau)\ldots$, and similarly for y, the *zero*th order balance is:

$$\frac{dx_0}{dt} = -y_0$$

$$\frac{dy_0}{dt} = x_0 \tag{14.30}$$

with solution $x_0 = X_0(\tau)\cos t$, $y_0 = Y_0(\tau)\sin t$ [cf. Eq. (14.25)]. At next order Eq. (14.28) is

$$\frac{dx_1}{dt} + y_1 = \left(X_0 - \frac{dX_0}{d\tau}\right)\cos t - \frac{1}{2}(X_0\cos t + Y_0\sin t)^2$$

$$- \frac{1}{3!}(X_0\cos t + Y_0\sin t)^3 \dots \qquad (14.31)$$

with a similar equation for $Y_0(\tau)$. To avoid secular terms in the solution for $x_1(t)$ we demand that all terms proportional to $\cos t$ or $\sin t$ on the right-hand side of Eq. (14.31) (or its y equivalent) vanish, again to avoid a resonance with the terms on the left-hand side. The quadratic terms do not project onto $\cos t$ or $\sin t$ but the cubic terms do. Evaluating the projection integrals (which are of the form $(2\pi)^{-1}\int_0^{2\pi}\cos^2 t\sin^2 t\,dt$) and restoring the dimensions (noting that $\tau = \varepsilon t$) we obtain, after a little algebra, the equation for the slowly varying amplitude

$$\frac{dr^2}{dt} = 2\varepsilon r^2 - \frac{1}{2}r^4, \qquad (14.32)$$

where $r^2(\tau) = (X_0^2(\tau) + X_0^2(\tau))/2$. Equation (14.32) is the Landau equation for this problem. The Landau constant is positive. The growth of the amplitude $x^2 + y^2$ is initially exponential, before equilibrating, and is at all times modified by a fast oscillation. Figure 14.4 shows the equilibrated amplitude from a direct numerical integration of Eqs. (14.19) and (14.20), along with the predicted time-averaged equilibrated amplitude $r^2 = 4\varepsilon$. Quantitative agreement is obtained for small ε values.

We still have not derived a Landau equation from the Navier–Stokes equations. For a real-world problem this is generally an arduous task. Indeed, it was 15 years after Landau first proposed the equation that Stuart and Watson [8] were first able to properly derive a Landau equation for a fluid stability problem in plane parallel flow.

If α is positive then it is clear from Eq. (14.18) that supercritical equilibration can readily be achieved. The solution of Eq. (14.18) is obtained by writing it as a linear equation in $|A|^2$, namely

$$\frac{d|A|^{-2}}{dt} + 2\sigma|A|^{-2} = \alpha, \qquad (14.33)$$

which leads to

$$|A|^2 = A_0^2 \left/ \left\{\frac{\alpha}{2\sigma}A_0^2 + \left(1 - \frac{\alpha}{2\sigma}A_0^2\right)e^{-2\sigma t}\right\}\right., \qquad (14.34)$$

where A_0 is its initial value. For large times $|A|^2$ tends to the limit

$$|A|^2 = 2\sigma/\alpha. \qquad (14.35)$$

The growth rate σ will be some function of the Reynolds number (or other controlling parameter), and because we have no reason to expect the first-order terms to vanish, we have that $\sigma = \text{constant} \times (R - R_c) + O(R - R_c)^2$. Hence, the amplitude of the equilibrated value of $|A|$ is proportional to $\sqrt{R - R_c}$. For $R < R_c$ the flow is stable. The bifurcation at $R = R_c$ is a supercritical Hopf bifurcation; it is a bifurcation from a stable fixed point to a limit cycle. For our model problem, Eq. (14.19), direct numerical solution of the original equations confirms the predictions of the weakly nonlinear theory. The averaged value of the equilibrated amplitude squared is, as predicted, a *linear function* of the growth rate ε. Furthermore, the dependence on the cubic term in the nonlinear function f_1 is much stronger than the dependence on the quadratic terms: if the quadratic terms are eliminated the equilibrated amplitude is very similar (Fig. 14.4), whereas if the cubic terms are eliminated, now retaining the second-order terms, the equilibrated values (not shown) are in face quite different.

If α is negative, then for $R < R_c$ there is a stable fixed point surrounded by an unstable limit cycle. The flow is thus metastable, because infinitesimal perturbations are damped back to the focus, but finite size perturbations (which exceed the radius of the limit cycle) are unstable. For $R > R_c$ there is no steady flow (at least to fourth order) because both terms on the right-hand side of Eq. (14.18) are positive and $|A|$ increases very rapidly; higher-order terms must be included for equilibration. In fact, we may hypothesize that after such a "subcritical" Hopf bifurcation the transition to turbulence is very rapid. (Indeed in the Lorenz equations the transition to chaos occurs after a single subcritical Hopf bifurcation [4].)

14.3   From Nonlinear Equilibration to Weak Turbulence

Suppose that the control parameter has been turned up past a first instability and equilibration has occurred, roughly obeying the Landau equation. (Actually, even at this stage, detailed equilibration mechanisms differ. See, for example, Pedlosky [8]). We now have a limit cycle. The question arises: what happens if the control parameter R is increased further. Common experience tells us that, even after a supercritical bifurcation, the limit cycle does not persist (if only turbulence were so simple). One possible scenario, now generally thought false, is the Landau–Hopf picture. After the first bifurcation, the flow is periodic with a frequency f_1 say, and hence is a single point on a Poincaré map. In principle it is possible to perform a stability analysis about this (time-dependent) flow, in much the same way as one performs a stability analysis about a stationary flow (except it is harder). Now, as R increases the periodic motion increases in amplitude, and, it may be supposed, the flow will eventually become unstable (say at R_2). If this instability is caused by a Hopf bifurcation to another limit cycle, then a second frequency, f_2, will be present. There is no reason that f_2 and f_1 be related, so for

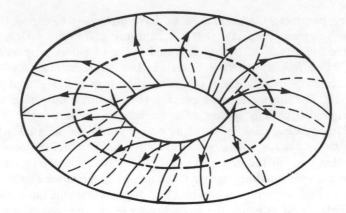

FIGURE 14.5. Motion on a torus. If the path round the torus does not close, the motion is quasiperiodic.

almost all cases f_1/f_2 will be irrational. The flow is then *quasiperiodic* on a torus (Fig. 14.5). Why quasiperiodic? If the flow due to the first frequency is $\phi \sim \exp[if_1t]$ then the phase of this component will return to its initial value after a time t such that $f_1t = 2\pi m$ where m is any integer. Similarly the phase of the second flow will return to its initial value after a time $t = 2\pi n/f_2$, where n is another integer. The flow will thus have an overall period $T = 2\pi m/f_1 = 2\pi n/f_2$, requiring $f_1/f_2 = m/n$. But if f_1 and f_2 are irrationally related, no such integers exist. Thus the flow has infinite period. But we can approximate f_1/f_2 as accurately as we wish by the ratio of two integers, so to any desired degree of accuracy the flow may be regarded as periodic.

A third bifurcation will produce another frequency, and so on. Eventually, Landau supposed, after many bifurcations a "turbulent flow," comprising many independent and irrationally related frequencies, arises. Such a picture is false, on both experimental and theoretical grounds. We shall now discuss why, and what in fact does happen. We will discuss only three transition sequences—although more may exist. We will pay most attention to the period-doubling route, partly in the interests of space and partly because the scaling properties are quite transparent here.

14.3.1   The Quasi-Periodic Sequence

The Landau sequence may be schematized as

$$\begin{pmatrix} \text{Steady} \\ \text{flow} \end{pmatrix} \Rightarrow \begin{pmatrix} \text{Periodic} \\ \text{Flow} \end{pmatrix} \Rightarrow \begin{pmatrix} \text{Periodic 2} \\ \text{flow} \end{pmatrix} \Rightarrow \begin{pmatrix} \text{Periodic 3} \\ \text{flow} \end{pmatrix} \Rightarrow \begin{pmatrix} \text{Periodic 4} \\ \text{flow} \end{pmatrix} \cdots,$$

where the arrows denote Hopf bifurcations. Now, a quasiperiodic flow is predictable, meaning it does not display sensitive dependence on initial conditions. This is because quasiperiodic flow is just flow on a torus, and although flow on a torus may be ergodic (i.e., it explores all of the available phase space) the flow is not mixing (two orbits initially close together explore the torus together) and is not unpredictable. However, turbulent flow is known to be unpredictable, and was known to be so before "chaos theory," excepting the work by Poincaré, was discovered [9]. (We discuss predictability properties later in this chapter.) However, it is fair to say that the reasons for its unpredictability were not at that time fully understood. Thus, the Landau picture is not supported by observation or experiment. Theoretically, too, it turns out that quasiperiodic motion is unstable to small perturbations of the equations of motion. Two phenomena occur. First, the addition of a small amount of nonlinearity may lead to frequency locking, in which the independent frequencies become truly rationally related. Second, for larger values of a nonlinearity parameter, the surface of the torus may crinkle, and a strange attractor may arise. This is the "Ruelle–Takens" or "quasiperiodic" route to chaos [10], and the sequence may be characterized as

$$\begin{pmatrix} \text{Steady} \\ \text{Flow} \end{pmatrix} \Rightarrow \begin{pmatrix} \text{Periodic} \\ \text{Flow} \end{pmatrix} \Rightarrow \begin{pmatrix} \text{Periodic 2} \\ \text{flow} \end{pmatrix} \left(\Rightarrow \begin{pmatrix} \text{Periodic 3} \\ \text{flow} \end{pmatrix} \right) \Rightarrow \begin{pmatrix} \text{Strange} \\ \text{Attractor} \end{pmatrix}.$$

The precise number of Hopf bifurcations before chaos emerges is probably not important. The important point is that after a small number of bifurcations a strange attractor generically emerges.

To examine these phenomena, and to keep the analysis tractable, it is unfortunately necessary to keep only the vestiges of the Navier–Stokes equations. The price is that the approach is less deductive than we like; the reward is universality. The simplest model displaying quasiperiodicity is probably the circle map—the map of the circle onto itself

$$\theta_{n+1} = \theta_n + \Omega - \frac{K}{2\pi} \sin 2\pi\theta_n \quad \text{(modulo 1)}. \tag{14.36}$$

If the nonlinearity parameter K is zero then the map displays only two kinds of behavior, periodic for Ω rational and quasiperiodic for Ω irrational. Rational numbers can by definition be expressed as the ratio of two integers, p/q. Although dense (i.e., we can approximate any number arbitrarily closely by a rational) they form a set of zero measure on the real line—which means, loosely, that if we pick a number at random its chances of being rational are zero. Thus, periodic behavior for this map is nongeneric. However, if we add a small amount of nonlinearity, then frequency locking occurs and for a set of Ω of nonzero measure (indicated by the shaded areas in Fig. 14.6) periodic behavior occurs. (Note that for some numbers the quasiperiodic regime persists longer than for other; these numbers are "more irrational," in the sense of the continued fraction representation. The golden mean $(\sqrt{5}-1)/2 = 0.618\ldots$ is the most irrational number in this sense, having entries that are all unity.) For values of K greater than unity chaotic motion may arise.

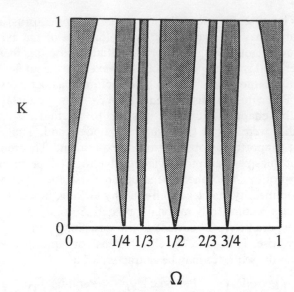

$$\Omega$$

FIGURE 14.6. Frequency locking in the circle map (schema). The shaded regions indicate periodic behavior, for which the parameter region grows as nonlinearity increases. For $K > 1$ chaotic motion may ensue.

14.3.2   The Period Doubling Sequence

This is based on the pitchfork bifurcation. In a pitchfork bifurcation a periodic orbit (which can be replaced by a fixed point using a Poincaré map) is replaced by a periodic orbit of twice the period. A great deal of progress can be made by studying one-dimensional maps of the form:

$$x_{n+1} = f(x_n), \tag{14.37}$$

where $f(x)$ is an analytic function of x. Perhaps the most well known of these is the logistic map

$$f(x) = rx_n(1 - x_n), \tag{14.38}$$

where r plays the role of control parameter. We will study this map is some detail, to get a feeling for the mechanisms of period doubling. After that we will study the universal scaling features of this and similar maps, and apply some simple renormalization group arguments. First we note some general properties of the map Eq. (14.37). The second iterate of the map yields

$$x_{n+2} = f(f(x_n)) = f_2(x_n). \tag{14.39}$$

The nth iterate will be denoted $f_n(x)$. Fixed points of Eq. (14.37) are found by solving

$$x^* = f(x^*).$$

Stability of these is then determined by examining small perturbations around the fixed point. Let

$$x_n = x^* + \varepsilon_n.$$

Then

$$\begin{aligned} x_{n+1} = f(x_n) &= f(x^* + \varepsilon_n) \\ &= f(x^*) + \varepsilon_n f'(x^*) = x^* + \varepsilon_n f'(x^*) \\ &= x^* + \varepsilon_{n+1} \quad \text{(say)}. \end{aligned} \tag{14.40}$$

The error grows if $|\varepsilon_{n+1}/\varepsilon_n| > 1$; thus, the condition for instability is that

$$|f'(x^*)| > 1. \tag{14.41}$$

If a fixed point at x_0 is unstable, then all higher iterates are also unstable at x_0. This is because

$$f_2'(x_0) = f'(x_0)f'(f(x_0)) = f'(x_0)f'(x_1), \tag{14.42}$$

and at the fixed point $x_1 = x_0$, and so

$$f_2'(x^*) = |f'(x^*)|^2. \tag{14.43}$$

In general the slope of the nth iterate is given by a simple extension of Eq. (14.42), namely

$$f_n' = \prod_{i=0}^{n-1} f'(x_i). \tag{14.44}$$

Now, for specificity, consider the logistic map Eq. (14.37). Graphically, the situation is illustrated in Fig. 14.7. The solid curve displays the function Eq. (14.38) and the dashed line is of unit slope (the identity line). To obtain successive iterates pick an initial value, x_0, along the abscissae and move parallel to the ordinate until the function curve is intersected; the ordinate value then gives x_1. Then move horizontally to meet the dashed line, and then vertically again to meet the solid curve, to obtain the next value in the iteration sequence and so on. If $0 < r < 4$ then the function always maps the interval $\{0, 1\}$ to itself. Because the map is quadratic, and therefore with single extremum, there are up to two fixed points, at the origin $x = 0$ and (for $r > 1$) at $x = 1 - 1/r$. These occur where the function crosses the identity line. Their stability is determined by the value of f', and we have

(i) Origin ($x^* = 0$); $f' = r$; and
(ii) $x^* = 1 - 1/r$; $f' = 2 - r$.

For $0 < r < 1$ the origin, being the only fixed point, is stable. Beyond $r = 1$ the fixed point loses its stability and the other fixed point emerges, and for all initial conditions the flow will eventually converge to this fixed point. This

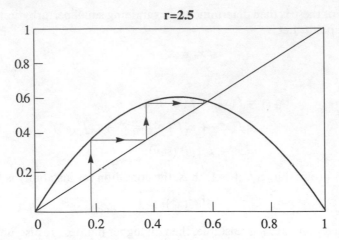

FIGURE 14.7. The logistic map for $r = 2 \cdot 5$. Note the convergence to the stable fixed point from any initial condition.

point itself becomes unstable when

$$|2 - r| > 1 \Rightarrow r = 3. \tag{14.45}$$

Thus for $r > 3$ there are no stable fixed points. What happens?

The answer to this is that the flow bifurcates from a fixed point to a periodic flow (a Period 1 flow). To see this consider the map obtained by iterating Eq. (14.38),

$$f_2(x) = f(f(x)) = r^2 x (1 - x)(1 - rx(1 - x))$$
$$= r^2 x - (1 + r)r^2 x^2 + 2r^3 x^3 - r^3 x^4. \tag{14.46}$$

Its precise form will not turn out to be too important. However, we note that it is a quartic, with three extrema, symmetrical about and with a minimum at $x = 1/2$. [The positions of the other two minima can be obtained by an application of Eq. (14.42). Because

$$f_2'(x_0) = f'(x_0)f'(x_1), \tag{14.47}$$

there is a zero at x_0 when the point to which it iterates has zero slope. Thus, the points that iterate to $x = 1/2$, that is, $f^{-1}(1/2)$, are maxima of f_2.]

Now, the fixed point of f loses its stability as r increases through $r = R_1 = 3$, with $x = 2/3$. The stability is lost in f_2 also [by Eq. (14.42) or (14.44)]. However, as can be seen from Fig. 14.8, two new (stable) fixed points in f_2 emerge. This is the first pitchfork bifurcation, so-called because the values of the fixed points as r is increased look like a pitchfork (Fig. 14.9). This bifurcation has given rise to the phenomena of period doubling, and the

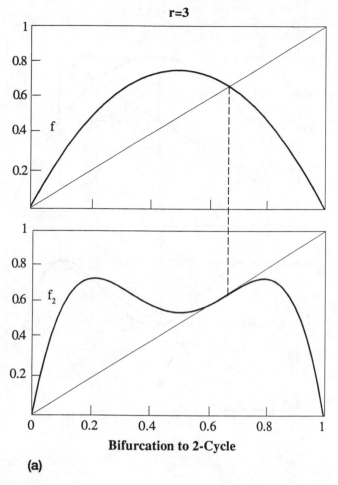

FIGURE 14.8. Period doubling in the logistic map. (a) For $r > 3$ the fixed point of f is unstable.

two new fixed points form a Period 2 flow, or a "2-cycle." The slopes of these two point are the same. We can see this because the actual iteration oscillates between them:

$$x_1 = f(x_0)$$

$$x_0 = f(x_1). \tag{14.48}$$

Thus, using Eq. (14.47),

$$f_2'(x_0) = f'(x_0)f'(x_1) = f_2'(x_1). \tag{14.49}$$

Indeed in general if the set of points $\{x_i\}$ forms an n-cycle such that for each i

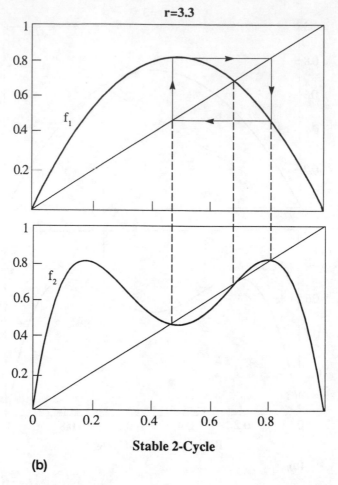

r=3.3

Stable 2-Cycle

(b)

FIGURE 14.8 *(continued)*. (b) For $r = 3.3$ the 2-cycle is stable.

$$x_i^* = f_n(x_i^*) \tag{14.50}$$

then each fixed point has the same slope in the f_n map, given by

$$f_n' = \prod_{i=1}^{n} f'(x_i^*). \tag{14.51}$$

For r marginally bigger than R_1, the two new fixed points are very close together, and consequently their slopes are less than unity and they are stable. However, as r increases they move further apart, passing through a so-called superstable cycle when the slopes of the slopes of the fixed points of f_2 are zero. At a particular value of r, in fact at $r = R_2 = (1 + \sqrt{6}) = 3.4495$, their slopes become greater in magnitude than -1 and another pitchfork bifurcation gives rise to four new stable fixed points, and Period 4 flow. Note that the

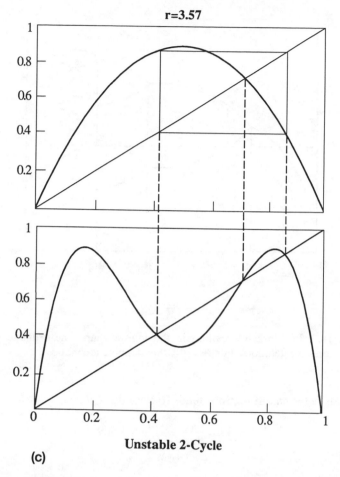

(c)

FIGURE 14.8 *(continued)*. (c) For $r > 1 + \sqrt{6}$ the 2-cycle is unstable. For $r = 3.57$ (illustrated) the absolute value of the slope of f_2 is clearly greater than unity.

pitchfork bifurcations occur simultaneously because it is a single trajectory that is being split, and the slopes at each of the fixed points are the same. These new fixed points eventually become unstable and bifurcate to a Period 8 flow, and so on in a period-doubling sequence. The successive bifurcations occur closer and closer together, in a geometric progression, and eventually accumulate at some particular value of r, denoted R_∞, whose value is 3.5699456 for the logistic map. For $r > R_\infty$ the iteration sequence is chaotic.

Scaling and Universality

If one numerically integrates the logistic map and notes the values of r at which successive bifurcations occur (R_1, R_2, etc.) then one finds that the ratio

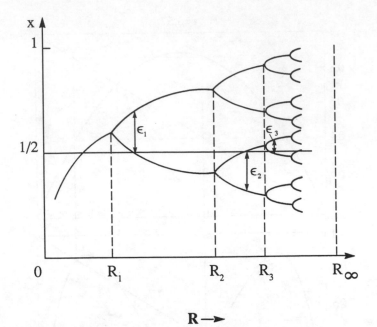

FIGURE 14.9. Pitchfork bifurcations in the logistic map. The ordinate gives the values of successive iterations. For $R > R_\infty$ the mapping is chaotic.

of intervals between bifurcations tends to a limit:

$$\delta = \lim_{n \to \infty} \frac{R_n - R_{n-1}}{R_{n+1} - R_n} = 4.6692. \qquad (14.52a)$$

Because

$$\lim_{n \to \infty} \frac{R_\infty - R_n}{R_\infty - R_{n+1}} = \lim_{n \to \infty} \frac{R_\infty - R_{n-1}}{R_\infty - R_n},$$

then Eq. (14.52a) is equivalent to

$$\delta = \lim_{n \to \infty} \frac{R_\infty - R_n}{R_\infty - R_{n+1}} = 4.6692. \qquad (14.52b)$$

Furthermore, the relative scale of branch splittings (see Fig. 14.9) is also universal:

$$\alpha = \lim_{n \to \infty} \frac{\varepsilon_n}{\varepsilon_{i+1}} = 2.503. \qquad (14.53)$$

Here α essentially measures the reduction in scale that follows each bifurcation. These parameters are *universal*, in the sense that their values do not depend on the particular map (except that it be quadratic) and the ratios hold (in the limit $n \to \infty$) for every bifurcation. Let us try to understand why this should be, before trying to deduce their values.

It is convenient to transform the logistic map Eq. (14.38) first by the shift $x \Rightarrow x + 1/2$ and then by $x \Rightarrow x(r/2 - 1)/2$. These then yield the map

$$x_{n+1} = 1 - \lambda x_n^2. \tag{14.54}$$

where

$$\lambda = \frac{r}{2}\left(\frac{r}{2} - 1\right). \tag{14.55}$$

The map is centered at $x = 0$ and is such that if $0 < \lambda < 2$ then $-1 < x < 1$. The fixed points of this are determined by $\lambda x^2 + x - 1 = 0$, which become unstable [by Eq. (14.41)] when $2\lambda x = 1$, giving $\lambda = 3/4$ (and $x = 2/3$). Thus the first bifurcation occurs at $\lambda = \Lambda_1 = 3/4$. Iterating the transformation gives

$$f_2 = x_{n+2} = 1 - \lambda + 2\lambda^2 x_n^2 - \lambda^3 x^4. \tag{14.56}$$

Neglecting the quartic term, and rescaling by $x \Rightarrow x/\alpha$ we obtain

$$x_{n+2} = 1 + 2\lambda^2(1 - \lambda)x_n^2 = 1 - \lambda_1 x_n^2, \tag{14.57}$$

which is of the same form as Eq. (14.54) but with

$$\lambda_1 = \phi(\lambda) = 2\lambda^2(\lambda - 1). \tag{14.58}$$

Successive iterations of this map then yield

$$x_{n+2^m} = 1 - \lambda_m x_n^2 \qquad \lambda_m = \phi(\lambda_{m-1}). \tag{14.59}$$

Now, the first bifurcation occurs when $\lambda = \Lambda_1 = 3/4$; the next bifurcation (denoted by $\lambda = \Lambda_2$) occurs when $\lambda_1 = 3/4$; that is, when $\phi(\Lambda_2) = \Lambda_1$. The bifurcation after that occurs when $\phi(\phi(\Lambda_3)) = \Lambda_1$; that is, when $\phi(\Lambda_3) = (\Lambda_2)$. Thus, the sequence of bifurcations is calculated by the sequence:

$$\Lambda_1 = \tfrac{3}{4}, \quad \phi(\Lambda_2) = \Lambda_1, \quad \phi(\Lambda_3) = \Lambda_2 \ldots \tag{14.60}$$

These are easily calculated to be at 0.75, 1.2428, 1.3440, 1.3622, 1.3654, 1.3659, 1.3660, ... The bifurcations *accumulate* at the fixed point given by $\Lambda_\infty = \phi(\Lambda_\infty)$ giving $\Lambda_\infty = (1 + \sqrt{3})/2 = 1.3660$. This is to be compared to the exact value 1.4011 obtained by a direct solution of Eq. (14.54). [Using Eq. (14.55), we obtain the accumulation point for the map in its more common form (14.38), $R_\infty = 3.54246$, whereas the exact value is 3.56994]. The value of Λ_∞ is not universal, being dependent on how the map is specifically defined. However, the scale factors α_m tend in the limit $n \to \infty$ to $\alpha_m = 1/(1 - \Lambda_\infty) = -2.73$ (cf. the exact value $\alpha = -2.5029$). We can also estimate the ratio of

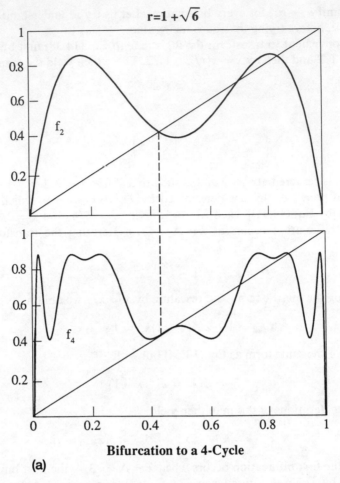

$$r = 1 + \sqrt{6}$$

Bifurcation to a 4-Cycle

(a)

FIGURE 14.10. From Period 2 to Period 4. (a) At $r = 1 + \sqrt{6}$ the 2-cycle becomes unstable.

bifurcation intervals, as follows. Using Eq. (14.60) in the limit $m \to \infty$ we have

$$
\begin{aligned}
\Lambda_\infty - \Lambda_m &= \Lambda_\infty - \phi(\Lambda_{m+1}) \\
&\approx \Lambda_\infty - \{\phi'(\Lambda_\infty)(\Lambda_{m+1} - \Lambda_\infty) + \phi(\Lambda_\infty)\} \\
&= (\Lambda_\infty - \Lambda_{m+1})\phi'(\Lambda_\infty)
\end{aligned}
\tag{14.61}
$$

Hence

$$
\delta = \phi'(\Lambda_\infty) = \lim_{m \to \infty} \frac{\Lambda_\infty - \Lambda_m}{\Lambda_\infty - \Lambda_{m+1}}.
\tag{14.62}
$$

For our simple approximations, this gives the value $\delta = 4 + \sqrt{3} = 5.732$, which is within 25% of the exact value. It is becoming clearer now why these

r=3.57

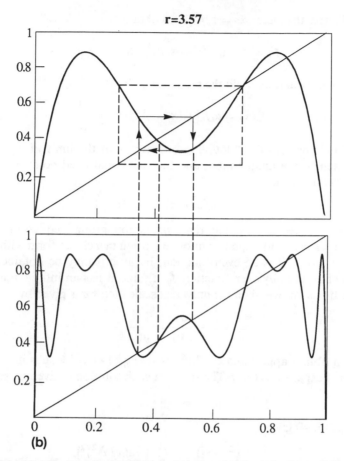

(b)

FIGURE 14.10 *(Continued)*. (b) The 4-cycle is unstable (as indeed are all higher iterates) at $r = 3.57$. Note that the inverse of the center portion (dashed) of (b) is very similar to the entirety of Fig. 14.8 (c). This is a characteristic of universality.

values are universal (see also Fig. 14.10). After each bifurcation the values of the iterated map f_m in the range $[-1, 1]$ are determined by the values of f_m in a part of the range smaller by a factor α. After many iterations, the determination of the iterated function is determined by the initial function closer and closer to its maximum. If the maximum of f_0 (the initial map) is quadratic, this fact alone suffices to determine α and δ.

The General Doubling Transformation

Rather than truncate the map after each bifurcation to keep its quadratic form, we will try now to be a little more general. Consider the mapping $f_1(x; \lambda)$ such that $f(0) = 1$. Equation (14.54) is an example of this. Then the iterated map is $f_2 = f(f(x; \lambda))$. However, the previous discussion suggests we rescale with the factor $\alpha_1 = 1/f_1(1)$ to obtain

$$f_2(x) = \alpha_1 f_1[f_1(x/\alpha_1)], \quad \alpha_1 = \frac{1}{f_1(1)}.$$

Further transformations yield the sequence

$$f_{n+1}(x) = Tf_n(x) = \alpha_n f_n[f_n(x/\alpha_n)], \quad \alpha_n = \frac{1}{f_n(1)}, \tag{14.63}$$

where T is known as the "doubling operator." In the limit of $n \to \infty$ this sequence tends to a unique limit, and the functional fixed point of the doubling operator satisfies

$$g(x) = \alpha g[g(x/\alpha)], \quad g(0) = 1. \tag{14.64}$$

This is a so-called "functional renormalisation group" equation. Such a function does exist, although it cannot be written in a closed form with a finite number of terms. It is an even function, because it may be obtained as an iteration of an initially even function $f_0(x)$, and it has an infinite number of extrema. If g is known, then the universal scale factor α is given by

$$\alpha = \frac{g(0)}{g[g(0)]} = \frac{1}{g(1)}. \tag{14.65}$$

We can obtain approximate solutions to Eq (14.64) by substituting even polynomial expansions for g. The simplest such solution is given by setting

$$g(x) = 1 + mx^2. \tag{14.66}$$

Then Eq. (14.64) gives

$$1 + mx^2 = \alpha[1 + m(1 + m(x/\alpha)^2)^2]. \tag{14.67}$$

Equating powers of x gives $\alpha = 1/(1 + m)$ and $\alpha = 2m$ yielding the numerical values

$$m = -(1 + \sqrt{3})/2 = -1.366, \quad \alpha = (1 + \sqrt{3}). \tag{14.68}$$

Going to the next order, we set $g(x) = 1 + mx^2 + nx^4$. Substituting into Eq. (14.64) we obtain

$$m = -1.52224, \quad n = 0.127613, \quad \alpha = -2.53404. \tag{14.69}$$

These are quite close to the exact values [11]:

$$g(x) = 1 - 1.52763x^2 + 0.104815x^4 - 0.0267057x^6 \dots$$
$$\alpha = -2.5029 \dots \tag{14.70}$$

[To obtain a polynomial expansion it is easiest just to iterate the sequence (14.63), rather than to assume a polynomial expansion with unknown coefficients and substitute in (14.69) to obtain their values.] Finally, we may obtain accurate values of δ by linearizing around the universal function $g(x)$. Close to the fixed point we may write

$$g_\varepsilon = g(x) + \varepsilon h(x). \tag{14.71}$$

Then

$$g_\varepsilon(g_\varepsilon(x)) = g_\varepsilon(g(x)) + \varepsilon h(x))$$
$$\approx g(g(x)) + \varepsilon g'(g(x))h(x) + h(gx))]. \tag{14.72}$$

And the universal equation for δ is [11]:

$$g'(g(x)h(x) + h(g(x)) = -\frac{\delta}{\alpha}h(\alpha x). \tag{14.73}$$

Because $g(x)$ and α are already known, we can solve this to any desired accuracy by substituting a polynomial expansion, to obtain $\delta = 4.66920\ldots$

Summarizing the period-doubling sequence we have then

$$\begin{pmatrix} \text{Steady} \\ \text{flow} \end{pmatrix} \Rightarrow \begin{pmatrix} \text{Periodic} \\ \text{Flow} \end{pmatrix} \rightarrow \begin{pmatrix} \text{Period 2} \\ \text{flow} \end{pmatrix} \rightarrow \begin{pmatrix} \text{Period 4} \\ \text{flow} \end{pmatrix} \rightarrow \begin{pmatrix} \text{Strange} \\ \text{Attractor} \end{pmatrix} \cdots,$$

where the first bifurcation is typically of the Hopf type, with subsequent bifurcations being pitchfork or period-doubling.

There is one other sequence we shall discuss, albeit rather briefly, before trying to figure out what all this has to do with fluids.

14.3.3   The Intermittent Sequence

For this sequence we return to the one dimensional map. Suppose that the an appropriate map (which governs the orbit) has, in some region, two intersections with the 45° line (Fig. 14.11). In this diagram the left-most intersection is an attracting fixed point and the right-most one is a repellor. As our control parameter varies, suppose the map effectively moves to the left, and undergoes a *tangent bifurcation*. When the map is just tangent, the intersection is an attracting fixed point that immediately disappears just after the bifurcation. Consider now some initial conditions far from x_0, just after the tangent bifurcation, but still in its prior "basin of attraction." The orbit will still converge toward x_0, because it does not realize yet that the fixed point is no more. The system then spends many iterations close to the pseudo-fixed point, before wandering away. The tangent bifurcation has replaced an attracting fixed point by a strange attractor, but one in which the system spends a lot of time near the pseudo-fixed point. This route is called the intermittent sequence because for many iterations the map is close to the pseudo-fixed point, undergoing motion that is very close to periodic—remember that at the fixed point motion is periodic, having been reduced to a point via a Poincaré map. The system will occasionally wander away from the pseudo-fixed point, undergoing truly chaotic (broad-band) motion, or intermittent

(a)

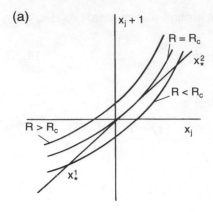

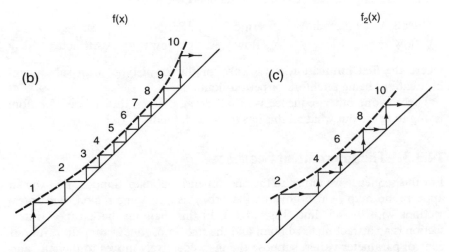

FIGURE 14.11. A tangent bifurcation. Two fixed points merge and then disappear, giving rise to intermittent chaos. The similarity of f_1 and f_2 give rise to universality.

bursts of turbulence. The bifurcation sequence may be summarized as

$$\begin{pmatrix} \text{Steady} \\ \text{flow} \end{pmatrix} \Rightarrow \begin{pmatrix} \text{Periodic} \\ \text{Flow} \end{pmatrix} \rightarrow \begin{pmatrix} \text{Intermittent} \\ \text{chaotic flow} \end{pmatrix},$$

where the bifurcation to the strange attractor is the tangent bifurcation.

This sequence too displays universal behavior [12]. Close to the tangent fixed point, the function $f(x)$ and its iterate $f(f(x))$ are rather similar, except that the steps in the iterated map are twice as long. Let us express this self-similarity by the equation

$$g(x) = \alpha g(g(x/\alpha)), \tag{14.74}$$

where $\alpha = 2$. This differs from Eq. (14.64) in that an extra tangency condition must be imposed; namely, $g'(0) = 1$ as well as $g(0) = 0$. We now find

$$g(x) = \frac{x}{1 - rx}, \tag{14.75}$$

where r is arbitrary. The parameter convergence ratio can be found in a similar way to that used in the period doubling case, yielding $\delta = 4$. Now, if the control parameter R increases such that we move $1/\delta = 1/4$ time closer to the fixed point $(R = R_c)$, the system stays twice as long (because $\alpha = 2$) near to the pseudo-fixed point. Thus, the average length of periodic motion τ_p scales as

$$\tau_p \sim \frac{1}{(R - R_c)^{1/2}}, \tag{14.76}$$

and therefore decreases as the supercriticality increases.

Another way to see this result is as follows. Near the critical point $R = R_c$ we may suppose that the mapping function can be approximated by

$$x_{n+1} = (R - R_c) + x_n + x_n^2, \tag{14.77}$$

which we approximate by

$$\frac{dx}{dt} = R - R_c + x^2. \tag{14.78}$$

Integrating this between two points on either side of the pseudo-fixed point yields the time taken for the passage between those two points, namely

$$\tau_p = \frac{1}{(R - R_c)^{1/2}} \left[\tan^{-1}(x/(R - R_c)^{1/2}]_{x_1}^{x_2} \right.$$

$$\sim \frac{1}{(R - R_c)^{1/2}}. \tag{14.79}$$

14.3.4   Fluid Relevance and Experimental Evidence

It is perhaps remarkable that anything in the above few sections has anything at all to do with fluids, yet a number of experiments and simulations have reproduced various of these sequences. Let's first try to figure out why this should be, before mentioning the actual experiments.

The obvious first question to ask is: When and why does a fluid behave like a one-dimensional map? Now, a fluid, although multidimensional, is dissipative; the Navier–Stokes equations are of the form

$$\frac{\partial \mathbf{u}}{\partial t} + \text{Nonlinear and Pressure Terms} = \nu \nabla^2 \mathbf{u}.$$

This means its phase space volume shrinks (for example, for each Fourier component of the velocity field, u_K, we have that $\partial \dot{u}_k / \partial u_k < 0$). If the flow is chaotic, the flow is stretching and folding in some directions, while shrinking in others. Thus, the dimensionality of the flow is constantly reduced until it is on its attractor. Because the attractor has been produced in this complex way it is both thin and complicated. Prior to chaos, its dimension may be quite small (order unity), and hence in some instances ideas from mapping theory may apply [13]. This qualitative, but probably essentially correct, argument nevertheless cannot be proven to hold (say by a series of rational approximations of the Navier–Stokes equation) in general for a given arbitrary fluid dynamical instability. Thus, there are many unknowns, and it is unclear precisely when any particular transition sequence will occur.

It is the *universality* of the sequences that gives rise to their robustness. For example, the period-doubling sequence of Feigenbaum is by no means unique to the logistic map. Most one-dimensional maps with a hump in the middle will give quantitatively the same behavior (this is why the behavior is called universal). What are the experimental signatures of a period-doubling cascade [14]? Suppose we heat a liquid from below in a low aspect ratio container. If the heating is too small, a balance between heating and heat diffusion is stable, and there is no motion. As the Rayleigh number increases (by making the heating more intense), convection begins, and if the geometry is appropriate two convective rolls appear. A probe measuring temperature at any given point would still, however, give a constant reading, and hence the flow corresponds to an attracting fixed point. As the temperature of the lower surface is slowly increased (by slowly we mean that at each value of the temperature the fluid is allowed to come to an equilibrium), then at some critical value the rolls become unstable to a wave propagating along the roll axis. At any particular Rayleigh number in this regime a probe in the fluid would reveal a single sinusoid; a power spectrum would yield a single frequency f, or single period T. The trajectory of the system in its phase space is a single closed loop (Fig. 14.12). Further increasing the temperature contrast, a new oscillatory mode (wave) appears, superimposed on the original one. Examining the phase space trajectory would reveal a double loop, which only exactly repeats after two cycles. A Fourier analysis would reveal the presence of a second sinusoid, at *double the original period*. This is the first period doubling. How does this look like a pitchfork? Well, for each value of the control parameter (e.g., the Rayleigh number) plot each maximum and minimum value of the temperature. For a single loop there are two such values, for a double loop two maxima and two minima and so on. When plotted this way, each period doubling bifurcation looks like a pitchfork. Note that the energy in the second mode is less than that in the original (note that from a distance the phase space trajectory still looks like a single loop), just as the amplitude of the branch splittings gets smaller with each successive bifurcation on the one-dimensional maps.

Increasing the temperature a little further, the trajectory splits again. Another period doubling bifurcation has occurred, and the phase space trajectory turns into a quadruple loop. The Fourier spectrum shows the domi-

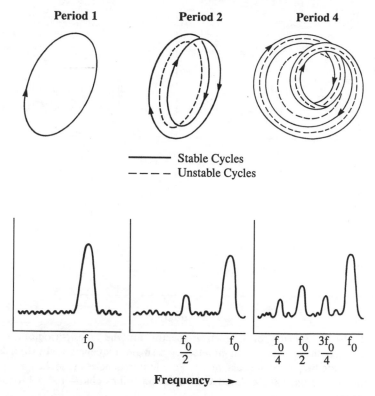

FIGURE 14.12. Period doubling. The upper row illustrates the trajectories in phase space, as the period doubles and then quadruples. The lower row is a schema of the corresponding frequency spectra.

nant scale is still at f, with a weaker contribution at $f/2$ (the doubled period) and a still weaker contribution at $f/4$ and a $3f/4$ harmonic. Ideally, we would be able to see still more period doublings, but it is very hard to discern them, as they they get closer and closer together as a function of Rayleigh number, just as the theory predicts. Period doubling has also been obtained in numerical experiments in two-dimensional convection [15].

The quasiperiodic route has also been observed in a number of hydro-dynamical experiments. The experimentalist's goal here is to increase the control parameter slowly, and at closely spaced values obtain a frequency spectra (Fig. 14.13). The signature of this route would then be the appearance of one, two, and perhaps even three or four independent frequencies, after which the flow becomes chaotic and the spectral signature is broad-band. In the experiments of Gollub and Benson [14] two apparently incommensurate frequencies arose, f_1 and f_2, followed by frequency locking with $f_1/f_2 \approx 9/4$. Broadband spectra, a typical signature (although not a proof) of chaos then appeared, with f_1 and f_2 still very visible. This was followed by strongly

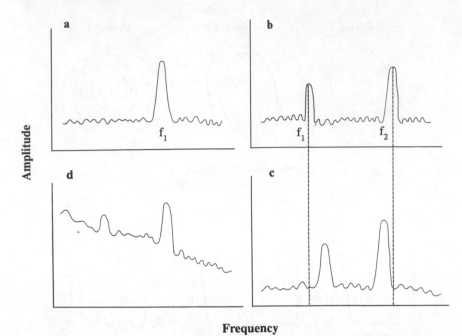

Frequency

FIGURE 14.13. Schema of frequency spectra for the quasiperiodic transition sequence. In (a) the spectrum is dominated by a single frequency, f_1. In (b) a second incommensurate frequency appears, and in Panel (c) freqencies are locked, by a small shift from their initial values. Another bifurcation brings chaos and a broad-band spectra, in (d).

chaotic flow, with the initial frequencies largely subsumed by the turbulence. Of course it may very difficult for the experimentalist to differentiate this sequence from the appearance of more and more independent frequencies (quasiperiodic flow), which in the presence of a small amount of experimental noise may also have a broad-band spectrum, and very careful experimentation is needed to overcome this. Finally, intermittency has also been observed in experiments [16, 17]. Figure 14.14 schematically illustrates a typical time-series. Note that the intermittency predicted by this Pomeau–Manneville mechanism has no obvious connection with the intermittency observed in strong turbulence, although of course this is not to say one does not exist.

The experimental support for these routes is fairly conclusive evidence for the existence of chaos and strange attractors in weak turbulence. It would be surprising indeed if, after passing through one of a number of fairly well-understood transition sequences, which all imply chaos with all its ramifications, the subsequent passage to strong turbulence occurred via some sequence through which the chaos were removed. Although this is not a proof that chaos exists in turbulence, the question now of much more interest to the physicist is: What good does it do me knowing there is a strange attractor in turbulence?

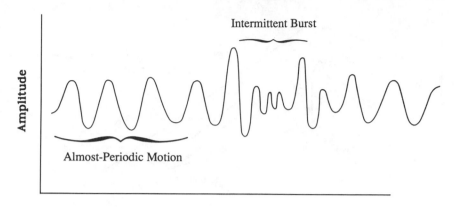

FIGURE 14.14. Schema of an intermittent time series. Periodic flows, when the flow orbits close to a pseudo-fixed point, are separated by bursts of turbulence.

14.4 Strong Turbulence

14.4.1 Scaling Arguments for Inertial Ranges

We will skip over the little-understood area of the transition from low-order chaos to fully developed turbulence, and for the remainder of the chapter focus on fully developed or strong turbulence. Much of the "modern" (say post-1940) theories of this area have foundations made of the scaling arguments of Kolmogorov [18]. For simplicity consider homogeneous, isotropic turbulence in a fluid of constant, unit, density. Suppose that energy is input into the fluid at some length scale L_I, by a stirring process, and a typical resulting velocity is U. The ratio of the inertial terms to the nonlinear terms is the Reynolds number $Re = UL_I/\nu$. If this is large (and just by stirring vigorously we can make it large), there is no effective means of removing energy at the input scale. We may nevertheless expect there to exist some much smaller scale, say $L_D \ll L_I$, at which the Reynolds number (based on L_D and the velocity at that scale) is close to unity, and hence for there to be energy removal at that scale. Thus, energy must be transferred from the larger scales to the smaller scales, for which nonlinearity is necessary. If the stirring is vigorous enough, the input and dissipation scales will be spectrally far removed, and there will exist a range of intermediate scales for which neither stirring or dissipation explicitly is important. This assumption, known as the locality hypothesis, depends on the nonlinear transfer of energy being sufficiently local (in spectral space). Given this, this intermediate range is known as the inertial range. If the stirring produces an energy source of magnitude ε then in a statistically steady state the flux of energy through the inertial range and the dissipation must both equal ε (see Fig. 14.15). All the dynamical fields may be

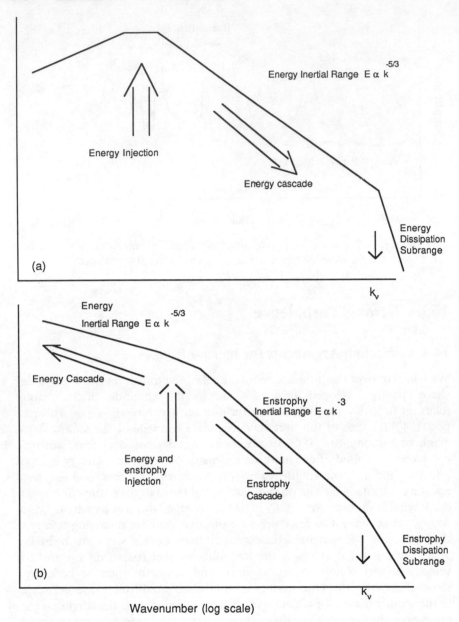

FIGURE 14.15. The putative energy and enstrophy cascades in three-dimensional (a) and (b) two-dimensional turbulence. The ordinate is energy (log scale) and abscissae wave number. The various subranges in reality blend smoothly together.

Fourier decomposed, and denoting the amplitude of the wavenumber by k, we may define an energy spectrum $\mathscr{E}(k)$ such that that the total energy $E = \frac{1}{2} \int \mathbf{u}^2 \, d\mathbf{x}$ (remember the density is unity) is given by

$$E = \int_0^\infty \mathscr{E}(k) dk. \tag{14.80}$$

We will also denote a velocity magnitude at a scale $l \sim 1/k$ by v_l or $v(k)$, so that $\mathscr{E}(k) \sim v(k)/k$. This is also a useful stage at which to note that Eqs. (14.1) and (14.2), in the absence of viscosity, conserve the total energy. Simple scaling arguments will now be used to give a relationship between $\mathscr{E}(k)$ and ε.

In the inertial range, the energy transfer is constant and cannot depend explicitly on wave number. The energy spectrum cannot depend on the particular stirring and dissipation processes, because the energy transfer is local. Thus, the energy spectrum $\mathscr{E}(k)$ is a universal function of ε and wavenumber (what else is there?). To obtain the actual functional relationship, it is convenient to define an eddy turnover time $\tau(k)$, which is the time taken for a parcel with energy $\mathscr{E}(k)$ to move a distance $1/k$. Thus,

$$\tau(k) = (k^3 \mathscr{E}(k))^{-1/2}. \tag{14.81}$$

Kolmogorov's assumptions are then equivalent to setting

$$\varepsilon \sim \frac{k\mathscr{E}(k)}{\tau(k)}, \tag{14.82}$$

which, because we demand that ε be constant, yields the famous law:

$$\mathscr{E}(k) = \mathscr{K}\varepsilon^{2/3}k^{-5/3}, \tag{14.83}$$

where $\mathscr{K}$ is a universal, hopefully order one, constant. This spectral form has been verified many times observationally, the first time using some very high Reynolds number oceanographic observations [19].

The scaling relationship [Eq. (14.83)], as well as some other useful scaling relationships, can be obtained in a slightly different, but essentially equivalent, way as follows. If we for the moment ignore viscosity, the equation of motion (14.1) is invariant under the following scaling transformation:

$$x \Rightarrow x\lambda \qquad v \Rightarrow v\lambda^r \qquad t \Rightarrow t\lambda^{1-r}, \tag{14.84}$$

where r is an arbitrary scaling exponent. So far there is no physics. Now make the following physical assumptions: First we make the locality hypothesis, namely that the energy flux through a wavenumber k depends only on local quantities [namely, the wavenumber itself and the energy $\mathscr{E}(k)$ or velocity $v(k)$]. Second, the flux of energy from large to small scales is assumed finite and constant. Third we assume that the scale invariance (14.84) holds, on a time-average, in the intermediate scales between the forcing scales and dissipation scales. This is likely to be strictly valid only in the limit of infinite

Reynolds number, but for finite Reynolds number it is made plausible by the locality hypothesis. [It is important to note that the infinite Reynolds number limit *is* a limit, and is different from simply neglecting the viscous term in Eq. (14.1), which gives the so-called Euler equations. This is because, as we shall see, this term contributes even in the zero-viscosity limit.] The time average in practice need be no longer than a few longest eddy turnover times, and depending on how local the energy transfer actually is we do not need an infinite Reynolds number for the scaling to be valid in the inertial range.

Dimensional analysis then tells us that the energy flux scales as

$$\varepsilon \sim \frac{v^3}{l} \sim \lambda^{3r-1}, \tag{14.85}$$

from which the assumed constancy of ε gives $r = 1/3$. This has a number of interesting consequences.

The velocity scales as $v \sim \varepsilon^{1/3} k^{-1/3}$. The velocity gradient scales as $\nabla v \sim \varepsilon^{1/3} k^{2/3}$, as does the vorticity $\omega = \nabla \times \mathbf{v}$. These quantities thus blow up (i.e., become infinite) at very small scales, but this is in fact avoided by a viscous cut-off.

We can now recover Eq. (14.83) easily, because dimensionally

$$\mathscr{E} \sim v^2 k^{-1} \sim \varepsilon^{2/3} k^{-2/3} k^{-1} \sim \varepsilon^{2/3} k^{-5/3},$$

which is Eq. (14.83).

The structure functions S_m of order m, which are the average of the m'th power of the velocity difference over distances $l \sim 1/k$, scale as $(\delta v_l)^m \sim \varepsilon^{m/3 l m/3} \sim \varepsilon^{m/3} k^{-m/3}$. In particular the second-order structure function, which is the fourier transform of the energy spectra, scales as $S_2 \sim \varepsilon^{2/3} k^{-2/3}$.

The viscous effects effects become important at a range given by equating the viscous and inertial terms in Eq. (14.1); that is

$$\nu k^2 v \sim k v^2,$$

which yields

$$k_v \sim \left(\frac{\varepsilon}{\nu^3}\right)^{1/4}. \tag{14.86}$$

The scale $l_v \sim k_v^{-1}$ is called the *Kolmogorov scale*. In the limit of viscosity tending to zero, l_v tends to zero, but the energy dissipation, perhaps amazingly, does not. The energy dissipation is given by

$$\dot{E} = \int \nu \mathbf{v} \cdot \nabla^2 \mathbf{v} \, d\mathbf{x}. \tag{14.87}$$

Because the length at which dissipation acts is the Kolmogorov scale, this expression scales as (for a box of unit volume)

$$\dot{E} \sim \nu k_v^2 v^2 \sim \nu \frac{\varepsilon^{2/3}}{k_v^{2/3}} k^2 \sim \varepsilon, \tag{14.88}$$

with $k = k_v$. Hence, energy dissipation apparently does not depend on the viscosity at all! This result is actually quite consistent with the whole picture. Energy is input at some large scales, and the magnitude of the stirring largely determines the energy input and cascade rate. The scale at which viscous effects become important is determined by the value of the molecular viscosity by Eq. (14.86). If viscosity tends to zero, this scale becomes smaller and smaller in such a way as to preserve the constancy of the energy dissipation.

Finally, the time scales as $t \sim \lambda^{2/3}$, implying that for smaller scales the "eddy-turnover time," on which structures at that scale deform, becomes smaller and smaller.

Two-Dimensional Turbulence

In two dimensions the situation is complicated by another quadratic invariant, the enstrophy. Taking the curl of Eq. (14.1) to give a vorticity equation, and restricting attention to two-dimensional flows, yields the vorticity equation:

$$\frac{\partial \zeta}{\partial t} + \mathbf{u} \cdot \nabla \zeta = \nu \nabla^2 \zeta, \qquad (14.89)$$

where $\mathbf{u} = u\mathbf{i} + v\mathbf{j}$ and $\zeta = \mathbf{k} \cdot \text{curl } \mathbf{u}$. It is easily verified that when $\nu = 0$, Eq. (14.89) conserves not only the energy but also the enstrophy $Z = \int \frac{1}{2}\zeta^2 \, d\mathbf{x} = \int k^2 \mathscr{E}(k) dk$.

We now ask, how does the distribution of energy and enstrophy change in a turbulent flow? The problem is analogous to that of rearranging mass on a lever while still preserving the moment of inertia, with energy playing the role of mass, enstrophy that of moment of inertia, and wave number the distance from the fulcrum. Any rearrangement of mass such that its distribution also becomes wider must be such that the center of mass moves toward the fulcrum. Thus, energy would move to *smaller* wave numbers and enstrophy to larger. Consistent with this, it is easy to show that energy dissipation goes to zero as Reynolds number rises. The total dissipation of energy is, from Eq. (14.89),

$$\frac{dE}{dt} = -\nu \int \zeta^2 \, d\mathbf{x}. \qquad (14.90)$$

Because vorticity itself is bounded from above [again using Eq. (14.89)] we see that energy dissipation goes to zero as viscosity goes to zero, and hence also in the infinite Reynolds number (but finite energy) limit. Thus, unlike the three dimensional case, there is no mechanism for the dissipation of energy at small scales in high Reynolds number two-dimensional turbulence. On the other hand, we do expect enstrophy to be dissipated at large wave numbers.

These arguments lead one to propose the following scenario in two-dimensional turbulence. Energy and enstrophy are input at some scale L_I and energy is transferred to larger scales (toward the fulcrum), and enstrophy is cascaded to small scales where ultimately it is dissipated. In the enstrophy inertial range the enstrophy cascade rate η is assumed constant. Using the dimensionally correct scaling

$$\eta \sim \frac{k^3 \mathscr{E}(k)}{\tau(k)} \tag{14.91}$$

yields the prediction

$$\mathscr{E}(k) = \mathscr{K}' \eta^{2/3} k^{-3}, \tag{14.92}$$

where $\mathscr{K}'$ is also, it is supposed, a universal, order one, constant. It is of course also quite possible to obtain Eq. (14.92) from scaling arguments identical to those following Eq. (14.84). The scaling transformation (14.84) still holds, but now instead of (14.85) we assume that the enstrophy flux is constant with wave number. Dimensionally we have

$$\eta \sim \frac{v^3}{l^3} \sim \lambda^{3r-3}, \tag{14.93}$$

which gives $\lambda = 1$. The exponent n determining the slope of the inertial range is given, as before, by $n = -(2r + 1)$ yielding the -3 spectra of Eq. (14.92). Thus, the velocity now scales as $v \sim \eta^{1/3} k^{-1}$, and the time scales with distance as $t \sim l/v \sim \eta^{-1/3}$. Thus, it is length-scale invariant. The appropriate Kolmogorov scale is given by equating the inertial and viscous term in Eq. (14.1) or (14.89), which gives, analogously to (14.85)

$$k_v \sim \left(\frac{\eta^{1/3}}{v} \right)^{1/2}. \tag{14.94}$$

The energy dissipation is easily calculated to go to zero as $v \to 0$. The enstrophy dissipation, analogously to Eq. (14.88), goes to a finite limit given by

$$\dot{Z} = \frac{d}{dt} \int \frac{1}{2} \zeta^2 \, dx = v \int \zeta \nabla^2 \zeta$$
$$\sim v k_v^4 v^2 \sim \eta. \tag{14.95}$$

So far so good. However, things in two dimensions are, unfortunately, not quite as simple as they appear. First, note that timescale given by Eq. (14.81) is apparently independent of scale. If the spectra were any steeper, then turnover times would actually increase with wave number. In fact the estimate Eq. (14.81) of an eddy turnover time is actually rather poor for the steep spectra found in two dimensional turbulence, and a useful refinement is:

$$\tau = \left\{ \int_{k_0}^{k} (p^2 \mathscr{E}(p)) \, dp \right\}^{-1/2}, \tag{14.96}$$

where k_0 is a lower wavenumber cut-off, recognizing the straining effects of all velocity scales larger than the scale of interest. Using this in Eq. (14.90) yields the log-corrected range

$$\mathscr{E}(k) = \mathscr{K}'\eta^{2/3}(\log(k/k_0))^{-1/3}k^{-3}. \tag{14.97}$$

This is likely to be observationally indistinguishable from the uncorrected range. (Generally speaking, we can safely leave out logarithmic corrections in final results, if not always in intermediate calculations. We will subsequently neglect them.)

However, this has not fixed the underlying problem with the two-dimensional phenomenology, which is as follows. The inertial range predictions are based on the assumption of locality, in spectral space, of energy and enstrophy transfers. Now, a useful measure of this locality is given by the straining at a particular wavenumber, say k, from other wavenumbers. The total strain $T(k)$ at k is given by

$$T(k) = \left\{ \int_0^k \mathscr{E}(p)p^3 d\log p \right\}^{1/2}. \tag{14.98}$$

The contributions to the integrand from each octave are given by

$$\mathscr{E}(p)p^3\Delta\log p. \tag{14.99}$$

In three dimensions, use of the $-5/3$ spectra indicates that the contributions from each octave below k increase with wave number, being a maximum close to k, implying locality and a posteriori being consistent with the locality hypothesis. However, in two dimensions each octave makes the same contribution. The strain, and possibly the enstrophy transfer, are hardly local after all! This very heuristic result implies that the two-dimensional phenomenology is on the verge of not being self-consistent, and suggests that the -3 spectral slope is the shallowest limit that is likely to be actually achieved in nature or in any particular computer simulation, rather than a very robust result. Why? Well, suppose the detailed dynamics attempt in some way to produce a shallower slope; using Eq. (14.99) the strain is then local and the shallow slope is forbidden by the Kolmogorovian scaling results. However, if the dynamics organizes itself into structures with a steeper slope (say k^{-4}), the strain is quite nonlocal. The fundamental assumption of Kolmogorov scaling is not satisfied, and there is no inconsistency. In fact numerical simulations do reveal a slope steeper than k^{-3}, often dominated by isolated vortices. However, the dynamical processes leading to their formation, and their precise relationship with the enstrophy cascade, are not at this time fully understood.

There is one other aspect of the phenomenology that has been sometimes thought to be a problem, but in fact is not. This is that in the limit of zero viscosity, Eq. (14.95) implies that enstrophy dissipation remains constant, whereas it has been shown rigorously that the inviscid equations—Eq. (14.89) with the right-hand side set to zero—have no singularities and enstrophy dissipation remains zero. This is not in fact a contradiction, first because we are concerned with the zero viscosity *limit* in Eq. (14.95). Even if we were to suddenly 'turn off' the viscosity in an infinitely high resolution simulation of Eq. (14.89), then the enstrophy inertial range (assuming it exists) would slowly spread to larger and larger wave numbers. During this period of adjustment the fluid indeed has zero enstrophy dissipation. It takes the fluid an infinite time to come to equilibrium with an infinitely long inertial range. Only then is the enstrophy dissipation nonzero, which is not an inconsistency with the rigorous results.

14.4.2 Predictability of Strong Turbulence

One of the central properties of turbulence is its unpredictability due to nonlinear interactions. Some authors will draw a distinction between "sensitive dependence on initial conditions" and "unpredictability." The former's meaning is unambiguous, and it is normally applied to deterministic systems. The latter is sometimes applied only to indeterminism arising out of stochasticity, when the equations of motion are not known. However, in this chapter we take them to be synonymous, and use the latter (because it is but one word) to mean unpredictability arising from chaos. Actually, the difference between chaos and stochasticity lies not so much in the underlying dynamics, but in our knowledge of them. Whereas chaos is essentially but a word for deterministic "randomness," stochasticity describes randomness arising from incomplete knowledge of the system, as for example in Brownian motion. Thus, in most cases the difference between stochasticity and chaos may be thought of as merely a difference in our knowledge of the dynamics. For example, most computers have "random number generators" built in, and these are often used in the simulation of stochastic systems. However, the algorithm producing the random numbers is completely deterministic, and if we regard that algorithm as part of the system, we have chaos, not stochasticity.

The modern ideas of nonlinear dynamics and chaos have not, interestingly enough, had at this time much impact on theories of, or ideas of how to cope with, strong turbulence. Even prior to the classical paper of Lorenz in 1963 and later Ruelle and Takens in 1971, it was believed that turbulence was truly unpredictable [9], notwithstanding the picture of Landau of turbulence as a large collection of periodic, and presumably predictable, motions. The unpredictability was thought to arise from the utter complexity of the flow. The reasons for the loss of predictability were probably only properly understood when it was realized that even systems with a small number of degrees of freedom could be unpredictable. Assuming that the dynamical systems arguments applicable to weak turbulence apply to strong turbulence, and hence that a turbulent fluid *is* in fact unpredictable, then just using the scaling laws we can heuristically obtain estimates of the predictability time for a turbulent fluid [20].

The physical space fields $\zeta(\mathbf{x})$ may be expressed as an infinite Fourier sum or integral, for example $\zeta = \sum \hat{\zeta}_{\mathbf{k}} \exp(i\mathbf{x} \cdot \mathbf{k})$ or $\zeta = \int \hat{\zeta}_{\mathbf{k}} \exp(i\mathbf{x} \cdot \mathbf{k}) \, dk$. The former is appropriate in a bounded domain (where the wave numbers are quantized), the latter in an infinite domain. We are usually concerned with a finite domain, but will nevertheless often replace sums by integrals where it will simplify things. In two dimensions (for simplicity) the inviscid vorticity equation may be written in spectral form

$$\frac{\partial \zeta_k}{\partial t} + \sum a_{kpq} \zeta_p \zeta_q = -vk^2 \zeta_k, \qquad (14.100)$$

where a_{kpq} are geometrical coupling coefficients which arise when Eq. (14.89) is Fourier transformed. The hats over transformed quantities have been dropped. At any given instant the equation of motion may be linearized about its current state, and the subsequent motion would then be described by an equation, valid for short times, of the form:

$$\frac{\partial \zeta_k'}{\partial t} + A_{kq} \zeta_q' = 0, \qquad (14.101)$$

and the eigenvalues of the matrix A_{kq} (whose explicit form does not concern us here) determine the short-term growth of errors in the system. Because the system is chaotic, A_{kq} has positive (growing) eigenvalues. If spectral inter-action in the inertial range are sufficiently local, it becomes meaningful to inquire as to the growth of errors at any particular scale k, for then the matrix A_{kq} is dominated by terms close to its diagonal. In particular, the rate of error growth at any particular scale is then given by the size of the appropriate coefficient of A_{kq}, which is ku_k where u_k is just a typical velocity at scale k. This of course is just the inverse of the eddy turnover time (14.81). After a time τ_k, errors will have grown sufficiently that a linear approximation is no longer valid; at that scale errors will saturate but at the same time will begin to contaminate the "next larger" (in a logarithmic sense) scale, and so on. Thus, errors initially confined to a scale k at $t = 0$ will contaminate the scale $2k$ after a time τ_k. The total time taken for errors to contaminate all scales from k' to the largest scale k_0 is then given by, treating the wavenumber spectrum as continuous,

$$T = \int_{k_0}^{k'} \frac{d(\ln k)}{\tau_k}$$

$$= \int_{k_0}^{k'} \frac{d(\ln k)}{\sqrt{k^3 E(k)}}. \qquad (14.102)$$

If the energy spectrum is a power law of the form $E = C'k^{-n}$ this becomes

$$T = [C'k^{(n-3)/2}]_{k_0}^{k'} \frac{2}{(n-3)}. \qquad (14.103)$$

As $k' \to \infty$ the estimate diverges for $n > 3$, but converges if $n < 3$.

What does this mean? First, we should point out that these arguments are at best heuristic, and do not account for the more esoteric phenomena, such as intermittency and coherent structures, believed by many to be important in strong turbulence. Nevertheless, taking them at face value they imply that two-dimensional turbulence is indefinitely predictable; if we can confine the initial error to smaller and smaller scales of motion, the payoff is that the "predictability time" (the time taken for errors to propagate to all scales of motion) can be made longer and longer, indeed infinite. This is consistent with what has been rigorously proven about the two-dimensional Navier–Stokes equations, with or without viscosity; namely, that they exhibit "global regularity," meaning they stay analytic for all time provided the initial conditions are sufficiently smooth (Rose and Sulem [4]). This does *not* mean that two-dimensional flow is in practice necessarily predictable. Two-dimensional turbulence is almost certainly chaotic, has positive Lyapunov exponents, and an arbitrarily small amount of noise will render a flow truly unpredictable sometime in the future. It is just that we can put off that time indefinitely if we know the initial conditions well enough, and can reduce the amount of external noise sufficiently.

In three dimensions, on the other hand, things are more worrisome. The predictability time estimate from Eq. (14.103) converges as $k' \to \infty$ so that even if we push our initial error out to smaller and smaller scales, the predictability time does not keep on increasing. The time it takes for errors initially confined to small scales to spread to the largest scales is simply a few *large* eddy turnover times (because the eddy turnover times of the small scales are so small). This is an indicator that something is badly wrong, either with our methodology or with the Euler equations; because the system is classical, we do not expect such finite time catastrophes. If one were able to prove global regularity for the three-dimensional Euler equations then we would know our analysis was wrong, but such a proof is lacking, and may not exist. So it is still an open question as to whether the equations are well posed or not. If the Euler equations were ill posed, it would mean that they are an incorrect description of zero-viscosity turbulent flow, which would perhaps not be so terrible anyway as no classical flow is inviscid, and we would be saved from having to throw away the Navier–Stokes equations by viscosity. No matter how small viscosity, if not zero, then at some small wave number the local Reynolds number will be small and viscous effects will start to dominate over inertial effects. Beyond the dissipation wave number, it is possible to show that the energy spectrum gets steeper, and as soon as the asymptotic spectra is steeper than -3 we are again assured of indefinite predictability. If the Navier–Stokes equations were shown to have singularities, it would be a more serious matter.

So what about the weather? Well, in the lower part of the atmosphere (below 10 km, where the weather is) the large-scale flow behaves more like a two-dimensional fluid than a three-dimensional fluid. This is because of the twin effects of rotation and stratification, but we shall not go into that here. At scales smaller than about 100 km, the atmosphere starts to behave three-dimensionally. Now the current atmospheric observing system is such that over continents the atmosphere is fairly well observed down to scales of a couple of hundred kilometers. If we knew the enstrophy cascade rate through the atmosphere we could evaluate the predictability time using the formulae derived above, but it is easier simply to do the sum manually, Fourier transforming in our heads, as it were. Suppose then we have no knowledge of the dynamical fields at scales smaller than 200 km. Aside from certain rather intense small-scale phenomena, the atmosphere is not especially energetic at these scales (hurricanes, for example, are rather larger as well as being intermittent) and we could estimate a typical velocity of about 1 m/s giving an eddy turnover time of about 2 days. So in 2 days motion at 400 km scales is unpredictable. The dynamics at these scales is a little more intense, say $U \sim 2$ m/s. Coincidentally (?), this also gives a 2-day eddy turnover time, so after 4 days motion at 800 km is unpredictable. Continuing the proccess, after about 12 days motion at 6000 km is completely unpredictable, and our weather forecasts are essentially useless. This is probably a little better than our experience suggests as to how good weather forecasts are in practice, but of course our models of the atmosphere are certainly not perfect. (Actually, I've fudged the numbers so they come out reasonable, having been rather cavalier about factors of 2, π, etc. More careful calculations, as well as computer simulations, do give similar results though.) In principle, we could make better forecasts if we could observe the atmosphere down to smaller scales of motion. Observing down to 100, 50, and 25 km would (if the atmosphere remained two dimensional) each add about a couple of days to our forecast times.

However, we can't go on forever, because at small scales of motion the atmosphere starts behaving three-dimensionally. As we have seen, because the energy spectrum for three-dimensional turbulence is shallow, the eddy turnover times decrease rapidly with scale and the predictability time is largely governed by the predictability time of the largest scale of motion. Thus, the *theoretical* limit to predictability is governed by the scale at which the atmosphere turns three dimensional, probably about 100 km. So we see that we can't increase the length of time we can make good weather forecasts for longer than about 2 weeks, no matter how good our models and no matter how good our observing system. This is the theoretical predictability limit of the atmosphere. The so-called butterfly effect has its origins in this argument: a butterfly flapping its wings over the Amazon is, so it goes, able to change the course of the weather a week or so later. How farfetched is this? Well, the affect of a lone butterfly are probably drowned both by viscous dissipation and more energetic eddies at larger scale. So although this argument may be an exaggeration, there is little doubt that small-scale phenomena will affect global weather some time later in an unpredictable manner.

One other point may be apposite. The predictability of a system is often characterized by its spectrum of Lyapunov exponents. In a turbulent system the largest Lyapunov exponent is likely be associated with the smallest scales of motion, and the error growth associated with this effectively saturates at small scales. The time scales of error growth affecting the larger scales, which are the time scales of most interest, are determined by slower, larger-scale processes whether or not the cascade-like growth of error described above is correct. This means that the largest Lyapunov exponents probably have nothing whatever to do with the growth of error at the larger scales in a turbulent fluid.

14.4.3   Renormalizing the Diffusivity

To obtain a predictability time for the large scales of turbulence, we successively summed the effects of the smaller scales. I'd now like to briefly discuss one other application of this kind of approach, the idea now being to successively average over the smaller scales to produce an effective eddy diffusivity for the large scales. This is the same idea used in renormalization group techniques in condensed matter physics and discussed earlier with reference to transition problems, but the arguments given here will be simple and self-contained. Nevertheless, the discussion may be a little brief for those with no background in turbulence theory. Let us first discuss the problem.

A turbulent fluid may have many decades of scales of motion. Indeed, in the most-observed fluid of all (the earth's atmosphere) the forcing scales are several thousands of kilometers and the dissipation scales are perhaps order millimeters. Nevertheless, we cannot hope to explicitly describe all these scales of motion, even with future generations of computers. (If we attempted to do so by constructing a numerical model of the atmosphere with grid points every millimeter, it would take a time longer than the current age of the universe to advance just one time step, even with computers 10 times as fast as today's.) If we just use the value of the molecular viscosity in a model resolving only the large scales, energy will not be removed correctly, if at all. This means that in an equation such as (14.89) we must use a much larger viscosity, appropriate to these resolved scales, which represents the effects that subgridscale, or unresolved, motions have on those resolved. Because models of the large scales *can* be constructed and run on the computer, the problem of turbulence lies, in a nontrivial sense, in deriving an expression for such an "eddy viscosity."

As a step toward that goal, we will discuss a simpler problem, that of the eddy diffusivity of a passive tracer. Such a tracer (ϕ) obeys the equation

$$\frac{\partial \phi}{\partial t} + (\mathbf{u} \cdot \nabla)\phi = \kappa \nabla^2 \phi, \tag{14.104}$$

where κ is the molecular diffusivity, akin to the viscosity. The Peclet number, UL/κ is analogous to the Reynolds number and is a measure of the size of the inertial terms to the diffusive terms. Again, the problem arises as to what to do if our model does not reach the small scales at which the diffusivity is effective; we must derive an effective eddy diffusivity appropriate for high Peclet number regimes. The problem is, or at least should be, simpler than the eddy viscosity problem because Eq. (14.104) is a linear equation.

Just as there is a cascade of energy (in three dimensional turbulence) to the small scales there is a cascade of tracer variance, ϕ^2-stuff, from large scales to small. Scaling arguments similarly give rise to inertial range predictions for the spectrum of tracer variance. These are:

$$\Phi(k) = C\chi k^{-5/3}\varepsilon^{-1/3} \qquad (14.105)$$

in three dimensions and

$$\Phi(k) = C'\chi k^{-1}\eta^{-1/3} \qquad (14.106)$$

in two dimensions, where C and C' are undetermined, dimensionaless, constants and χ is the rate of cascade of tracer variance, $\Phi(k)$.

At sufficiently small scales the Peclet number becomes of order unity and dissipation of ϕ^2-stuff occurs. If the diffusivity is sufficiently large, diffusion occurs before (i.e., at larger wave numbers) than dissipation of energy by viscosity. Because Eq. (14.36) is linear, it turns out that it is possible (after a great deal of algebra [21]) to derive an expression for the eddy diffusivity, in a certain *low Peclet number limit*. The expression is

$$\mathscr{D}(k) = \frac{2}{3}\int_k^\infty \frac{\mathscr{E}(p)}{\kappa p^2}\, dp. \qquad (14.107)$$

This is a sensible and unsurprising result, because it merely says that the eddy diffusivity is determined by the combined motion of small eddies from a size $1/k$ and smaller. The result is of no immediate help, because it is valid only in a low Peclet number limit.

Note that Eq. (14.107) contains the actual molecular diffusivity in the denominator. Now, suppose that Eq. (14.107), or an expression very similar to it, is accurate at some wavenumber k, in the dissipation regime, and we wish to obtain an expression for the eddy diffusivity at some slightly larger scale, or smaller wave number $k - \Delta k$. Then at $k - \Delta k$ the correct expression for the eddy diffusivity will be the value at k plus a small contribution from the wavenumber interval between k and $k - \Delta k$. But in this interval, the diffusivity appearing in the denominator should not only be the molecular diffusivity, but should include the eddy diffusivity appropriate at that wavenumber. That is:

$$D(k - \Delta k) = D(k) + \frac{2\mathscr{E}(k)}{3(\kappa + D(k))}\frac{\Delta k}{k^2}. \qquad (14.108)$$

Thus, given the result valid for low Peclet number, we are able by successive applications of Eq. (14.108) to bootstrap ourselves to a result valid in a low wave number, high Peclet number regime (see [22]). From Eq. (14.108) we obtain the differential equation

$$\frac{\partial D}{\partial k} = -\frac{2}{3(D + \kappa)}\frac{\mathscr{E}(k)}{k^2}, \qquad (14.109)$$

which integrates, with the boundary condition of $D(\infty) = 0$, to give

$$D(k) = -\kappa + \left[k^2 + \frac{4}{3}\int_k^\infty \frac{\mathscr{E}(p)}{p^2}\,dp\right]^{1/2}. \qquad (14.110)$$

In the low Peclet number limit, the expression is strongly dependent on the molecular value. In the high Peclet number limit Eq. (14.110) simply reduces to

$$D(k) = \left[\frac{4}{3}\int_k^\infty \frac{\mathscr{E}(p)}{p^2}\,dp\right]^{1/2} \qquad (14.111)$$

and there is no explicit dependence on the molecular diffusivity. We have apparently succeeded, then, in obtaining a "renormalized" value of the diffusivity, which would be appropriate to use in calculations that do not explicitly resolve the diffusive subrange.

It is possible to use Eq. (14.111) to make testable predictions about the values of $\mathscr{K}$ and $\mathscr{K}_t$ appearing in the inertial range expressions (14.83) and (14.105), and (14.92) and (14.106), for energy and passive tracer variance in three and two dimensions, respectively. To do this first evaluate Eq. (14.111) using Eq. (14.83) or (14.92) to give

$$D_3(k) = \left(\frac{\mathscr{K}}{2}\right)^{1/2}\varepsilon^{1/3}k^{-4/3} \qquad (14.112)$$

in three dimensions and

$$D_2(k) = \left(\frac{\mathscr{K}'}{3}\right)^{1/2}\eta^{1/3}k^{-2} \qquad (14.113)$$

in two. Now, the dissipation of tracer variance (ϕ^2-stuff) in reality is given by

$$\chi = \kappa \int_0^\infty p^2\Phi(p)\,dp. \qquad (14.114)$$

Similarly, it is also equal to

$$\chi = D(k)\int_0^k p^2\Phi(p)\,dp, \qquad (14.115)$$

where the upper limit may now lie in the inertial range. For the three-dimensional case, substituting Eq. (14.105) for $\Phi(p)$ and Eq. (14.112) for $D(k)$, and performing the integration, gives a relationship between $\mathscr{K}$ and C, namely

$$C = \frac{4}{3}\left(\frac{2}{\mathscr{K}}\right)^{1/2}. \qquad (14.116)$$

Experimentally, C and $\mathscr{K}$ are both approximately 1.5, which is consistent with the prediction.

In two dimensions, a similar calculation yields

$$C' = \left(\frac{12}{\mathscr{K}'}\right)^{1/2}. \qquad (14.117)$$

There is currently no experimental confirmation or falsification of this prediction. However, because the Kolmogorov scaling itself in two dimensions remains unconfirmed, this may be moot.

It is tempting, but unjustified on any a priori basis, to suppose that in two-dimensions the above ideas apply not only to a passive tracer but also to vorticity—for note that Eq. (14.89) is the same form as Eq. (14.104). Then from Eq. (14.116) we obtain a prediction for the value of the Kolmogorov constant by setting $C' = \mathscr{K}'$, to give $\mathscr{K}' = 12^{1/3}$. We emphasize that this last step is very speculative, although there are a couple of indicators that it may not be completely nonsensical [23]. First, numerical simulations do indicate that at the small scales of two-dimensional turbulence the vorticity field is passively advected by the large field. Second, a subsequent but independent full renormalization group treatment gave, perhaps remarkably, precisely the same numerical number for $\mathscr{K}'$. It may be that such a treatment is implicitly treating the vorticity as passive, although it is difficult to disentangle the assumptions from the very elaborate calculations. In any case, as this stage our reach has now far exceeded our grasp (but then what's a heaven for?) and we should now sum up.

14.5   Remarks

In this chapter we first discussed some basic notions of linear instability and nonlinear equilibration. Then, temporarily leaving the Navier–Stokes equations behind, we discussed various sequences to do with the transition to chaos. Much progress has undeniably been made in understanding the transition problem using tools from "nonlinear science" or dynamical systems theory. However, for the physicist these would be little more than mental gymnastics were it not for the fact that a number of the sequences have actually been observed in experiment. The experimental verification is important because at the time of writing there is something of a gap between the scaling theories and rigorous mathematical hydrodynamics. Ideally, one would like to be able to do the following: take the Navier–Stokes equations for some fairly generic geometry (e.g., three-dimensional convection in a box of arbitrary aspect ratio, or a rotating cylinder) and show that a series of rational approximations for some particular range values of control parameter (say Rayleigh number or Reynolds number) leads to a low-dimensional map or system of ordinary differential equations for which the transition to instability and then chaos is well understood. Although in some circumstances this goal can be approached, in general it is a very difficult proposition—it is even difficult to rigorously obtain the appropriate Landau equation for a given instability, and this is a simpler task than elucidating the full transition sequence. In the absence of this it is very difficult to predict ahead of time what particular route a given flow will take, or even to say with confidence that the transition to turbulence is understood. Numerical simulations and algebraic manipulation languages, such as Maple, will undoubtedly help, as will using any symmetry in the geometry.

The astute reader will also have noticed that there was no section entitled "From Chaos to Turbulence." This lamentable lack is not solely due to laziness or ignorance on the part of the author, although both may be applicable. It is because little is quantitatively understood about the phenomenon. It is certain that turbulence is chaotic, although it is not certain how the transition from low-dimensional chaos to higher-dimensional turbulence occurs, whether it is in any sense universal, even whether it is a sensible question, and so on.

Bypassing this difficult topic, we then turned to strong turbulence. Although scaling arguments prove useful in understanding the basic phenomenology, further progress at a fundamental level has here too been slow. However, a lot of useful practical information, for example the predictability limits of the atmosphere, can be estimated using no more than these arguments. We then discussed a rather more recondite topic, that of renormalizing the eddy diffusivity in a turbulent flow using a successive averaging approach. Such arguments have proved successful in a number of problems involving many scales of motion, and their application to turbulence is attractive because they apparently afford a means of beginning with an expression or idea valid in a rather restrictive domain and bootstrapping to a regime of greater validity. Nevertheless, more sophisticated application of renormalization group theory has been a little controversial and there currently is probably no a priori reason to prefer them over the more established renormalized perturbation theories.

It is a truism to say that fluid mechanics in general is difficult because it is nonlinear. Turbulence in particular is made even more difficult by the fact that it involves many degrees of freedom, or put another way, its dimensionality is large. The impact of ideas in nonlinear dynamics, which have typically dealt only with systems of low dimensionality, on strong turbulence is yet to be seen. It may not be too important to understand the detailed structure of the strange attractor in turbulence (presuming that the attractor is strange); because the flow is so complex, any interesting properties manifest themselves before the attractor has been fully explored. However, it is probably true to say that the generic lack of predictability of fluid motion, as dictated and explained by the presence of strange attractors in turbulence, actually *underscores the necessity* for a statistical theory rather than a deterministic theory of turbulence. Whether the texture of what is traditionally meant by nonlinear dynamics (chaos, strange attractors, Lyapunov numbers, etc.) will play a direct role in turbulence is unclear; rather the ideas may provide theoretical foundation for statistical assumptions. In any case, the field (as it has been since Horace Lamb, in the 1920s, expressed more faith in the Almighty being able to explain quantum electrodynamics than turbulence) is ripe for progress.

References

[1] A.S. Monin and A.M. Yaglom, *Statistical Fluid Mechanics: Mechanics of Turbulence*, Vols. I and II. (MIT, Cambridge, 1971).

[2] P. Drazin and W.H. Reid, *Hydrodynamic Stability* (Cambridge University, Cambridge, 1981); *Hydrodynamic Instabilities and the Transition to Turbulence*, 2nd ed., edited by H.L. Swinney and J.P Gollub (Springer-Verlag, New York, 1985).

[3] B. Hu, Phys. Rep. **91**, 233 (1982); L.D. Landau and E.M. Lifshitz, *Fluid Mechanics*, 2nd ed. (Pergamon, Oxford, 1987).

[4] J. Miles, Adv. Appl. Mech. **24**, 189 (1984); J-P. Eckmann, Rev. Mod. Phys. **53**, 643 (1981); H. Rose and P.-L. Sulem, J. Phys. **39**, 441 (1978).

[5] R. Kraichnan, J. Fluid Mech. **5**, 497 (1959).

[6] F.M. Ludlam, *Clouds and Storms: the behavior and effect of water in the atmosphere.* (Pennsylvania State University, Philadelphia, 1980).

[7] C. Bender and S. Orszag, *Advanced Mathematical Methods for Scientists and Engineers* (McGraw-Hill, New York, 1978).

[8] T. Stuart, J. Fluid Mech **9**, 353 (1960); J. Watson, J. Fluid Mech. **9**, 370 (1960); J. Pedlosky, J. Atmos. Sci. **27**, 15 (1970).

[9] P.D. Thompson, Tellus **9**, 275 (1957); Ye.A. Novikov, On the predictability of synoptic processes. Izv. An. SSSR. Ser. Geophys **11**, (1959).

[10] D. Ruelle and F. Takens, Comm. Math. Phys. **20**, 167 (1971).

[11] M.J. Feigenbaum, J. Stat. Phys. **19**, 25 (1978); M.J. Feigenbaum, J. Stat. Phys. **21**, 669 (1979).

[12] Y. Pomeau and P. Manneville, Comm. Math. Phys. **74**, 189 (1980); P. Manneville and Y. Pomeau, Physica D **1**, 219 (1980).

[13] J. Collet and J.-P. Eckmann, *Iterated Maps on the Interval as Dynamical Systems* (Birkhauser, Boston, 1980).

[14] J.P. Gollub, S.V. Benson, and J.A. Steinman, Ann. NY Acad. Sci. **357**, 22 (1980); M. Giglio, S. Musazzi, and U. Perini, Phys. Rev. Lett. **47**, 243 (1981); A. Libchaber, C. Laroche, and S. Fauve, J. Phys. Lett. **43**, 211 (1982).

[15] D.R. Moore, J. Toomre, E. Knobloch, and N.O. Weiss, Nature **303**, 663 (1983).

[16] J. Gollub and S.V. Benson, J. Fluid Mech. **100**, 449 (1980).

[17] A. Libchaber and J. Maurer, J. Phys. Colloq. **41**, 51 (1980).

[18] A.N. Kolmogorov, Doklady AN SSSR **30**, 299 (1941).

[19] H.L. Grant, R.W. Stewart, and A. Moilliet, J. Fluid Mech. **12**, 241 (1962).

[20] C.E. Leith and R.H. Kraichnan, J. Atmos. Sci., **29**, 1041 (1972); G.K. Vallis, Q.J. Roy. Meteor. Soc. **111**, 1039 (1985); G.K. Vallis, in *Nonlinear Phenomena in the Atmospheric and Oceanic Sciences*, edited by G. Carnevale and R. Pierre-humbert (Springer-Verlag, New York, 1992).

[21] G.K. Batchelor, I.D. Howells, and A.A. Townsend, J. Fluid Mech. **5**, 134 (1959); I.D. Howells, J. Fluid Mech. **9**, 104 (1960).

[22] H.K. Moffatt, J. Fluid Mech. **106**, 27 (1981).

[23] M.E. Maltrud and G.K. Vallis, Phys. Fluids A **5**, 1760 (1993); P. Olla, Int. J. Mod. Phys. B **8**, 581 (1994).

15

Active Walks: Pattern Formation, Self-Organization, and Complex Systems

Lui Lam

15.1 Introduction

A nonequilibrium system is one in which there is an exchange of energy or matter between the system and its environment. Nonequilibrium systems are very common and are studied in problems in physics, engineering, chemistry, biology, geology, computer sciences, economy, ecology, sociology, and so forth [1, 2]. Examples include the heating from below of a horizontal layer of liquid (the Rayleigh–Bénard problem), crystal growth, the coherent emission of a laser, the breakdown of bridges, spatiotemporal patterns in chemical reactions, ant swarms, biological evolutions, structures in geological deposits, the stock market, pollution control, traffic flows, and the growth of cities.

Most nonequilibrium system of interest are made up of a large number of "simple" components such as molecules, cells, neurons, animals, or humans. In some cases, all the components are identical to each other (e.g., the molecules of a simple liquid in a Rayleigh–Bénard cell). In other cases, there may be more than one type of component (e.g., the different species coexisting in an ecological system). A common feature of these constituent components in a nonequilibrium system is that, under suitable conditions, they may interact and *self-organize* themselves to give the system a collective behavior or state that is not dictated from a central command—an *emerging* behavior [3]. For example, when the temperature gradient exceeds a critical value, a series of convection rolls appear in a Rayleigh–Bénard cell. In the process of food searching and gathering, ants organize themselves to form trails between their nest and the food sources without a central command from the queen ant.

And in many circumstances, the emerging state or behavior—be it the stock market or the spatial (and temporal) patterns resulting from chemical reactions or ant movements—can be very complex in nature or in appearance [1, 4, 5]. The study of these *complex phenomena* [6] is now in the forefront of research. It provides a real opportunity for scientists to break out from their inherited mode of training, talk to each other, and explore subjects of *any* discipline of their liking from both natural and social sciences. It is a very exciting, unifying science in the making.

Although a universal principle governing self-organization of complex systems is still lacking [7], it seems that many self-organizing processes in nature can be described by the *principle of active walks* [8].

Active walk (AW) is a new paradigm proposed recently by Lam and co-workers [9, 10], to treat problems in pattern formation [11, 12], self-organization [7], and the dynamics of complex systems [13, 14]. In an AW the walker changes the landscape as it walks, and its next step is influenced by the changed landscape. Many adaptive, self-organizing systems in natural and social sciences—simple or complex—can be described by such an active walker model. The idea is that each component of the system, being an active walker, communicates with and thus influences others through a landscape (a potential) set up in space. In principle, the active walk model (AWM) is applicable to river formation, heat-seeking missiles, ant swarms, bacteria colonies, retinal neurons and vessels, electrodeposit and dielectric breakdown patterns, percolation in soft materials, ion transport in glasses, polymer reptation, rough surfaces, biological evolution, population dynamics, mass communication, and so forth [15].

In the rest of this chapter, the basic concepts and formulations, some exemplary applications, recent developments, and open problems of the AW will be presented.

15.2   Basic Concepts

Two masses attract each other. This can be described as a direct interaction through Newton's law of gravity. Or, one can say that one mass sets up a gravitational potential around it, and the second mass feels this potential and walks/moves from high to low potential. In doing this, the second mass also changes the potential surrounding itself. In other words, the two masses change and interact indirectly through the potential they share. This is well known in physics.

Similarly, charged particles interact indirectly through the electric potential, which is the sum of the potentials set up by each charge. In this case, depending on whether the charge is positive or negative, the particle will walk from high to low, or from low to high potential, respectively. For a charged particle that is small enough (e.g., an electron), with or without the presence of other charges, it may even interact with its own potential—the well known self-energy effect detected by the Lamb shift. As history shows, this description of interaction through a potential is more fundamental than the original direct-interaction scenario. And as demonstrated by the Aharonov–Bohm effect [16, 17], the potential itself is a real entity to be reckoned with.

Let us now consider systems made up of macroscopic or mesoscopic particles/bodies. A common macroscopic body is a human body, and we are talking about social science. Human beings, except in rare occasions, do not interact with their bodies in contact; they interact through a medium. A nice example is a large group of people interacting with each other in cyberspace, through the Internet, say. Something noticeable usually results from this interaction (e.g., they may pair off and get married or form a political party, which is an emergent property). Another example is a crowded swimming pool: people in the pool influence each other's swimming path or motion by disturbing the water they share.

How about systems in natural science? Electrons in a solid disturb the lattice as they move and may attract each other *through* their interaction with phonons (vibrations of the lattice), resulting in superconductivity! A heat-seeking missile releases heat as it moves and rushes toward a high-temperature region. It is a "walker," changing the temperature "potential" as it walks, and is influenced by the potential in selecting its next step. With the presence of random forces due perhaps to a bad weather, even in the absence of other moving heat sources, the track of the missile could become very complex. The track may be extended or localized, terminate abruptly in midair (in the absence of gravity), or develop into a closed path—a periodic motion with the missile chasing its own tail. Imagine what will happen if there are a number of these heat-seeking missiles sharing a limited space in the sky, assuming they are moving slowly; the situation will be more complex and can even be unpredictable. This complexity arises because the system is *adaptive* (heat-seeking), *noisy*, and has a *feedback* (heat-sensing) mechanism. For an ant swarm the picture is the same, except that heat is now replaced by the scent released by the ants and the potential is the distribution of the scent concentration.

We thus see that numerous examples in natural and social systems can be described in a unified and simple manner by a model of active walkers. By definition, an active walker—a term coined by Lam [9]—is one that changes the landscape (the potential) as it walks, and its steps are influenced by the changing landscape (Fig. 15.1); coexisting active walkers interact with each other through the shared landscape.

15.3   Continuum Description

The coupled Langevin equations for a single walker look like this [15]:

$$md^2\mathbf{R}/dt^2 = -\eta\, d\mathbf{R}/dt - \int d\mathbf{r}\,\delta(\mathbf{r}-\mathbf{R})\nabla V(\mathbf{r},t) + \mathbf{F}(t) \tag{15.1}$$

and

$$\partial V(\mathbf{r},t)/\partial t = D\nabla^2 V(\mathbf{r},t) + t_0^{-1}W(\mathbf{r}-\mathbf{R}(t)), \tag{15.2}$$

where m, the mass of the walker, represents the inertial effect; $\mathbf{R}$ is the position of the walker; η is a damping coefficient; t_0 is a relaxation time constant; V is the landscape/potential; $\mathbf{F}$ is a fluctuating force; D is a diffusion constant; W, the *landscaping function*, represents how the walker changes the landscape. As an example, for the heat-seeking missile, $-V$ is the temperature

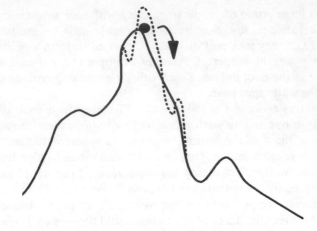

FIGURE 15.1. Sketch of an active walk. The walker (solid dot) changes the land-scape, a potential V, around itself according to a landscaping rule (from the solid line to the broken line). It then chooses its next step according to a stepping rule based on the new landscape. The landscaping rule and the stepping rule depend on the system under study. In a computer algorithm these two rules have to be specified, which could be fixed or genetic (i.e., evolving in time).

distribution; for the ants, the scent concentration. The W function should be determined by the system under study (see section 15.5). Computation-ally, two forms of W (both isotropic), as shown in Fig. 15.2 have been used [18].

Equations (15.1) and (15.2) are not easy to solve, and in fact have not been solved. The Fokker–Planck equations,

$$\partial p(\mathbf{r}, t)/\partial t = \nabla \cdot [p(\mathbf{r}, t)\nabla V(\mathbf{r}, t)] + D_w \nabla^2 p(\mathbf{r}, t) \qquad (15.3)$$

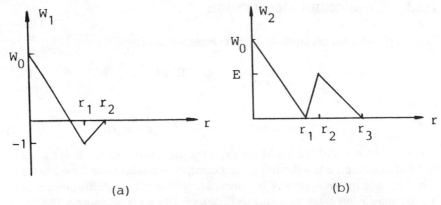

FIGURE 15.2. Two types of isotropic landscaping function, W_1 (a) and W_2 (b).

and

$$\partial V(\mathbf{r}, t)/\partial t = D\nabla^2 V(\mathbf{r}, t) + t_0^{-1} \int d\mathbf{r}' W(\mathbf{r} - \mathbf{r}') p(\mathbf{r}', t), \qquad (15.4)$$

are not necessarily simpler, where $p(\mathbf{r}, t)$ and D_w are the probability density and the diffusion constant of the walker, respectively. Moreover these equations, being stochastic, can only give the statistical properties of the walker's track or the potential, but not their exact topologies. Yet in the application to pattern formation, it is precisely the track pattern that is of interest. We therefore turn to the computer models of active walks.

15.4   Computer Models

The basic AWM [9, 10] is described here, starting from a single active walker without branching. Then the branching rule and the possible updating schemes of multiwalkers are presented. Some examples of the track pattern from the AWM, corresponding to observed filamentary patterns, are given. For the simulation results presented in sections 15.4 and 15.5.1 a square lattice is used.

Compact patterns, obtained from a variation of the AWM described in this section, are shown in section 15.6. Discussion of the physical origin of the landscaping function W and other assumptions adopted in the AWM is deferred to section 15.5. The time variation of the landscape, a rough surface resulted from the AW, is given in section 15.7.

15.4.1   A Single Walker

In an AW, a single-valued potential $V(i, n)$ is defined at every site i on a lattice at time n, where $n = 0, 1, 2, \ldots$ (The unit of time is taken to be t_0.) Furthermore,

$$V(i, n) = V_0(i, n) + V_1(i, n), \qquad (15.5)$$

where V_0 is the external background; V_1 evolves in time due to the action of the walker. At $n = 0$, a walker is placed on the initial landscape $V_0(i, 0)$. V_1 is updated according to the *landscaping rule*

$$V_1(i, n+1) = V_1(i, n) + W(\mathbf{r}_i - \mathbf{R}(n)), \qquad (15.6)$$

with $V_1(i, 0) = 0$; $\mathbf{r}_i$ is the position vector of site i, $\mathbf{R}(n)$ the position of the walker at time n. After the landscape is updated, the walker takes a step according to the *stepping rule* specified by P_{ij}, the probability for the walker to step from its present site i to a site j, with j belonging to A_i, the set of *available sites* surrounding site i. For example, for a self-avoiding walk A_i could be taken to be all the nearest-neighbor sites not yet visited by the walker. (An

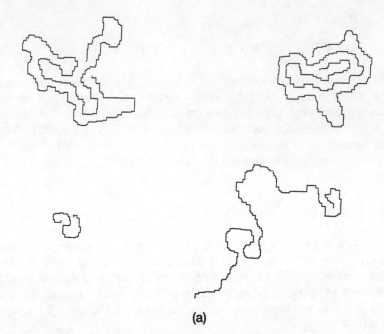

(a)

FIGURE 15.3. Track patterns obtained from the PAW model of a single active walker [9]. $\eta = 1$, $V_0 = 0$, and $W = W_1$. (a) Four different walks from four different runs of the algorithm. The same set of parameters are used, but the random number sequence in each run is different. Each track terminates naturally when the walker reaches a local minimum in the landscape.

alternative procedure is to allow the walker to walk before the landscape is updated. The two versions are equivalent to each other if self-crossing is allowed, but not in a self-avoiding walk.)

Four forms of P_{ij} have been studied [10, 19]:

(i) The deterministic active walk (DAW). The j site in A_i with the lowest potential below that at i is stepped onto.
(ii) The probabilistic active walk (PAW). This walk is specified by

$$P_{ij} \propto \begin{cases} [V(i,n) - V(j,n)]^{\eta}, & \text{if } V(i,n) > V(j,n), \\ 0, & \text{otherwise,} \end{cases} \tag{15.7}$$

where η is a parameter.
(iii) The Boltzmann active walk (BAW). This walk is given by

$$P_{ij} \propto \exp\{[V(i,n) - V(j,n)]/T\}, \tag{15.8}$$

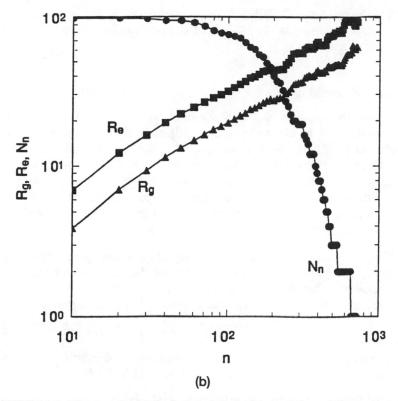

(b)

FIGURE 15.3 *(Continued)*. (b) Statistics of single filaments from 100 repeated runs [18]. The ensemble averaged end-to-end length R_e and the radius of gyration R_g obeys roughly $R_e \sim n^v \sim R_g$, with $v = 0.63$. v is not universal but depends on the parameters in W_1. N_n is the number of filaments with length greater than or equal to n.

where the parameter T may be interpreted as the "temperature."
(iv) The fuzzy active walk (FAW). Some fuzzy logic and control [20] is incorporated into the active walk [19] (see section 15.8).

If all the available sites have the same potential as $\mathbf{R}(n)$, an available site is chosen at random. More generally, the W function could be time dependent, time delayed, or have a finite lifetime [10, 15]. A diffusion term could be added to Eqs. (15.5) and (15.6) if needed.

Note that in the DAW and PAW, the walk terminates if the potential at $\mathbf{R}(n)$ is a local minimum, while the walker can go uphill and downhill in the BAW. The DAW is deterministic; the PAW and BAW are stochastic: different runs of the same algorithm (corresponding to different sequences of random number) lead to different results (Fig. 15.3). Also, as $T \rightarrow 0$, the

BAW reduces to the DAW; as $T \to \infty$, the BAW is just a random walk, albeit active, resulting in the absence of the second (first) term on the right-hand side of Eq. (15.1) [Eq. (15.3)] [21].

15.4.2 Branching

Branching, the splitting of a walker's track into two, is often observed in experiments such as electrodeposition [22] and dielectric breakdown of a liquid layer [23]. Also, if $-V$ is interpreted as the fitness landscape in evolutionary biology [24] and the walker's movement as the evolutionary path of a species, then branching will correspond to the emergence of two species sharing the same family tree. Although the exact mechanism of branching varies from system to system, a phenomenological description within the AWM can be constructed [18]. The *branching rule* consists of two steps:

(i) If the chosen site, j say, for the next step of the walker presently at site i is *not* the site with the lowest potential among the set A_i, then no branching is allowed; otherwise, go to step (ii).
(ii) The site in A_i with the next lowest potential, k say, is examined. If

$$[V(i,n) - V(k,n)] \geq \gamma[V(i,n) - V(j,n)], \qquad (15.9)$$

then a new walker is placed on site k, resulting in the splitting of one track into two; otherwise, no branching is allowed (Fig. 15.4).

Here the *branching factor* γ is a parameter satisfying $0 \leq \gamma \leq 1$. Note that $\gamma = 1$ implies no branching and $\gamma = 0$, sure branching.

15.4.3 Multiwalkers and Updating Rules

Once branching occurs, or when more than one walker is initially placed on the lattice, we have a situation of multiwalkers sharing a common landscape. A rule is obviously needed in updating the multiwalkers. The multiwalkers could be updated in parallel, but then a decision has to be made on what to do when two walkers try to step onto the same available site. This problem is avoided in the alternative scheme of updating the multiwalkers sequentially, and there are several ways of doing this. Here are some examples.

S1: Each walker is labeled with a fixed seniority number, for example, according to their "age," the order the walker is created. The walkers appearing simultaneously at time $n = 0$ could have their seniority number assigned randomly or by a certain well-defined scheme; the newly added one due to branching assumes a new number while the original walker from which branching occurs continues to use the old number. In each round of updating, each walker changes the landscape and walks one step in turn according to their seniority.

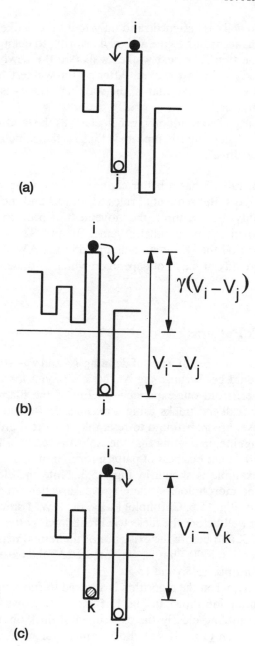

FIGURE 15.4. The branching rule. $V_i \equiv V(i, n)$, etc. (a) No branching, because the chosen site j is not the one with the lowest potential. (b) No branching, because the condition in Eq. (15.9) is not satisfied. (c) Branching occurs. The shaded circle represents the new walker added to the site with the second lowest potential.

S2: A priority to walk is assigned randomly to all the walkers. Each walker takes its turn, according to the assigned priority, to change the landscape and walk one step. The next walker walks on the newly changed landscape effected by the last walker. After all the walkers have their turns, the procedure is repeated; that is, the walking priority is usually not the same in each round.

S3: One walker is picked randomly, which changes the landscape and moves one step. The procedure is repeated; that is, the same walker could be chosen consecutively.

The sequential rule S3 has a better chance of producing walker tracks of very different lengths. But sequential rules S1 and S2 and the parallel scheme are closer to reality. For example, the movement of pedestrians in the street should be governed by the parallel scheme. S1 and S2 have been used in [9, 18]. Depending on the W function assumed in the AWM, the track morphology obtained may or may not depend on which of these two rules is used [18].

15.4.4   Track Patterns

There is an infinite number of ways of defining W and P_{ij}—the AWM is very flexible, as it should be, because the AWM is a model for various complex systems. However, from our experience, many of the filamentary patterns produced by the walkers' tracks using a reasonable W and one of the four forms of P_{ij} above, are *later* found to resemble some real growth morphologies. This is intriguing, indicating that the AWM is really in the bag of tricks of Mother Nature in her business of pattern formation.

A dramatic example is shown in Fig. 15.5. Note the close resemblance between the three morphologies; there is even excellent quantitative agreement, as shown in Fig. 15.6. Chronologically, the AWM simulation was produced by us [18] well before the dielectric breakdown pattern was created in the laboratory [25] and before the existence of the retinal neuron pattern [27] came to our attention [26]. That is, we "predicted" the existence of the two patterns shown in Figs. 15.5b and 15.5c!

More amazingly, when the algorithm is allowed to run longer (Fig. 15.7a–c), the AWM simulation shown in Fig. 15.5a actually grows to a dense radial morphology, resembling closely the experimental ones observed in electrodeposit pattern of Zn (Fig. 15.7d) and in other systems [29]. These results imply that even though the microscopic mechanisms of the three real systems are very different—the dielectric breakdown and electrodeposit ones are physical/chemical and the retinal neuron is biological—their morphologies can nevertheless be described by the unifying principle of active walks. In other words, the AW description is at a level higher than the microscopic

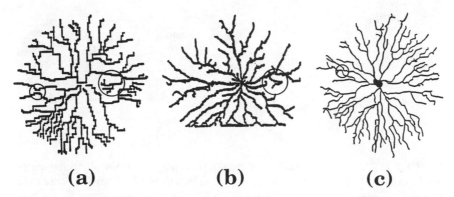

(a)   **(b)**   **(c)**

FIGURE 15.5. (a) A PAW track pattern formed by four initial walkers with branching [18]. The sequential updating rule S1 is used. $\eta = 1$; V_0 is a cone pointing upward, and $W = W_1$. (b) A dielectric breakdown pattern of olive oil put between two parallel plate electrodes [25]. (c) A retinal neuron [26]. Note the similarity in details highlighted by the circles.

description, in the sense that the Navier–Stokes equation in fluid mechanics is higher in level than the Hamiltonian description of a system of interacting molecules.

Spirals can be obtained with a time-delayed W [9, 10]. An example is given in Fig. 15.8a, which compares well with experiments shown in Fig. 15.8b. See

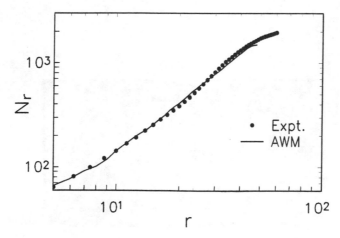

FIGURE 15.6. A quantitative comparison between the morphologies in Fig. 15.5a and c [26]. N_r is the number of pixels within a circle of radius r.

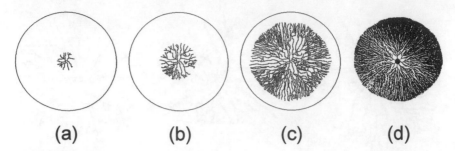

FIGURE 15.7. (a–c) Time sequence of growth from the same PAW algorithm used in producing Fig. 15.5a, which in fact is taken from the (b) here [18]. (d) The dense radial morphology observed in electrodeposit of Zn [28], in close resemblance to the simulation in (c).

[31, 32] for spirals in other systems. In Fig. 15.9, an interesting pattern formed by *one* initial walker with branching is presented. Note that the morphology changes *spontaneously* from compact to filamentary as it grows outward, without the change of any parameter in the model (see also section 15.6). More track patterns can be found in [8–10, 18, 26] and Fig. 15.11 to Fig. 15.13, with many of them similar to the filamentary patterns found in biological, physical, and chemical systems.

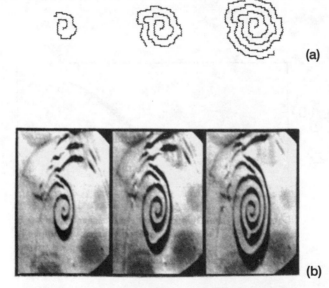

FIGURE 15.8. Growth of a spiral. (a) Time sequence from a PAW simulation. See [9]. (b) Experimental result observed in the oxidation of CO on a Pt(110) surface [30].

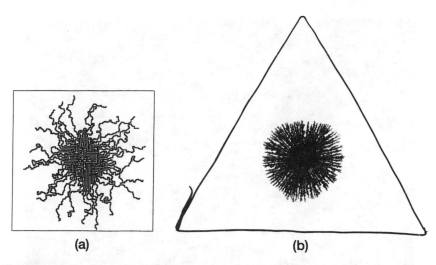

FIGURE 15.9. Spontaneous morphology transition. In the cases shown here, the pattern grows from a compact core to filamentary at larger radius. (a) A computer example from the PAW, with one initial walker and branching allowed [15]. (b) An experimental example from electrodeposits (in the presence of a triangular anode) [22].

15.5   Three Applications

The physical bases for the landscaping function, the landscaping rule, and the stepping rule—the central ingredients in the AWM—will be discussed in the context of three real systems of different origins. The dielectric breakdown of liquid is chemical/physical in origin, the ion transport in glasses is physical, and the ant trails are biological.

15.5.1   Dielectric Breakdown in a Thin Layer of Liquid

Lightning is an example of dielectric breakdown (DB) occurring in nature. In a more controlled quasi-two-dimensional experiment first carried out by Lam et al. [23], a thin layer of liquid is sandwiched between two glass plates coated with conductive transparent material on the inner surfaces. When the DC electric potential V across the plates is higher than a threshold value, DB of the liquid occurs and a filamentary pattern is left on the inner surfaces of the two plates. Examples of these DB track patterns are given in Figs. 15.5b, 15.10, 15.11a, 15.12a, and 15.13a. Essentially, when V is high, the DB process is very fast and one obtains the random pattern with little branching [23] (Fig. 15.11a). When V is of an intermediate value, one has a fast process and a radial wiggling pattern with more branching results [18] (Fig. 15.10a). When V is low, the process is slow and dense winding is observed [23] (Fig. 15.10b).

FIGURE 15.10. Experimental DB patterns from a thin layer of mineral oil, similar to those observed in [18, 23]. The cell is 3.5×4.0 cm^2 in size and of thickness 6 μm. The images are from the negative potential plate. (a) The radial wiggling pattern. $V = 310$ V. (b) The dense winding pattern. $V = 75$ V. (Courtesy of Ru-Pin Pan.)

The DB tracks correspond to the paths of the charge carriers (ions and electrons) moving between the two cell plates and are due to chemical reactions occurring at the coated inner surfaces of the plates. Assuming that the chemical reaction is favored at high temperature, the chance for the next chemical reaction to occur (i.e., where the growing tip of the DB track will go) will be proportional to the product of T and ρ, where T is the temperature field and ρ is the surface density of unconsumed chemicals on the inner

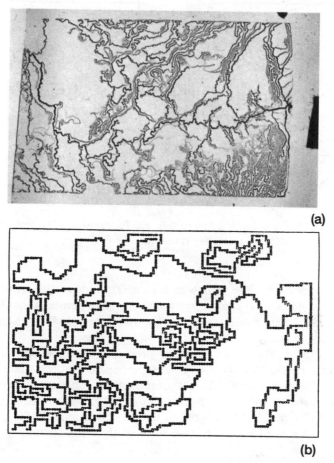

(a)

(b)

FIGURE 15.11. The random pattern. (a) Experimental DB result from a thin layer of mineral oil [23]. (b) Computer result from the PAW [9].

plates. The distribution of T is governed by a diffusion process so that at a time t_0 shortly after the chemical reaction takes place at a point in space, it will look like that in Fig. 15.14. The $T\rho$ curve has a peak and a finite range. In the AWM, because by design the walker prefers to go from high V to low V, the value of V at a point actually represents the probability that the walker does *not* want to go there and thus

$$W \sim -T\rho, \tag{15.10}$$

say. To guarantee that the walker does not cross its own track, as oberved in the DB experiment, one should assign a high W value at $r = 0$. The W_1 function assumed for W in Fig. 15.2a is a straightforward realization of all

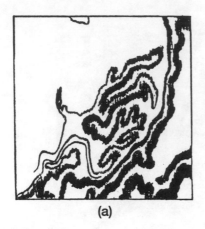 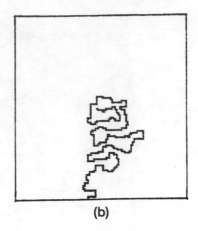

(a)   (b)

FIGURE 15.12. The dense winding pattern. (a) Experimental DB result from a thin layer of mineral oil [9]. (b) Computer result from the PAW.

these considerations. The trough in W_1 corresponds to the peak in the $T\rho$ curve; the finite range r_2 comes from the short range nature of the Guassian T function; the linear nature of W_1 in the two intervals $[0, r_1]$ and $[r_1, r_2]$ is assumed for simplicity.

The fact that heat diffusion is a linear process and the physical nature of W as described by Eq. (15.10), justify the superposition process assumed in Eq. (15.6). Realistically, one should let the landscaping function W created by the walker at each step to relax. However, in the PAW used to create Figs. 15.5a, 15.11a, 15.12a, and 15.13a this relaxation process is replaced by the use of a fixed W_1, approximating the relaxed form of W at a later time t_0. This sim-

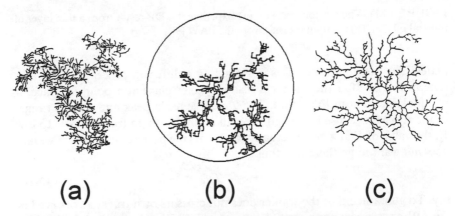

(a)   (b)   (c)

FIGURE 15.13. Patterns from the AWM and two different kinds of experiments. (a) Experimental DB result from a thin layer of nematic liquid crystal [26]. (b) Computer result from the PAW [18]. (c) Experimental electrodeposit pattern from a thin cell [33].

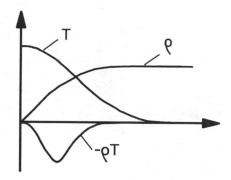

FIGURE 15.14. Sketch of T and ρ, and the product $T\rho$ [15]. r is the distance from the point where the chemical reaction takes place, t_0 time before.

plification can be physically justified for each of three DB processes described above [15]. Good agreement between the AWM simulations and experimental results can be seen in Figs. 15.5 and 15.11–15.13. Furthermore, as shown in Fig. 15.13, patterns from two different kinds of experiments agree with the same AWM simulation, offering more evidence (in addition to Fig. 15.5) of the universal power of the AW.

Under some reasonable assumptions, the two parameters r_1 and r_2 in W_1 can even be related to the physical characteristics of the DB experiment [15]; namely,

$$\sinh(2r_1/w) = 4\chi t_0/(wr_1) \tag{15.11}$$

and

$$r_2 = 3(\chi t_0)^{1/2}, \tag{15.12}$$

where w is the observed DB track width and χ is the thermal diffusivity within the cell. To determine the value of W_0 in W_1, more detailed knowledge about the DB process is needed.

15.5.2 Ion Transport in Glasses

To explain, for example, the existence of a deep minimum in the ionic conductivity as a function of the ratio of Na to Li in mixed Na/Li silicate glasses (Fig. 15.15), and the drastic increase of the DC conductivity in single ionic glasses with the concentration of mobile ions, a dynamic structure model was proposed by Bunde and co-workers [34–36]. The model is physically motivated [36] and is based on the assumption that ions maintain their distinct local environments, leading to the formation of fluctuating structural pathways. This model is in fact a BAW model [15]. To put it in our language, the model described below [15] is a version differing slightly from that in [34–36], but this difference is not believed to be crucial.

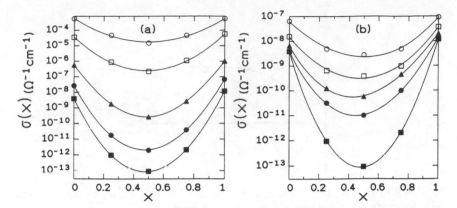

FIGURE 15.15. Experimental conductivity $\sigma(x)$ as a function of x in $x\mathrm{Na_2O}(1\text{-}x)\mathrm{Li_2O}$-$3B_2\mathrm{O}$. (a) Fixed frequency $\omega = 0$; temperature T increases from 25°C to 300°C for the bottom to top curve. (b) Fixed temperature $T = 25$°C; ω increases from 0 to 100 kHz for the bottom to top curve. The data are from M. Tomozawa and M. Yoshiyagawa [Glastech. Ber. **56k**, 939 (1983)], as redrawn in [35].

For simplicity of description, consider the case of a single type of mobile ions, A say. The model assumes two types of vacant interstitial sites: $\bar{C}$ and $\bar{A}$. $\bar{A}$ is a vacant site previously occupied by an A ion, and $\bar{C}$ is a vacant site not adjusted to an A ion. We assume that a vacant $\bar{C}$ ($\bar{A}$) site has potential $V_{\bar{C}}$ ($V_{\bar{A}}$); *any* site occupied by an A ion will have its potential changed to V_A instantaneously, with $V_{\bar{C}} > V_A > V_{\bar{A}}$; a vacant $\bar{A}$ site will change into a vacant $\bar{C}$ site in time τ; a site left vacant by an A ion will have its potential changed to $V_{\bar{A}}$ in time τ. Because ion jumping between sites is an activated process, the BAW stepping rule is used so that $w_{A\bar{A}} = \exp[(V_A - V_{\bar{A}})/T]$ and $w_{A\bar{C}} = \exp[(V_A - V_{\bar{C}})/T]$. Here $w_{A\bar{A}}(w_{A\bar{C}})$ is the probability for an A ion to jump to a nearest neighbor site $\bar{A}(\bar{C})$. Note that $w_{A\bar{C}} = w_{A\bar{A}} \exp(-T_1/T)$ with $T_1 = V_{\bar{C}} - V_{\bar{A}}$. In this BAW model of ion transport, there are three types of sites: vacant $\bar{C}$ and $\bar{A}$, and occupied A sites, each having its own characteristic potential. There are two W functions according to whether the site the walker moves in is $\bar{C}$ or $\bar{A}$; both of which are delta functions—the former has a negative height $V_A - V_{\bar{C}}$, the latter has a positive height $V_A - V_{\bar{A}}$. The case of two types of mobile ion can be handled similarly. Presumably the parameters involved in this model can be determined experimentally.

Numerical results from the original model [35] are in excellent agreement with experiments. An example is shown in Fig. 15.16. In these simulations, multiwalkers without branching are used.

15.5.3   Ant Trails in Food Collection

As mentioned in section 15.2 ants are natural active walkers, with $-V$ representing the scent density emitted by the ants. Indeed, an active walk simu-

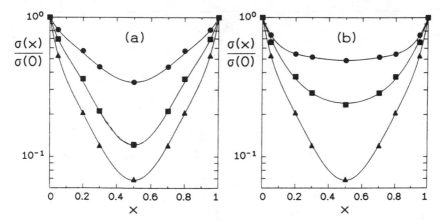

FIGURE 15.16. Normalized real part of the AC conductivity $\sigma(x)/\sigma(0)$ as a function of x, from computer simulations [35]. Note the close agreement between Figs. 15.15 and 15.16. (a) Fixed frequency; temperature increases from bottom to top. (b) Fixed temperature; frequency increases from bottom to top.

lation of the behavior of an ant swarm in food search and collection is provided by Schweitzer et al. [37]. In this work, the ants come out from a nest at the center of a triangular lattice. Two different food source distributions are considered: a continuous food distribution at the top and bottom lines of the lattice, and a random distribution of five separated food sources.

Motivated by real ant behaviors [38], two kinds of scent are used: Scent A (B) is dropped by those ants which do not (do) have food in their "hands." The simulation is divided into two stages. In the beginning, a group of ants, the "scouts," came out of the nest and are guided in their motion by Scent A before and after they find food. Once a food-carrying ant returns to the nest spontaneously, guided by the high concentration of Scent A near the nest and not by any external direction, a new group of ants, the "recruits," come out of the nest. Each recruit follows the trail of Scent B (A) before (after) it finds the food. Therefore, two landscape potentials (corresponding to Scents A and B) are used in this model.

Each ant acts as an active walker and moves toward high-scent regions. The inertial effect of the walker is included by giving a preference to the forward direction in the stepping rule. Roughly speaking, it is a kind of PAW with a delta-function W (see [37] for more details).

The agreement between this active walk simulation and real ant behavior is, at least qualitatively, amazingly good. For the case of food distribution along the top and bottom lines, a dendritic foraging track pattern of the ants (as shown in Fig. 15.17) is produced that compares very well with what is actually observed from desert ants [39]. For the case of five random food sources, the ants essentially attack and exhaust the food sources one at a time, a behavior conforming to reality (see, e.g., p. 234 of [1]). In short, the active

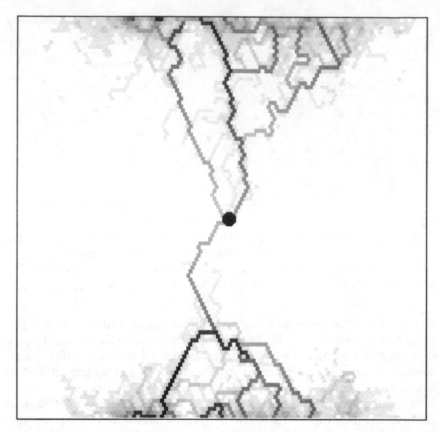

FIGURE 15.17. Trails of ants carrying food home, simulated with an active walk model [37]. The trails exist as a concentration of phermone/Scent B as dropped by food-carrying ants. The grey scale indicate the concentration. Note the branching of the trails.

walk model is able to describe the self-organized behavior of ants in food collection, for the two cases considered here.

15.6   Intrinsic Abnormal Growth

Abnormal growth represents exceptional growth behaviors or patterns [10, 40, 41]. Some are harmless; for example, the appearance of grey hairs when a person ages. Others, such as the cancerous growth of cells in a human body, could be deadly. Two types of abnormal growth are of particular interest because they are often observed experimentally. The first type is *transformational* growth, the occurrence of qualitative change in behavior during a growth process. The second type is *irreproducible* growth, the case that distinctively different growth forms are obtained from presumably identical samples.

The cause for both types of growth is usually attributed to an extrinsic mechanism. For transformational growth, it is assumed that an external control parameter changes during the growth process; for example, a mother's intake of drugs can cause brain damage to the growing fetus. We might blame irreproducible growth on unnoticed differences in the preparation of the samples, or perhaps the carelessness of the student who performs the experiment. Although extrinsic mechanisms are certainly responsible in many cases, for the nonequilibrium systems under consideration, is it possible that some growth processes are intrinsically abnormal?

A nonequilibrium system is an open system in which noise or some form of stochasticity is ever-present. Although noise was considered to be a nuisance in the past, in recent years noise has been recognized as a source of order and complex behavior in its own right, such as in noise-induced transitions, stochastic resonance, and stochastic chaos [42]. For example, noise is certainly present in the dynamics of a sandpile, a disordered system in which friction and voids between the granular particles play important roles. Indeed, it is found that as a semicylindrical drum of sand is rotated slowly, when the slope of the sandpile surface θ is below a certain angle θ_r, the angle of repose, the pile is always stable and is stationary. For $\theta > \theta_m$, an angle about 2° above θ_r, the pile is unstable and a global avalanche occurs. Within the range of $\theta_r < \theta < \theta_m$, however, the pile can either flow or be stationary and the result is irreproducible [43]. Furthermore, the use of sandpiles to check the power laws predicted by self-organized criticality (SOC) [44–46] gives differing experimental results [47, 48]. These and other examples [15, 40, 41] show that irreproducibility is a real issue in nonequilibrium systems.

It turns out that, in principle, noise can indeed cause transformational and irreproducible growths, as demonstrated by the AWM. For this purpose, two versions of the AWM are used. The first version is the PAW model, showing transformational growth as depicted in Fig. 15.9a. The second version is the so-called boundary probabilistic active walker (BPAW) model [10, 40].

In the BPAW, starting from a seed particle from the center, one of the perimeter sites is chosen with a rule similar to the PAW rule. A new particle is placed at this chosen site, and the process is repeated. Each particle, including the seed, changes the potential according to the landscaping rule like an active walker does. Note that the particles do not actually move; the pattern is formed by an aggregation process. Examples of transformational and irreproducible growths produced by the BPAW are presented in [10, 40, 41]. All these abnormal growths occur spontaneously due to noise—as represented by the random numbers used in the algorithm—and are thus intrinsic in nature.

Similar to chaos and SOC, intrinsic abnormal growth (IAG) is a new paradigm concerning exceptional phenomena [40]. All three paradigms provide unexpected, intrinsic mechanisms for explaining phenomena that were thought to have extrinsic origins. The existence of chaos reminds us that some apparently random behavior can be explained by deterministic nonlinear equations; the signature is strange attractors and sensitive dependence on initial conditions. The existence of SOC suggests that some complex, dissipative systems can adjust by themselves toward a critical state; the signature is power laws in space and time. The IAG shows the sensitive dependence of growth and form on noise; the signature is the existence of sensitive zone(s) in the morphogram [40], the "phase" diagram in a parameter space for non-equilibrium systems.

In a sensitive zone more than one type of growth morphology is generated from different computer runs, due to the differing sequence of random numbers employed in each run. This is best illustrated by the BPAW model. With the choice of $W = W_1$ and by varying the two parameters η and ρ ($\equiv r_1/r_2$; see Fig. 15.2a), the BPAW is capable of producing morphologies ranging from compact to filamentary, which can be classfied into five classes: blob, jellyfish, diamond, lollipop, and needle (Fig. 15.18). As can be seen in Fig. 15.19, there is a region in the middle bounded by the two solid lines—the sensitive zone—within which, for the same set of model parameters, more than one type of morphology is generated from different computer runs. In other words, if experiments are performed and repeated with fixed parameters inside the sensitive zone, different growth morphologies will be obtained—the result is irreproducible. A corollary is that two (or more) morphologies may coexist in the sample [40], a situation that seems to have been observed by Ben-Jacob et al. in the growth of bacterial colonies [49].

For some reason, noise plays a more important role in the AWM than in other stochastic models such as the diffusion-limited aggregation model and the dielectric breakdown model (see part III). In contrary to the AWM, in the latter two cases, statistically similar patterns are always obtained in repeated runs of the algorithms for the same set of parameters. To make the time-consuming construction of the morphogram more efficient, a neural network classifier for active walk patterns, with a success rate of about 99% and based on the use of radial basis functions, has been constructed by Castillo et al. [50].

It should be emphasized that IAG and sensitive zones are believed to exist objectively in nature, independent of the AWM. The latter is used merely as a computer model to demonstrate the feasibility of the former.

In short, some interesting and important abnormal growth in nature (such as cancer!?) could just be intrinsic in nature. A search for sensitive zones in experimental systems (including electrodeposit systems [41]) as suggested by Lam et al. [40] would be welcomed.

15.7   Landscapes and Rough Surfaces

The study of the formation and dynamics of rough surfaces, some of which are self-affine fractals [51], is important in basic and applied researches.

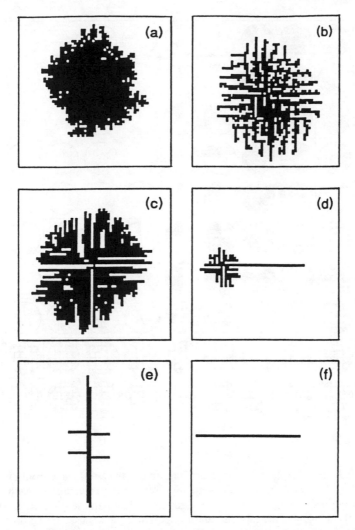

FIGURE 15.18. The five classes of patterns obtained from the BPAW model [40]. (a) blob; (b) jellyfish; (c) diamond; (d) lollipop; (e) and (f) needle.

Examples of rough surfaces include crystal growth, vapor deposition, electroplating, biological growth, corrosion, and photoablation. Although there is an undeniable need to construct realistic models in modeling physical rough surfaces, especially from the industrial point of view, the study of simplified models delineating each aspect of the physical mechanisms is equally important, as evidenced by the study of the Ising and other models in relation to phase transition and ferromagnetism.

The landscape (resulting from the action of the walker) in the AWM is a fractal surface [18], the properties of which are described in this section.

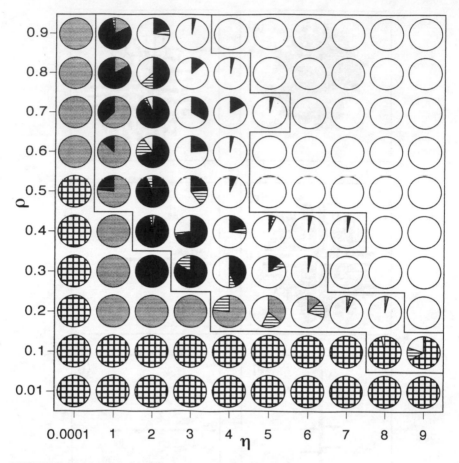

FIGURE 15.19. Morphogram, or phase diagram, in the (η, ρ) plane from the BPAW model [40]. The piechart at each point represents the percentage of each kind of pattern obtained from 30 runs of the algorithm. The parameters used in these 30 runs are the same, but the random number sequence (noise) used in the probabilistic rule varies from run to run. The grid corresponds to blob, grey to jellyfish, black to diamond, horizontal shades to lollipop, and white to needle.

15.7.1   Groove States

Many physical rough surfaces are formed by the combined effects of two processes: deposition and relaxation. Both these two processes are included in the PAW model in $(1 + 1)$ [18] and $(2 + 1)$ dimensions [26]; power laws are indeed found and rough exponents are calculated. In [18] the $W(i)$ function, for simplicity, is assumed to be given by $W(0) = W_0$, $W(\pm 1) = -1$, and $W = 0$ otherwise. $W_0 > 2 \, (<2)$ corresponds to deposition (evaporation), and $W_0 = 2$ is the case with the average height of the surface conserved. For $W_0 = 1$, a surface consisting of a "periodic" structure of deep grooves is obtained by Lam et al. [18] (Fig. 15.20a), which resembles the experimental Si growth on Si(100) surfaces [52] (Fig. 15.20c). In this model probabilistic active walkers are dropped randomly on the surface in sequence; each walker is allowed to walk until it dies and is removed before the next walker is deposited. The exponent β (called α in [18]) is 3/4 before it changes to 1 at large time. Here the exponent β is defined by $\sigma_T \sim n^\beta$ and, for later use, α by $\sigma_\infty \sim L^\alpha$ [51], where the root-mean-square surface height σ_T is given by

$$\sigma_T(L, n) \equiv \left\{ L^{-1} \left\langle \sum_i [V(i, n) - V_{av}(n)]^2 \right\rangle \right\}^{1/2}, \tag{15.13}$$

where $\langle \dots \rangle$ represents an average over N_R runs, $V_{av}(n)$ is the height averaged over special sites is, and L is the size of the initially flat surface.

To pinpoint the origin of the grooves, a model with relaxation alone is studied by Pochy et al. [21] and Kayser et al. [26]. In this BAW model [21, 26] a *single* Boltzmann walker walking forever is used. A small number (one or two) of grooves remain in the surface when $W_0 < 2$ (at time $n = 10^4$, as shown in Fig. 15.20b), indicating that relaxation, and not the particular stepping rule, is essential for the formation of grooves.

15.7.2   Localization–Delocalization Transition

However, relaxation alone is not sufficient for grooves to form. As shown by Kayser et al. [26] for the BAW model in $(1 + 1)$ dimensions with integral values of W_0, for fixed temperature T and L, the behavior of σ_T undergoes a sudden change as W_0 is varied across 2. Specifically, σ_T saturates at large n if $W_0 \geq 2$; for $W_0 < 2$, σ_T seems to diverge with a power law such that $\sigma_T \sim n^\beta$ with $\beta = 1$ [26] (Fig. 15.21). Apparently, one has a first-order–like phase transition if the inverse of σ_∞ [$\equiv \sigma_T(L, \infty)$] is plotted against W_0, which shows a peak at $W_0 = 2$ (Fig. 15.22c). (Computationally, $n = 10^4$ is used as ∞ in σ_T here.) This phase transition is in fact a transition between a grooved surface (when $W_0 < 2$) and a rough surface (when $W_0 \geq 2$) [15]. This result can be understood as follows: For $W_0 < 2$, more surface is decreased around the walker than is being added to the walker's site; the walker does not stand high enough to be able to eventually escape from the trough it creates, and a groove of ever-increasing depth is formed. As far as the movement of the walker is concerned, the transition at $W_0 = 2$ is thus a localization–delocalization transition. In the localization regime the walker is self-trapped, leading to a grooved surface; in the delocalization regime the movement of the walker is extended, resulting in a rough surface.

If nonintegral value of W_0 is allowed, for fixed T and L, the $1/\sigma_\infty$ versus W_0 curve in fact peaks at $W_0 = W^* < 2$. Moreover, W^* depends on L with $W^* \to 2$ as $L \to \infty$ [15].

Note that the movement of a probabilistic active walker is guided by the

384   L. Lam

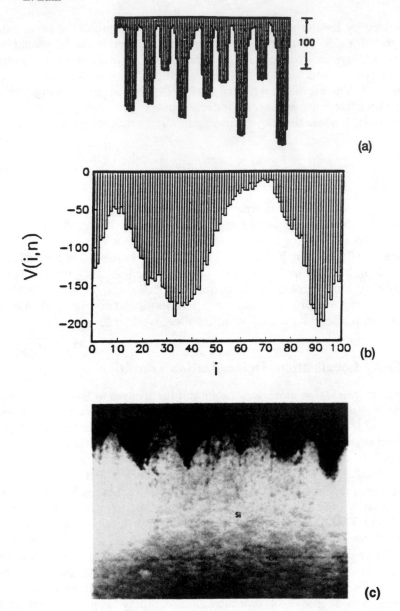

(a)

(b)

(c)

FIGURE 15.20. Groove surfaces. (a) A simulation from the PAW in $(1 + 1)$ dimensions, with active walkers dropped in sequence [18]. $W_0 = 1$. (b) A simulation from the BAW in $(1 + 1)$ dimensions, with a single active walker walking forever [10]. $W_0 = 1$; $T = 1000$. (c) Experimental Si film grown by MBE [52].

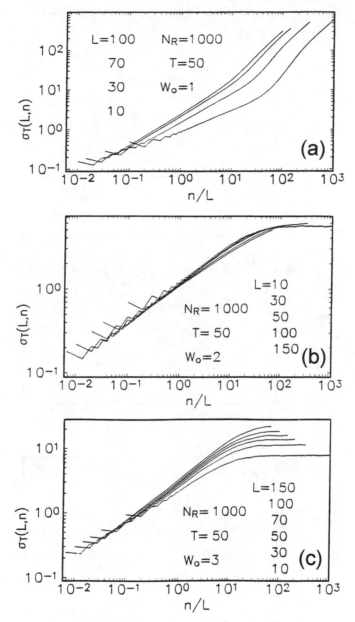

FIGURE 15.21. Dependences of σ_T on n, L, and W_0 from the BAW model. $T = 50$; $N_R = 1000$. (a) $W_0 = 1$. σ_T does not saturate at large time n. (b) $W_0 = 2$. σ_∞ is independent of L. (c) $W_0 = 3$. σ_∞ does depend on L.

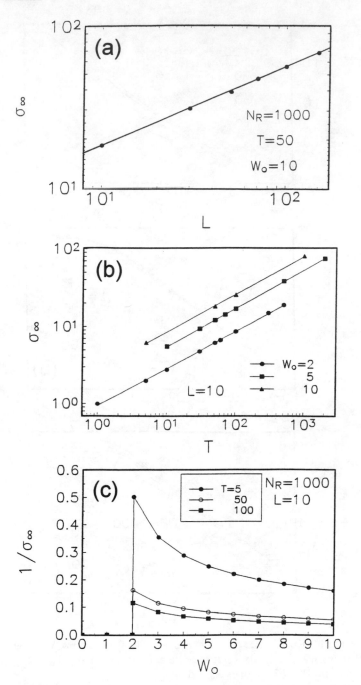

FIGURE 15.22. Dependences of σ_∞ on L, T, and W_0 from the BAW model.

potential gradient, and a Boltzmann active walker moves by an activated process. But because, as pointed out above, the formation of grooved surfaces is insensitive to the stepping rule of the walker, one may use Eqs. (15.3) and (15.4) with $D = 0$ to study this situation. For the particular W function used in our computer studies [21, 26], Eq. (15.4) reduces to [15]

$$\partial V(r,t)/\partial t = (W_0 - 2)p(r,t) - \partial^2 p(r,t)/\partial r^2, \qquad (15.14)$$

whereas on the right-hand side, the first term is a source (sink) for V if $W_0 > 2$ (< 2), and the second term is a negative diffusion term that helps to raise the height V if the walker spreads out. Equation (15.14) can be integrated with time and combined with Eq. (15.3), resulting in an integral equation for $p(r,t)$,

$$\partial p(r,t)/\partial t = (\partial/\partial r)\left\{ p(r,t)(\partial/\partial r)\int dt' [(W_0 - 2)p(r,t') - \partial^2 p(r,t')/\partial r^2] \right\}$$
$$+ D_W \partial^2 p(r,t)/\partial r^2. \qquad (15.15)$$

Equation (15.15) can be solved, at least numerically, and gives insight on the localization–delocalization problem. (See [53, 54] for related discussions on groove formation.) Note that in the simulation in $(2+1)$ dimensions [55], three-dimensional grooves do form at the locations where the active walkers cluster; clustering and groove-forming are two sides of the same coin in this particular problem.

15.7.3 Scaling Properties

The dependences of σ_T on L, n, T and W_0 from the BAW model are shown in Figs. 15.21 and 15.22. The scaling properties of these surfaces [21, 26] differ from those of other thermally activated models [56, 57]. For example, the exponent β increases monotonically from zero at $T = 0$ and saturates at high T in our BAW [21] (Fig. 15.23), while β decreases monotonically with increasing T in [57] or possesses a dip at a finite temperature in [56]. The temperature dependence of α is also different; in the BAW $\alpha = 0$ when $W_0 = 2$ [21], but increases with T and W_0 for $W_0 > 2$ (Fig. 15.24).

In the BAW model with $W_0 = 2$, the surface height is conserved and the scaling is particularly interesting: one finds

$$\sigma_T(L,n) \sim T^\gamma g(n/T^{\gamma/\beta(T)}) \qquad (15.16)$$

where $\gamma \approx 0.48$; $g(x) = $ const for $x \gg 1$, and $g(x) = x^{\beta(T)}$ for $x \ll 1$. Note that Eq. (15.16) has the same form as that in most other models [51], except that on the right-hand side of Eq. (15.16), T replaces L and β is temperature dependent: the spatial fluctuations are limited by T instead of L in the BAW when $W_0 = 2$. Equation (15.16) breaks down when $W_0 > 2$, the case without a conservation law in height, because α no longer vanishes (see Fig. 15.24). A

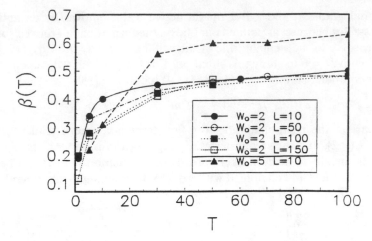

FIGURE 15.23. Dependences of the exponent β on T and W_0 from the BAW model.

theoretical understanding of these results and the new universality classes is needed.

Note the qualitative agreement between the landscapes from the AWM and real rough surfaces, as shown in Fig. 15.20 and Fig. 15.25. Consequently, the AWM may have practical as well as theoretical significance in the study of the formation of real surfaces. Furthermore, the BAW model with a single walker [21, 26] is related to two physical systems of much current interest. The first system is the motion of a single flux line in a superconducting sand-pile [58], in which the scaling law of Eq. (15, 16) also appears. The second system concerns solitons and $1/f$ noise in a molecular chain [59].

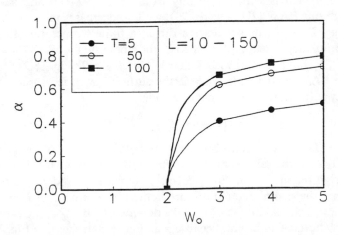

FIGURE 15.24. Dependences of the exponent α on W_0 and T from the BAW model.

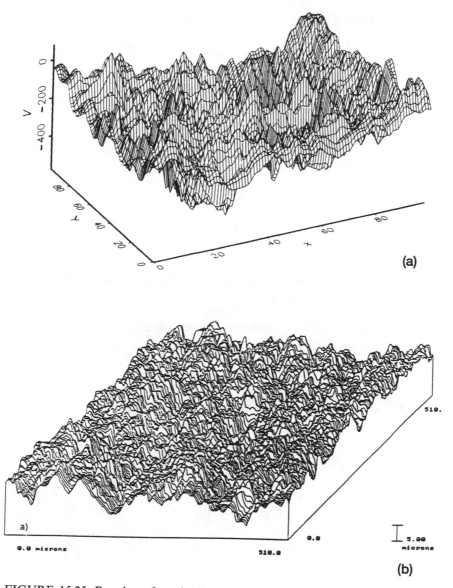

(a)

(b)

FIGURE 15.25. Rough surfaces in $(2 + 1)$ dimensions. (a) A simulation from the PAW, with active walkers dropped in sequence [26]. $W_0 = 15$. (b) An experimental sand-blasted brass surface [60].

15.8   Fuzzy Walks

When the active walker in the AW is identified with a real-life walker such as an ant or a human being, the first three kinds of mathematical stepping rule (i–iii in section 15.4.1) assumed for the active walker are not very realistic. It is hard to believe that an ant, or even a human being, will measure precisely the landscape heights around itself and make a mathematical calculation to determine P_{ij} before it actually proceeds with the next step. Ants, like humans, do it in a fuzzy way in their walks: the landscape is surveyed roughly and an estimate of some kind is made in deciding on the next step.

The theory of fuzzy systems invented by Zadeh in 1965 [61], now enjoying a wide range of applications in science and engineering [20, 62], seems to be a suitable vehicle in describing fuzzy walks [19]. Because fuzzy control theory is not familiar to most physicists, we give here a description of the basics via the simple example of a fuzzy walk, before we embark on the fuzzy active walk.

To be concrete, consider a fuzzy walk representing a partially drunk walker in one dimension. Assume that signal S_R (S_L) from the right (left) is received by the walker. A sober walker (SW) will move in the direction of the stronger signal; a drunk walker cannot differentiate the signals and will move at random (RW). A partially drunk walker is governed by a set of linguistic fuzzy rules as follows.

Rule 1: If $S_R \gg S_L$, step to the right.
Rule 2: If S_R is comparable to S_L, step to left or right at random.
Rule 3: If $S_R \ll S_L$, step to the left.

For simplicity, $S_R - S_L$ is assumed to be a linear function of x, the position of the walker, and normalized to be between -1 and 1 (Fig. 15. 26). To model the linquistic rules, three membership functions m_B, m_C, and m_S—representing the degrees of big, comparable, and small, the languages used in the fuzzy rules—are used for the input $(S_R - S_L)$ (Fig. 15.27). A similar scheme is used for the output membership functions m_R, m_N, and m_L (Fig. 15. 28).

The membership function associated with each rule is defined by

$$m_1(W) = m_B \wedge m_R(W)$$

$$m_2(W) = m_C \wedge m_N(W) \tag{15.17}$$

$$m_3(W) = m_S \wedge m_L(W),$$

where W is the output and $F \wedge G \equiv \min\{F, G\}$, where F and G are any two real numbers. The membership function for W is given by

$$m(W) = m_1(W) \vee m_2(W) \vee m_3(W), \tag{15.18}$$

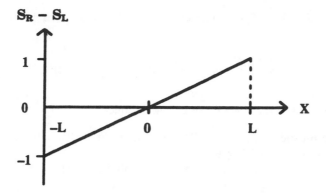

FIGURE 15.26. The assumed input function, $S_R(x) - S_L(x)$ versus x. This function is normalized and L is half the lattice size.

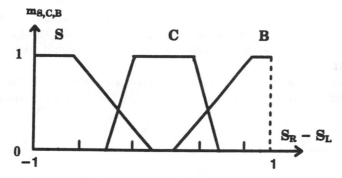

FIGURE 15.27. Membership functions m_S, m_C, and m_R for the input $S_R - S_L$; the latter is obtained from Fig. 15.26 for each x. Here S represents small, C comparable, and B big.

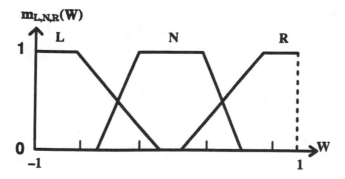

FIGURE 15.28. Membership function m_L, m_N, and m_R for the output W. Here L represents left direction, N (neutral) either left or right, R right direction.

where $F \vee G \equiv \max\{F, G\}$. The crisp output W_0 is given by the centroid rule,

$$W_0 = \int_{-1}^{1} m(W)W \, dW \Big/ \int_{-1}^{1} m(W) \, dW. \qquad (15.19)$$

Two versions of fuzzy walk can be formulated:

(i) Deterministic fuzzy walk (FW/D): Go right if $W_0 > 0$; go random if $W_0 = 0$; go left if $W_0 < 0$.
(ii) Probabilistic fuzzy walk (FW/P): Probability of going right $P_R = (W_0 + 1)/2$; $P_L = 1 - P_R$.

Computer results for $\langle x^2 \rangle$ as a function of time t are shown in Fig. 15.29, where $\langle \ldots \rangle$ represents average over N_R runs. In each run, the walker starts at the center, $x = 0$. In both FW/D and FW/P, the $\langle x^2 \rangle$ curve has an initial linear region where the walker moves randomly near the center, like an RW, when the left and right signals are comparable; the curve then turns up toward the SW line as the difference between the left and right signals becomes more distinguishable.

For the BAW, as given by Eq. (15.8), $P_{ij}(t) \propto \exp\{[V(i,t) - V(j,t)]/T\} \equiv U_{ij}(t)$, where n is rewritten as t, and $j = i \pm 1$ in one dimension with nearest-

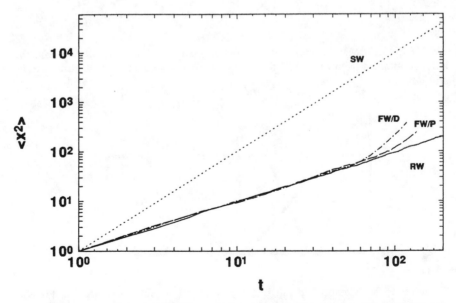

FIGURE 15.29. $\langle x^2 \rangle$ versus time t for different walks. (See Fig. 4 of [19] for parameters used.) FW/P: fuzzy walk (probabilistic version); FW/D: deterministic version SW: sober walk; RW: random walk.

neighbor jumps. In the fuzzy BAW, one may define the signal difference by

$$S_R(i) - S_L(i) \equiv (U_{iR} - U_{iL})/(U_{iR} + U_{iL}), \qquad (15.20)$$

where R and L correspond to the $i + 1$ and $i - 1$ sites, respectively. The rest of the model for determining the next step then proceeds as in the fuzzy (passive) walk described above, except that when the walker is at the new site it changes the landscape as in the BAW, and the whole process is repeated. The σ_T versus t curve from the landscape of the fuzzy BAW has been calculated, which shows the same qualitative behavior as for the nonfuzzy version [19].

The work described above is preliminary; there remain a lot of unknowns. As with any fuzzy system, the membership functions and the control scheme are not uniquely defined by the problem. Can one tell whether the walker is fuzzy or simply biased but probabilistic by looking at the $\langle x^2 \rangle$ curve, or any other curve? Does the fuzziness of ants show up in their trail patterns, or anywhere? Note that the study of fuzzy systems is rather new in the realm of physics. The study of fuzzy walks [19], simple and basic as it is, represents one of the early attempts in this direction.

15.9   Related Developments and Open Problems

In section 15.4.1 the DAW corresponds to $F = 0$ in Eq. (15.1), and the stepping rule is the steepest descent. In both the DAW and PAW, the meaning of the potential is very clear. However, in the BAW, while the potential V looks like the "energy level" at a site, because of the stepping rule $P_{ij} \propto \exp[(V_i - V_j)/T]$ [with $V_i \equiv V(i,n)$], it is more accurate to interpret $-(V_i - V_j)$ as the activation energy between the sites i and j.

Although the landscaping function W could be time dependent, the special case of a constant W is found to be quite sufficient in many important applications (see Figs. 15.5, 15.7, 15.11, and 15.13). In this connection, it is interesting to note that a constant potential—like the $1/r$ gravitational or Coulomb potential associated with a mass or a charged particle, respectively—is not uncommon in the realm of physics.

Track statistics for the PAW are studied by Freimuth and Lam [9], and Lam et al. [18]. The ensemble averaged end-to-end length R_e and the radius of gyration R_g are found to vary approximately as n^ν for the track of a single walker without branching. The exponent varies with the parameters used in the W function. In the so-called Red Queen's walk—a variation of the DAW—studied by Freund and Grassberger [63] the W function is assumed to be $W(i) = -\delta_{i0}$, where δ is the Kronecker delta, empty sites are allowed to relax monotonically, and self-crossing is rare but allowed. Furthermore, if more than one neighboring site has the same lowest V, a deterministic rule of preference (e.g., straight, left, right) in choosing them is assumed. For this model, the function R_n, defined by $R_n \equiv \langle |\mathbf{R}(t + n) - \mathbf{R}(t)|^2 \rangle^{1/2}$ is found to obey a power law, with logarithmic corrections at large n. Such a logarithmic correction appears also in the "true" self-avoiding walk proposed by Amit et al. [64], which in fact is a BAW with $W(i) = \delta_{i0}$, self-crossing allowed [21].

In the PAW the walker is attracted more or less toward the potential minimum and is not allowed to go uphill. However, if the latter rule is relaxed the walker will be able to walk forever. And if a number of such walkers are placed together on the same plane, they may attract each other indirectly through their individual interactions with the common potential, assuming that the W function is negative at the center. Clustering of the active walkers may occur unless the probabilistic noise built into the stepping rule is large enough to offset the effect of the attractive interaction. Depending on the relative strength of these two counteracting effects, a phase transition between a homogeneous state and a clustering, inhomogenous state may result. A very interesting study of such a phase transition is carried out computationally and formulated analytically by Schweitzer and Schimansky-Geier [55] (see also [15]).

Active walk is a new field of research and there are still many loose ends and open problems. On the theoretical side, many of the computer results need to be understood; the Langevin or Fokker–Planck equations corresponding to the computer models need to be *derived* and solved. Although there are some theoretical studies of active walkers in modeling real systems, notably the study of ant swarms by Millonas [65], analysis beyond the mean field approximation is very difficult. Perhaps new techniques have to be invented, and modeling of more real systems is definitely welcome.

As pointed out elsewhere, many natural phenomena could be modeled by the AWM, including river formation, movement of worms or bacterial cells, polymer reptation, excitable media, sandpiles, and evolutionary biology [10]. All these problems remain to be tackled. In addition, percolation in a soft material is a perfect system to be studied by the AWM. As a liquid flows through a soft porous medium, it changes the structure of the medium (e.g., the size and shape of the cavities) which then influences the subsequent flow. Another example is the microscopic modeling of a real sandpile as suggested by Mehta [66]. In this model, the free surface of a sandpile is divided into a collection of local clusters, and each cluster is represented by a multiparticle potential well. If a particle is ejected from its well, it will change the landscape of its own well and that of the one in which it lands. Obviously, the particle is an active walker.

Population distribution and flow is an example in social science which can be modeled by an AWM. For example, the population in a city can be represented by the density of active walkers; the walkers change the "potential"— for example, the index of living—of that city, and may move from city to city according to the gradient, or some more complicated function, of this potential. An ecological system can be modeled similarly. Recently, the AWM has been used by Helbing et al. in modeling pedestrian dynamics [67].

The active walker described here can be thought of as an "entity" equipped with a sensor to measure the potential around it, a simple calculator/brain to do a calculation from the measurements, and rotor/legs to move itself according to the calculation. It is not difficult to imagine that the "intelligence" of the active walker can be improved; for example, by assigning it memory and internal states, and the capacity to evolve its strategy of walking or modifying the landscape (using, perhaps, genetic or evolutionary algorithms [68]). In fact, in the study of pattern formation in biological systems, walkers with internal states have been adopted as the "bions" by Kessler and Levine [69], and in the bacteria modeling by Ben-Jacob et al. [70]. In some sense, the entities in these models can be considered as active walkers.

In the future, micromachines or nanorobots may become practical. For example, hundreds or thousands of these micromachines could be released into the veins of a human body to do a repairing job. For practical reasons, it would be preferred to build the micromachines as simply as possible and take advantage of their self-organized, collective behavior to get the job done in the most efficient way—something like the ants do in their food collecting effort. This type of potential but important application provides yet another impetus to study simple active walkers.

15.10   Conclusions

To summarize, an active walk is one in which the walker changes its environment as it walks, and its next step is influenced by the changed environment. In its simplest form, the landscaping action of the walker remains the same at every step, and the AWM is a template-laying model. More generally, the landscaping action of the walker and/or the environment itself may be time dependent. An active walk is certainly an adaptive walk [46], but not vice versa.

As shown above, in particular, through the three diverse examples in section 15.5, the AWM is very simple but powerful. The AWM can produce many patterns—both filamentary and compact—and rough surfaces, many of which are observed in nature. The AWM can produce spontaneous morphological changes and other intrinsic abnormal growth, which are believed to be general phenomena existing in nature. The landscapes produced in the AWM are fractal surfaces with unique properties. Furthermore, the AWM can simulate self-organized, emergent behavior of complex systems, such as ant swarms, urban growth, and increasing returns in economics [71]. Research in this exciting, new field of active walks has progressed very quickly in the last few years, but many fundamental and applied problems remain to be solved.

Finally, we note that the search for a unifying principle of self-organization is an ongoing endeavor [7, 46, 72]. Our feeling is that such a unifying principle should and has to exist. But there may be more than one unifying principle governing nonequilibrium, open systems. (After all, *three* laws are required to describe the thermodynamics of equilibrium, closed systems.) And, we believe, the *principle of active walks* is one of them.

References

[1] G. Nicolis and I. Prigogine, *Exploring Complexity* (Freeman, New York, 1989).

[2] H. Haken, *Advanced Synergetics* (Springer-Verlag, New York, 1983).

[3] *Emerging Syntheses in Science*, edited by D. Pines (Addison-Wesley, Menlo Park, 1988).

[4] M. Gell-Mann, *The Quark and the Jaguar* (Freeman, New York, 1994).

[5] M.M. Waldrop, *Complexity* (Simon & Schuster, New York, 1992).

[6] *Modeling Complex Phenomena*, edited by L. Lam and V. Naroditsky (Springer-Verlag, New York, 1992).

[7] *On Self-Organization*, edited by R.K. Mishra, D. Maaß, and E. Zwierlein (Springer-Verlag, New York, 1994).

[8] L. Lam, in *Lectures on Thermodynamics and Statistical Mechanics*, edited by M. Costas, R. Rodriquez, and A.L. Benavides (World Scientific, River Edge, 1994).

[9] R.D. Freimuth and L. Lam, in [6].

[10] L. Lam and R.D. Pochy, Comput. Phys. **7**, 534 (1993).

[11] *Nonlinear Structures in Physical Systems: Pattern Formation, Chaos, and Waves*, edited by L. Lam and H.C. Morris (Springer-Verlag, New York, 1990).

[12] *Spatio-Temporal Patterns in Nonequilibrium Complex Systems*, edited by P.E. Cladis and P. Palffy-Muhoray (Addison-Weslely, Menlo Park, 1995).

[13] L. Lam, *Nonlinear Physics for Beginners* (World Scientific, River Edge, 1996).

[14] *Lectures in the Sciences of Complexity*, edited by D.L. Stein (Addison-Wesley, Menlo Park, 1989).

[15] L. Lam, *Chaos Solitons Fractals* **6**, 267 (1995).

[16] Y. Aharonov and D. Bohm, Phys. Rev. **115**, 465 (1959).

[17] See, e.g., K. Huang, *Statistical Mechanics* (Wiley, New York, 1987).

[18] L. Lam, R.D. Freimuth, M.K. Pon, D.R. Kayser, J.T. Fredrick, and R.D. Pochy, in *Pattern Formation in Complex Dissipative Systems*, edited by S. Kai (World Scientific, River Edge, 1992).

[19] L. Lam, R.W. Koepcke, and T.Y. Lin, in *Soft Computing*, edited by T.Y. Lin and A.W. Wildberger (Society for Computer Simulation, San Diego, 1995).

[20] H.-J. Zimmermann, *Fuzzy Set Theory and Its Applications* (Kluwer Academic, Boston, 1991).

[21] R.D. Pochy, D.R. Kayser, L.K. Aberle, and L. Lam, Physica D **66**, 166 (1993).

[22] L. Lam, in *Defect Structure, Morphology and Properties of Deposits*, edited by H.D. Merchant (Minerals, Metals and Materials Society, Warrendale, PA, 1995); L. Lam, R.D. Pochy, and V.M. Castillo, in [11].

[23] L. Lam, R.D. Freimuth, and H.S. Lakkaraju, Mol. Cryst. Liq. Cryst. **199**, 249 (1991).

[24] S. Kauffman, *The Origins of Order: Self-Organization and Selection in Evolution* (Oxford University, New York, 1993).

[25] R.P. Pan, C.R. Sheu, and L. Lam, *Chaos Solitons Fractals* **6**, 495 (1995).

[26] D.R. Kayser, L.K. Aberle, R.D. Pochy, and L. Lam, Physica A **191**, 17 (1992).

[27] R.H. Masland, in *Fractals and Disordered Systems*, edited by A. Bunde and S. Havlin (Springer-Verlag, New York, 1991), Fig. 1.0.

[28] Y. Sawada, A. Dougherty, and J.P. Gollub, Phys. Rev. Lett. **56**, 1260 (1986).

[29] E. Ben-Jacob and P. Garik, Nature **343**, 523 (1990).

[30] S. Jakubith, H.H. Rotermund, W. Engel, A. von Oertzen, and G. Ertl, Phys. Rev. Lett. **65**, 3013 (1990).

[31] *Spiral Symmetry*, edited by I. Hargittai and C.A. Pickover (World Scientific, River Edge, 1992).

[32] T.A. Cook, *The Curves of Life* (Dover, New York, 1979).

[33] W.Y. Tam and J.J. Chae, Phys. Rev. A **43**, 4528 (1991).

[34] P. Maass, A. Bunde, and M.D. Ingram, Phys. Rev. Lett. **68**, 3064 (1992); A. Bunde, M.D. Ingram, P. Maass, and K.L. Ngai, J. Non-Cryst. Solids **131–133**, 1109 (1991).

[35] A. Bunde and P. Maass, Physica A **191**, 415 (1992).

[36] A. Bunde, M.D. Ingram, and P. Maass, J. Non-Cryst. Solids **172–174**, 1222 (1994). See also G.N. Greaves and K.L. Ngai, Phys. Rev. B **52**, 6358 (1995).

[37] F. Schweitzer, K. Lao, and F. Family, "Active Random Walkers Simulate Trunk Trail Formation by Ants," Biosystems (in press, 1996).

[38] B. Hölldobler and E.O. Wilson, *The Ants* (Belknap, Cambridge, MA, 1990).

[39] B. Hölldobler and M. Möglich, Insects Sociaux **27**, 237 (1980).

[40] L. Lam, M.C. Veinott, and R.D. Pochy, in [12].

[41] L. Lam, M.C. Veinott, D.A. Ratoff, and R.D. Pochy, in [42].

[42] *Fluctuations and Order: The New Synthesis*, edited by M.M. Millonas (Springer-Verlag, New York, 1996).

[43] H.M. Jaeger, C.H. Liu, and S.R. Nagel, Phys. Rev. Lett. **62**, 40 (1989).

[44] P. Bak, C. Tang, and K. Wiesenfeld, Phys. Rev. Lett. **59**, 381 (1987).

[45] P. Bak, in [46].

[46] *Complexity: Metaphors, Models, and Reality*, edited by G.A. Cowan, D. Pines, and D. Meltzer (Addison Wesley, Menlo Park, 1994).

398   L. Lam

[47] S.K. Grumbacher, K.M. McEwen, D.A. Halvorson, D.T. Jacobs, and J. Lindler, Am. J. Phys. **61**, 329 (1993).

[48] G.A. Held, D.H. Solina, II, D.T. Keane, W.J. Haag, P.M. Horn, and G. Grinstein, Phys. Rev. Lett. **65**, 1120 (1990).

[49] E. Ben-Jacob, O. Shochet, A. Tenenbaum, and O. Avidan, in [12], color plate 12.

[50] V.M. Castillo, M.C. Veinott, and L. Lam, *Chaos Solitons Fractals* **6**, 67 (1995).

[51] A.L. Barabási and H.E. Stanley, *Fractal Concepts in Surface Growth* (Cambridge University, New York, 1995).

[52] D.J. Eaglesham, H.-J. Gossmann, and M. Cerullo, Phys. Rev. Lett. **65**, 1227 (1984).

[53] M. Siegert and M. Plishke, Phys. Rev. Lett. **68**, 2035 (1992).

[54] J.G. Amar, P.-M. Lam, and F. Family, Phys. Rev. E **47**, 3242 (1993); F. Family and J.G. Amar, Fractals **1**, 753 (1993).

[55] F. Schweitzer and L. Schimansky-Geier, Physica A **206**, 359 (1994): See also L. Schimansky-Geier, M. Mieth, H. Rosé, and H. Malchow, Phys. Lett. A **207**, 140 (1995).

[56] J.G. Amar and F. Family, Phys. Rev. Lett. **64**, 236 (1990).

[57] S. Das Sarma and P. Tamborenea, Phys. Rev. Lett. **66**, 258 (1991).

[58] C. Tang, Physica A **194**, 315 (1993).

[59] H. Rosu and E. Canessa, Phys. Rev. E **47**, 3818 (1993).

[60] C.R. Carme, in *Fractals and Disordered Systems*, edited by A. Bunde and S. Havlin (Springer-Verlag, New York, 1991), Fig. 7.2.

[61] L.A. Zadeh, Inf. Contr. **8**, 338 (1965).

[62] T. Munakata and Y. Jani, Commun. ACM **37**(3), 69 (1994).

[63] H. Freund and P. Grassberger, Physica A **190**, 218 (1992).

[64] D.J. Amit, G. Parisi, and L. Peliti, Phys. Rev. B **27**, 1635 (1983).

[65] M.M. Millonas, in [12]. See also E.M. Rauch, M.M. Millionas, and D.R. Chialvo, Phys. Lett. A **207**, 185 (1995).

[66] A. Mehta, *in Correlations and Connectivity*, edited by H.E. Stanley and N. Ostrowsdy (Kluwer Academic, Boston, 1990); A. Mehta, *Granular Matter* (Springer-Verlag, New York, 1994).

[67] D. Helbing, P. Molnar, and F. Schweitzer, in *Evolution of Natural Structures: Principles, Strategies, and Models in Architecture and Nature*, Proceedings of the 3rd International Symposium of the Sonderforschungsbereich 230, Stuggart, 1994.

[68] D.E. Goldberg, Commun. ACM **37**(3), 113 (1994).

[69] D.A. Kessler and H. Levine, Phys. Rev. E **48**, 4801 (1993).

[70] E. Ben-Jacob, O. Schochet, A. Tenenbaum, I. Cohen, A. Czirok, and T. Vicsek, Nature **368**, 46 (1994).

[71] L. Lam and C.Q. Shu, "Active Walks and Increasing Returns in Economics" (preprint, 1996).

[72] *Evolution of Dynamical Structures in Complex Systems*, edited by R. Friedrich and A. Wunderlin (Springer-Verlag, New York, 1992); *Self-Organization of Complex Structures: From Individual to Collective Dynamics*, edited by F. Schweitzer (Gordon and Breach, London, 1996).

Appendix: Historical Remarks on Chaos

Michael Nauenberg

During the past decade there has been virtually an explosion of activity in a new field that has become commonly known as *Nonlinear Physics*. There are numerous journal articles [1, 2], textbooks [3, 4], conferences [5], and workshops on such diverse subjects as chaotic dynamics, strange attractors, fractal measures, and many others discussed in this book, which were virtually unknown to most physicists more than 20 years ago. Although some of the basic concepts underlying these new developments had been known as early as the turn of the century, these concepts did not enter into the mainstream of physics until rather recently. The main reason for the rapid development of these subjects during the past years has been the introduction of increasingly powerful computers and graphics workstations, which have enabled scientists to visualize very complex phenomena. While these developments have affected practically every branch of the *sciences*, we will confine our discussion here primarily to *physics*, and in particular to chaotic dynamics.

To many generations of students, *classical mechanics* has been taught as a perfect example of a causal and deterministic theory, at least for systems with only a few degrees of freedom. Given initial conditions in phase space, the evolution of a mechanical system obeying Hamilton's equation is determined uniquely for all future time. The behavior of such a system is predictable. Furthermore, this evolution is reversible in time. Numerous examples are given in textbooks to illustrate the validity of these fundamental principles. The most famous and historically important case is the motion of a planet around the sun which, in the absence of perturbations, follows an elliptic orbit in accordance with Kepler laws.

In retrospect, it is remarkable that there is no discussion in these textbooks about every day experiences that might have given rise to doubts about the general validity of these principles. For example, why is it apparently impossible to determine in practice the outcome of flipping a coin or rolling dice? Afterall, a coin and a die are rigid bodies that obey Hamiltonian dynamics, yet apparently we cannot predict their evolution. The same is true of other gambling devices, like roulette wheels and pinball machines, which are mechanical systems where the outcome is not *predictable* but *random*. Indeed, the theory of probability and statistics, which gives a precise meaning to the concept of a random process, was originally developed to give a mathematical description of these devices.

The basic reason for the uncertainty in the outcome of flipping a coin or rolling dice is that the evolution is very sensitive to the initial conditions. An infinitesimally small change in the initial impulse given to the coin or dice can lead to a totally different final state. Basically this sensitivity can be traced to the occurrence of unstable equilibria when the coin or dice hits the ground on edge. Many nonlinear dynamical systems exhibit this sensitivity to initial conditions, and the long time evolution of these systems cannot be predicted, because in practice there are always errors in these initial conditions which grow exponentially in time. The resulting motion has been called chaotic or deterministic chaos.

The effect of sensitivity to initial conditions in classical mechanics was emphasized clearly in recent times by the German physicist Max Born in an apparently little known article entitled "Is Classical Mechanics in Fact Deterministic?" written in 1955 [6]. He considered a simple model of a particle bouncing from fixed obstacles (the Lorentz model of a gas or a pinball machine). It is clear that the smallest change in initial direction will lead to large changes in the resulting zigzag motion of the particle. Born, who developed the probabilistic interpretation in quantum mechanics, was deeply concerned with the reluctance of leading physicists, including Einstein, to accept the fact that this fundamental theory of nature was not deterministic. Ironically, Einstein defended his views with the remark that *"God does not play dice."* Born concluded that the determinism of classical physics *turns out to be a false appearance* [7] due to the neglect of physical limitations in the initial conditions.

For systems with very many degrees of freedom, such as a gas of particles or a fluid, classical mechanics is supplemented by statistical methods, because one is interested in the average macroscopic properties of such systems. We are not concerned here with the fluctuations or noise of a statistical ensemble, but it is worthwhile to point out that the lack of time reversibility—the arrow of time—in the evolution of macroscopically averaged systems is again due to the sensitivity to initial conditions of the fundamental equations of classical mechanics, which are time reversal invariant. In the case of fluids we have the well-known phenomenon of turbulence, which is one of the most dramatic manifestations of chaotic behavior. Nevertheless, we believe that fluids satisfy by and large the Navier–Stokes equations, which are partial differential equations describing the average behavior of the motion of the particles composing the fluid. This is supported by a great deal of numerical work, although it should be remembered that solutions in the turbulent regime for three-dimensional fluid flow are barely obtainable even on the most powerful computers.

In the early 1960s, an MIT meteorologist, E. Lorenz, who had developed a simplified set of three ordinary differential equations to describe fluid flow under a thermal gradient, was solving these equations numerically on a computer, when he stumbled on the fact that for certain parameter values the solutions were very sensitive to initial conditions. Contrary to conventional wisdom, Lorenz found that a slight change of initial values of his variables (originally an unplanned change in roundoff in the numerical value), led to totally different solutions at later times. At the time only a few scientists paid much attention to this development, which was published in a journal of meteorology [8]. Among them was a mathematician at UC Berkeley, S. Smale, who had made fundamental contributions in understanding the evolution of dynamical systems, and J. Yorke from Maryland who circulated the work of Lorenz. An independent but related development occurred about 10 years later with the discovery by three mathematicians at Los Alamos, M. Metropolis, M.L. Stein, and P.R. Stein, of universal properties of iterates of certain nonlinear maps. Such maps give the simplest possible mathematical models for dynamical systems that exhibit a transition from regular or periodic motion to complex, irregular, or chaotic motion. This work [9] also remained unnoticed until the publication of the classic article on period doubling bifurcations of the logistic map by the biologist, R. May, which appeared in *Nature* in 1976 [10]. At about this time a physicist at Los Alamos, M. Feigenbaum, who was familiar with the seminal work that had been done there, discovered the existence of universal critical exponents characterizing the transition from periodic to chaotic motion for maps exhibiting period doubling bifurcations [11]. Remarkably, the same discovery was made independently by P. Coullet and C. Tresser [12], two graduate students at the University of Nice in France, and by S. Grossmann and S. Thomae, two physicists working at the University of Marburg in Germany [13]. Originally, the concepts of universality and critical exponents had been developed in statistical mechanics to described phase transitions, and the underlying theory of scaling and the renormalization group were applied by Feigenbaum and Cvitanovic and by Coullet and Tresser to develop a mathematical theory for period doubling bifurcations.

It is interesting to note that up to this time all the developments had been purely theoretical, supported by numerical explorations on computers. However, there was hope that these discoveries might be relevant to physics and in particular to the understanding of the transition to turbulence in fluid flow. Meanwhile, two physicists at the Ecole Normale Superiore in Paris, Libchaber and Maurer [14], unaware of these theoretical developments, had set up the famous Rayleigh–Bérnard cell for a fluid in a thermal gradient to study this transition, and unexpectedly discovered a period doubling bifurcation experimentally. There are now many experiments that have demonstrated this route to turbulence, in excellent agreement with theory. Perhaps one of the simplest cases for this transition, which can be observed in everyday life, is the dripping faucet, well known as a household headache. At a conference in Bavaria, Germany, Otto Rössler speculated that a leaky faucet might exhibit deterministic chaotic behavior in the transition from regular dripping to continuous flow, and Rob Shaw and his collaborators at UC Santa Cruz demonstrated the occurrence of period doubling *experimentally* [15].

It should be emphasized that there are other routes for a transition from laminar to turbulent flow in a fluid. A famous example is the pioneering experiment on Taylor–Couette flow of J. Gollub and H. Swinney [16], where the appearance of a few oscillating modes of the fluid that become successively unstable marks the transition to turbulence. The theory of L. Landau [17] and E. Hopf [18] proposed the occurrence of an infinite set of incommensurate frequencies to explain turbulence, but had been criticized earlier on theoretical grounds by D. Ruelle and F. Takens [19]. They introduced the fundamental concept of a strange attractor to explain chaotic behavior in a dissipative dynamical system [20]. Many simple mathematical models have been invented that give rise to a strange attractor, which is very complex and can be partly characterized by nonintegral or Hausdorff dimensions, popularly known as fractal dimensions [21]. The development of fractal geometry initiated by B. Mandelbrot has played an important role in elucidating complex behavior in general nonlinear systems.

We now return to a brief description of the parallel developments that occurred in Hamiltonian or conservative dynamical systems. In the early 1960s the Russian mathematician V.I. Arnold [22], and the German mathematician J. Moser [23] proved an important theorem of Kolmogorov [24], known as the KAM theorem, concerning the stability of invariant tori in Hamiltonian mechanics for integrable systems, under the influence of small perturbations. These results have been properly hailed as one of the greatest milestones in celestial mechanics since the pionering work of Poincaré [25]. Already around the turn of the century, Poincaré had called attention to the complexity of orbits in the three body problem, and progress toward understanding this behavior culminated in the KAM theorem. The motion of planets in the solar system has always been regarded in elementary texts as a canonical example of deterministic motion. The approximation usually made is that the orbit of a planet is in a Kepler elliptic orbit, neglecting the effects of other nearby planets as small perturbations. However, it is another matter to show that these effects are unimportant over long times, and indeed the question of the stability of the solar system has been a central problem of celestial mechanics. The KAM theorem proves that under certain rather stringent mathematical conditions orbits satisfying these conditions remain confined to tori. Actually, numerical solutions in many examples show that this is the case under much less restricted conditions than required by the theorem. Very recently, G.J. Sussman and J. Wisdom [26] at MIT have shown by an extensive numerical calculation on a specially dedicated computer, the Digital Orrery, that the orbit of one of the planets in the solar system, Pluto, is chaotic. A general property of the Hamiltonian dynamics for nonintegrable systems is that there exists regions of phase space where chaotic trajectories are interspersed, often in extraordinary complex fashion, with regular domains. In practice these regions can only be determined by direct numerical integration of the equations of motion.

In conclusion, we note that in spite of the considerable progress that has been made in understanding the nature of the onset of chaos or turbulence in physical systems, there still remain some fundamental unsolved problems. We have confined our discussion to systems that are well described by classical mechanics, but an important question concerns the applications of these ideas in the realm of atomic physics where quantum mechanics must be applied. The Schrödinger equation is a *linear* partial differential equation, and it is not apparent how it might exhibit any sensitive dependence on initial conditions. Nevertheless, we must face the fact that quantum mechanics should correspond to classical mechanics in the limit of large quantum numbers. Recently there has been a great deal of work in attempting to understand quantum properties of systems that have chaotic behavior in the classical regime [27]. The most striking example is the hydrogen atom in the presence of strong electromagnetic fields, and much progress has been made toward elucidating the behavior of this system both experimentally and theoretically [27, 28]. The interested student is urged to read the relevant references.

References

[1] P. Cvitanovic, *Universality in Chaos* (Adam Hilger, Bristol, 1989).

[2] Hao Bai-Lin, *Chaos II* (World Scientific, Teaneck, 1990).

[3] M. Tabor, *Chaos and Integrability in Nonlinear Dynamics* (Wiley, New York, 1989).

[4] G.L. Baker and J.P. Gollub, *Chaotic Dynamics* (Cambridge University, Cambridge, 1990).

[5] *Dynamical Chaos*, edited by M.V. Berry, I.C. Percival and N.O. Weiss (Princeton University, Princeton, 1987).

[6] M. Born, Physicalische Blatter **11** (9), 49 (1955).

[7] M. Born, Science **122**, 675 (1955).

[8] E. Lorenz, J. Atmos. Sci. **20**, 130 (1963).

[9] N. Metropolis, M.L. Stein, and P.R. Stein, J. Combinatorial Theor **15**, 25 (1973).

[10] R. May, Nature **261**, 459 (1976).

[11] M. Feigenbaum, J. Stat. Phys. **19**, 25 (1978); J. Stat. Phys. **21**, 669 (1979); Los Alamos Science **1**, 4 (1980).

[12] P. Coullet and C. Tresser, C.R. Acad. Sci. Paris **287**, 577 (1978); J. Phys. (Paris) **39** Coll., C 5–25 (1978).

[13] S. Grossmann and S. Thomae, Z. Naturforsch. **32a**, 1353 (1977).

[14] J. Maurer and A. Libchaber, J. Phys. Lett. (Paris), **40**, L4l9 (1979); in *Nonlinear Phenomena at Phase Transitions and Instabilities*, edited by R. Riste (Plenum, New York, 1982).

[15] R. Shaw, *The Dripping Fauces as a Model Chaotic System* (Aerial, Santa Cruz, 1984); P. Martien, S.C. Pope, P.L. Scott and R.S. Shaw, Phy. Lett. **110A**, 399 (1985).

[16] J.P. Gollub and H.L. Swinney, Phys. Rev. Lett. **35**, 927 (1975).

[17] L.D. Landau, C.R. Dokl. Acad. Sci. USSR **44**, 311, (1944); in *Collected Papers of L.D. Landau*, edited by D. ter Har (1965).

[18] E. Hopf, Commun. Pure Appl. Math. **1**, 303 (1948).

[19] D. Ruelle and F. Takens, Commun. Math. Phys. **20**, 167 (1971); **23**, 343 (1971).

[20] D. Ruelle, *Chaotic Evolution and Strange Attractors* (Cambridge University, Cambridge, 1989).

[21] B.B. Mandelbrot, *Fractal Geometry of Nature* (Freeman, New York, 1982).

[22] V.I. Arnold, Russ. Math. Surveys **18**, 9 (1963).

[23] J. Moser, Nachr. Akad. Wiss. Göttingen Math. **2**, 1 (1962).

[24] A.N. Kolmogorov, Akad. Nauk. USSR Doklady **98**, 527 (1954); Lecture Notes Phys. **93**, 51 (1979).

[25] H. Poincaré, *Methodes Nouvelles de la Mécanique Céleste*, three vols (Gauthier-Villars, Paris, 1892–1899). Reprinted by Dover (New York, 1957).

[26] G.J. Sussman and J. Wisdom, Science **241**, 433 (1988).

[27] B. Eckhardt, Phys. Reports **163**, 205 (1988).

[28] M.C. Gutzwiller, *Chaos in Classical and Quantum Mechanics* (Springer-Verlag, New York, 1990).

Contributors

Ralph H. Abraham received his Ph.D. in mathematics at the University of Michigan in 1960. During his career at UC Berkeley, Columbia, Princeton, and UC Santa Cruz, he worked in dynamical systems theory and applications, and is the author of a number of books on dynamics, chaos, and the history of mathematics. *Address: Department of Mathematics, University of California, Santa Cruz, CA 95064, USA. Email: rha@cats.ucsc.edu.*

Stephen G. Eubank received his B.A. from Swarthmore College and Ph.D. from the University of Texas at Austin in theoretical particle physics. After post-docs in fluid dynamics at La Jolla Institute and nonlinear dynamics at Los Alamos National Laboratory, he helped found Prediction Company in Santa Fe. He has since consulted for Biosphere 2 and for the TRANSIMS transportation systems simulation project at Los Alamos National Laboratory. His scientific interests are in the mechanisms by which small scale interactions generate large scale phenomena. He pays the bills by doing computational statistics and modeling. He is currently building models for machine parsing of English text at Advanced Telecommunications Research International (ATR) in Kyoto, Japan. *Email: eubank@santafe.edu.*

J. Doyne Farmer is Senior Scientist at Prediction Company, which specializes in the application of prediction methods to financial markets. Prior to founding Prediction Company, he obtained a B.S. in physics from Stanford University and a Ph.D. in physics from the University of California, Santa Cruz. In 1981 he went to work at Los Alamos National Laboratory, where he was first an Oppenheimer Fellow with the Center for Nonlinear Studies, and then became founder and leader of the Complex Systems Group in the Theoretical Division. Dr. Farmer has published more than fifty papers in the fields of chaos and complex systems, including contributions to problems in dynamical systems theory, fractals, time series modeling, theoretical immunology, neurophysiology, machine learning, and the origin of life. *Address: Prediction Company, 320 Aztec Street, Suite B, Santa Fe, NM 87501, USA.*

Thomas C. Halsey received a Ph.D. in physics from Harvard University in 1984. He did post-doctoral work at the University of Chicago and at CEA-Saclay, and joined the faculty at Chicago in 1987. In 1994 he became staff physicist at Exxon Research and Engineering, his current position. His research interests include pattern formation, nonlinear dynamics, and smart materials. He is the recipient of a Presidential Young Investigator Award and an Alfred P. Sloan Junior Fellowship. *Address: Exxon Research and Engineering, Route 22 East, Annandale, NJ 08801, USA. Email: tchalse@erenj.com.*

David E. Hiebeler received his B.S. in computer science from Rensselaer Polytechnic Institute, and his M.S. in applied mathematics from Harvard University. He has worked at Thinking Machines Corporation and the Santa Fe Institute, developing agent-based simulation environments for the exploration of complex systems. He is currently a graduate student in applied mathematics and ecology at Cornell University, studying stochastic spatial models. *Address: Center for Applied Mathematics, 657 Rhodes Hall, Cornell University, Ithaca, NY 14853, USA. Email: hiebeler@cam.cornell.edu.*

Lui Lam did his thesis at Bell Laboratories, received his Ph.D. from Columbia University, and is now a professor at San Jose State University. He is the originator of active walks, bowlics, the Springer book series *Partially Ordered Systems*, and the International Liquid Crystal Society. Professor Lam is also noted for his contributions to Compton profiles, the dissipation function formulation of hydrodynamics and irreversible thermodynamics, and solitons in liquid crystals. *Address: Department of Physics, San Jose State University, San Jose, CA 95192-0106, USA. Email: luilam@email.sjsu.edu.*

Michael Nauenberg received his Ph.D. from Cornell University, and he has taught at Columbia University and at the University of California, Santa Cruz, where since 1966 he helped developed the physics department. He was cofounder and director of the Institute for Nonlinear Science at the University of California, Santa Cruz. He has been awarded fellowships from the A.P. Sloan, J.S. Guggenheim, and A. von Humboldt foundations, and during the 1995–1996 academic year he was the Van der Waals Professor of Physics at the University of Amsterdam. He has had a broad range of interest in physics and has made contributions in quantum field theory (KNL theorem for massless charged leptons with T.D. Lee), astrophysics (mass limit of neutron stars with G. Chapline), statistical mechanics (renormalization fixed points for first order phase transitions with B. Nienhuis), chaotic dynamics (scaling theory of noise with J. Rudnick and J. Crutchfield), and foundations of quantum mechanics (coherent wavepackets in Rydberg Atoms). He also worked to debunk cold fusion (with S. Koonin), and recently he has elucidated some long-standing historical problems in the development of celestial mechanics in the seventeenth century. *Address: Department of Physics, University of California, Santa Cruz, CA 95064, USA. Email: michael@mike.ucsc.edu.*

Leonard M. Sander got his Ph.D. in theoretical physics at the University of California at Berkeley. He is now Professor of Physics at the University of Michigan in Ann Arbor. He worked in several areas of solid state physics before turning to his present interest, the statistical physics of systems far from equilibrium. His contributions to the theory of aggregation and fractals (the subject of Part III in this volume) include inventing, with T. Witten, the diffusion-limited aggregation (DLA) model. His more recent work is on thin film growth, non-classical chemical reaction kinetics, and pattern formation. Professor Sander is a fellow of the American Physical Society. *Address: H. M. Randall Laboratory of Physics, University of Michigan, Ann Arbor, MI 48109-1120, USA. Email: LSANDER@umich.edu.*

Robert Tatar received his Ph.D. from the University of Pennsylvania in theoretical solid state physics. He spent several years at the GE Research and Development laboratory in Schenectady, NY as project leader of a new medical diagnostic instrument. His interests in computers and computational physics led him to cellular automata and the founding of Automatrix, Inc. Today the company specializes in commercial applications of parallel computers and networks (see http://calendar.com/concerts). *Address: Automatrix, Inc., P. O. Box 196, Rexford, NY 12148-0196, USA. Email: rct@automatrix.com.*

Geoffrey K. Vallis received his B.A. from University of Oxford and his Ph.D. from Imperial College London, both in physics. He spent a number of years at Scripps Institution of Oceanography, and is now Professor of Ocean Science and Director of the Institute of Nonlinear Science at the University of California, Santa Cruz. His research interests encompass geophysical fluid mechanics, turbulence theory, ocean-atmosphere dynamics and climate. Recent interest in problems of climate change has led to a much-heightened awareness of the need for an understanding of fluid dynamics and turbulence, as well as the myriad interactions and feedbacks in the complex ocean-atmosphere system, and has led to the realization that these are some of the most fascinating and difficult problems in science. *Address: Institute of Nonlinear Science, Department of Ocean Science, University of California, Santa Cruz, CA 95064, USA. Email: vallis@cascade.ucsc.edu.*

Index

A

Active walk, 359
  basic concept, 360
  Boltzmann, 365, 383, 387, 393
  branching, 366
  computer model, 363
  continuum description, 361
  deterministic, 364
  fuzzy, 365, 390
  principle of, 9, 360, 395
  probabilistic, 364
Active walk model, 360, 394
  BPAW, 379
  track pattern, 368
  track statistics, 393
Active walker, 360
  clustering, 394
  coexisting, 361
  intelligent, 395
  single, 363
Adaptive system, 361
Aerosol, 180, 190
Aggregation, 4, 6, 179
  ballistic, 194
  ballistic cluster-cluster, 206
  cluster, 180
  cluster-cluster, 190, 205
  irreversible, 179
  particle, 179
  reaction-limited, 206
Aharonov-Bohm effect, 360
Anisotropy, 186, 202
Annealed averaging, 41
Antikink, 220

Ants, 9, 359, 376, 394, 395
Approximation, local, 159
Arnold tonque, 81
Artificial life, 8
Attractor, 74
  chaotic, 94, 102
  Lorenz, 97
  regular, 75
  Rössler, 94, 156, 306
  strange, 5, 51, 94
Autocorrelation, 125
Autoregressive process, 127
Averaging, 115
  ensemble, 116
  method of, 316
  time, 116
AW. *See* Active walk
AWM. *See* Active walk model

B

Bäcklund transformation, 225
Bacteria, 189
  cell colony, 196, 394, 395
Basin of attraction, 5, 74
Bifurcation, 82, 102
  catastrophic, 88
  diagram, 88
  period doubling, 84
  pitchfork, 326
  subtle, 88
  surface, 88
Biology, 234, 359
Bond-orientation order, 252

Bond percolation, 16
Branching, 366
  factor, 366
  rule, 366
Breather, 220
Butler-Volmer equation, 25

C

Cantor set, 139, 146
  two-scale, 33
Cellular automata, 1, 7, 266, 275, 278, 296
  hardware, 291
Cellular dynamata, 296, 297
  visualization technique, 296
Central limit theorem, 110, 123
Chaos, 1, 5, 8, 55, 89, 153, 379
  controlling, 6
  higher dimensions, 101
  history, 401
  soliton, 266
  spatiotemporal, 310
Chemical reaction, 289
Chemistry, 234, 359
Coarse graining, 57, 108, 196
  resolution of, 109
Colloid, 4, 180, 190
Commensurate-incommensurate phase
    transition, 255
Complex adaptive systems, 9
Complex Ginzburg-Landau equation, 267
Complex phenomenon, 359
Complex system, 1, 8, 359
Computational capacity, 279
Computer science, 359
Conduction polymer, 7
Configuration space, 60
Conservation law, 218
Control space, 61
Convection, 24
Coordinate
  Broomhead-King, 157
  delay, 155
  derivative, 155
Correlation, 125
  function, 20, 124, 198
  integral, 143
Covariance, 124
  matrix, 128

Crystallization, 185
Cycle, 72

D

Damped θ^4 equation, 239
Damped driven θ^4 equation, 241
D'Arcy's law, 188
Degree of freedom, 138, 214, 218
Dendrite, 186
Dendritic crystal, 180, 187
Dendritic pattern, 204
Dense radial morphology, 203, 204, 370
Determinism, 55, 62
Dielectric breakdown, 9, 189, 200, 368,
    371
Diffusion, 282
  constant, 184, 361
  equation, 184, 234
  length, 185, 202
Diffusion-limited aggregation, 15, 17,
    179, 183, 187, 189, 206
  growth probability, 40
  physical application, 22
  sticking probability, 26
Diffusivity, 352
Dilatational symmetry, 15
Dimension, 138, 148
  correlation, 141
  embedding, 144
  fractal, 15, 140, 185, 197
  information, 44, 140
  Lyapunov, 145
  pointwise, 139
  topological, 138
Dirac delta function, 108
Discommensuration, 255
Discrete physics, 275
Dislocation, 228
Dispersion, 215
Dissipative system, 5, 67
Distribution, 122
  Gaussian, 123
DLA. *See* Diffusion-limited aggregation
Domain wall, 227, 249
Double layer, 25
Double sine-Gordon equation, 237, 267
Doubling operator, 334
Doubling transformation, 333

Dynamical system, 55, 296
  autonomous, 62
  characterization, 162
  complex, 296, 297
  conservative, 67
  deterministic, 58, 113
  dissipative, 67
  invertible, 62
  linear, 76
  reversible, 62
  stochastic, 113

E

Ecology, 359
Economic system, 4, 359, 395
Eden model, 22, 196
Edge diffusion, 189
Edwards-Wilkinson model, 194
Electrochemical deposition, 187, 203
Electroconvection, 258
Electrodeposition, 23, 368, 374, 380
  diffusive, 27
Electrostatic scaling law, 44, 47
Elementary particle, 213
Elliptic point, 72
Emerging behavior, 359
Encoding, 157
End-to-end length, 393
Engineering, 359, 390
Enstrophy, 345
Entrained oscillators, 81
Entropy, 133, 306
Ergodicity, 116
η model, 22
Event, 109
Evolution equation, 234
Excitable medium, 394

F

Feedback, 361
Fermi-Pasta-Ulam problem, 217
Film growth, 182, 384
Fisher equation, 238, 243, 244
Fix point, 75
Floquet number, 72
Flow, 63, 92
  laminar, 308, 311

population, 394
suspension, 66
Fluid flow, 213
Fluxon, 227
Fokker-Planck equation, 362, 387, 394
Folding, 92
Forecast, 160
Fourier transform, 3, 122
Fractal, 1, 4, 5, 15, 94, 138, 179, 183, 197
  fat, 144
  geometry, 181
  measure, 30
  self-affine, 182, 192, 380
  surface, 381
Free electron laser, 267
Frenkel-Kontorova model, 228, 255
Fuzzy
  active walk, 393
  control, 365, 392
  logic, 365
  walk, 390

G

Galaxy, 4, 213
Game of life, 7, 277
Geology, 359
Glasses, ion transport, 9, 375
Granular flow, 267
Groove state, 382
Growth
  abnormal, 378
  ballistic, 181, 192
  city, 359, 395
  diffusion-limited, 183, 197
  diffusive, 6, 15
  disorderly, 201
  instability, 200
  irreproducible, 378
  irreversible, 180
  power law, 194
  reaction-limited, 196
  transformational, 378
Growth velocity, 186

H

Hamiltonian, 230, 236, 263
Hamiltonian system, 218

Hausdorff dimension, 15
Heat-seeking missile, 361
Hele-Shaw cell, 3, 6, 188
Hirota method, 226
Homoclinic
    intersection, 100
    point, 100
    tangle, 100
Hopf bifurcation, 83, 314
Hyperbolic point, 72
Hysteresis, 89

I

Increasing return, 395
Inertial range, 341
Information, 133
Integrability, 7, 218
Integrable system, 213, 267
Interface, 3
Intermittent sequence, 335
Internet, 361
Interval, 119
Intrinsic abnormal growth, 378
Invariant circle, 81
Invariant set, 71
Inverse scattering method, 4, 224, 244

J

Josephson junction, 213, 237
Jupiter, 213

K

Kadomtsev-Petviashvili Equation, 223
Kaplan-Yorke conjecture, 146
Kardar-Parisi-Zhang model, 197
Kelvin-Helmholtz instability, 312
Kernel density estimation, 110
Kernel function, 110
Kink, 220
Kolmogorov scale, 344
Kowteweg-de Vries equation, 215, 218,
    219, 223, 225, 231, 244, 246
Kullback information distance, 136

L

Lamb shift, 360
Landau constant, 316

Landau equation, 314, 319
Landscape, 361, 380
Landscaping function, 361, 373, 393
Langevin equation, 361, 394
Laplace equation, 3, 21
Laser pulse, 213, 265
Lattice anisotropy, 200
Lattice gas, 277, 280, 287
    FHP, 284
    HPP, 283
Lattice gas automata, 8, 277
Legendre transform, 31
Limit cycle, 78, 101
Linearization, 68
Linear stability, 200, 311
Linear system, 4, 61, 76
Liouville's theorem, 67
Liquid crystal, 213, 234, 252
    cholesteric, 252, 255
    nematic, 252, 254, 258
    smectic A, 252
Localization-delocalization transition,
    383
Lyapunov exponent, 64, 70, 89, 101, 145,
    148, 162, 306, 351
Lyapunov number, 71
Lyapunov spectrum, 71

M

Mackey-Glass equation, 102
Macroscopic system, 1, 360
Magnetic system, 213, 266
Makarov theorem, 45
Manifold
    orientable, 62
    stable, 98
    unstable, 98
Map, 63
    binary shift, 90, 106, 147
    Hénon, 66, 69
    logistic, 76, 84, 119, 123, 325
    Poincaré, 64, 321
    tent, 92, 119
Marginal stability, 245
Markov process, 113
Maximal entropy principle, 110
Mean, 122
Measure, 107
    dynamical, 116

invariant, 117
  Lebesgue, 108, 144
  natural, 118
  singular, 108
  support of, 108
Membership function, 390
Mesoscopic system, 360
Metric entropy, 146, 148
Micromachine, 395
Mixed alkali effect, 375
Mixing, 120
Molecular system, 213
Moment, 122
Morphogenesis, 296
Morphogram, 380
Morphology, 369
Moving average process, 127
Mullins-Sekerka instability, 200, 203
Multifractal, 30, 201
  correlation, 37
Multifractality, 30
Multiple time scales, 316
Multiply periodic motion, 74
Multisoliton, 220, 222
Multiwalkers, 367
  updating rule, 367
Mutual information, 135

N

Navier-Stokes equation, 7, 152, 309, 369
Neuron, retinal, 369
Noise, 113, 148
  dynamical, 114
  observational, 115
  pure white, 115
  reduction, 163, 170, 172
Nondeterminstic dynamics, 113
Nonequilibrium system, 359, 379
Nonintegrable system, 7, 230, 234, 267
Nonlinear diffusion equation, 234
Nonlinear equilibration, 311
Nonlinear evolution equation, 237
Nonlinearity, 2, 9
Nonlinear model, 56
Nonlinear Schrödinger equation, 219, 230
Nonlinear science, 1, 214
Nonlinear system, 2, 4, 215
Nyquist frequency, 131

O

1/f noise, 5, 388
Operator product expansion, 38
Optical communication, 7, 265
Optical fiber, 7, 264
Optics, 287
Orbit, 58

P

Parameter, 61
  space, 61
Partition, 108
Pattern
  compact, 6, 370, 381
  dielectric breakdown, 9, 369, 371
  filamentary, 6, 370, 381
Pattern formation, 1, 6, 180, 245, 252, 266, 359
Peclet number, 352
Pedestrian dynamics, 394
Percolation, 16
  cluster, 15
  soft medium, 394
Period doubling sequence, 324
Periodic orbit, 72
Personal computer, 4, 8
Pesin's identity, 148
Phase coherence, 104
Phase-locked oscillators, 81
Phase space, 58
Phase transition, 279, 381, 394
Phonon, 249
Plasma, 213
Poincaré-Bendixson theorem, 94
Poincaré section, 64
Pollution control, 359
Polyacetylene, 261
Polymer, 213, 394
Popular books, 9
Population distribution, 394
Porous medium, 188
Power law, 5
Power spectrum, 129
Prediction, 5, 153
  error, 160
  time, 106
Probability, 106, 107
  conditional, 109
  distribution, 109

function, 109
joint, 109
marginal, 109
self moment of, 134
Probability density function, 107
Propagation-collision cycle, 283

Q

Quantum mechanics, 1
Quasiparticle, 250, 280, 282, 287
Quasiperiodic motion, 74
Quasiperiodic sequence, 322
Quenched averaging, 41

R

Radius of gyration, 19, 393
Random process, 55, 58, 10
    Gaussian, 127
    linear, 127
Random variable, 107
    identically distributed, 107
Rayleigh-Bénard convection, 6, 81, 258,
    359
Reaction-diffusion equation, 234
Recurrence, 74
Relativity, 1
Renormalization, 352
Renormalization group, 324
Repellor, 72
Revolution, 1
River formation, 394
Rössler equations, 94
Rough surface, 380, 389
Roughness, 182, 194

S

Saddle point, 72
Sandpile, 5, 394
Scaling, 44, 160, 181, 192, 194, 197, 329,
    341, 387
Scaling function, 194, 196
Screening, 199
Self-focusing, 229
Self-organization, 359
Self-organized criticality, 5, 9, 15, 379
Self-similarity, 4
Sensitive zone, 380

Separatrix, 75
Shadowing, 164, 167
Shallow water wave, 227, 267
Shear, 254
Signal, 163
Simple pendulum, 3, 222
Simulation, 197, 281, 291
Sine-Gordon equation, 219, 222, 225,
    228, 231, 236, 251, 258
Social system, 5, 359
Sociology, 359
Solidification, 6
Soliton, 1, 6, 213, 388
    A, 239, 240, 242
    B, 239, 240, 242
    C, 239, 240, 242
    D, 240
    elastic collision, 213, 217, 234
    equation, 219
    formation, 244
    generation of, 217
    history, 215
    magnetic, 266
    mechanical analogue, 220, 235, 238,
        245
    method of construction, 243
    method of solution, 224
    nonrigorous, 234
    optical, 265
    origin, 215
    perturbation, 230, 246
    physical systems 227, 252
    statistical mechanics, 247
Soliton chaos, 266
Sound wave, 285
Spiral, 369
Spring, 3
Stability, 68, 82
    absolute 68
    local, 69
    marginal 68, 245
Standard deviation, 122
State 58
State space, 58
    reconstruction, 154
Statistic, 122
Statistical independence, 109
Statistical property, 122
Sticking probability, 199

Stochastic process, 58, 107
Stock market, 4, 5, 359
Stretching, 92
Structural phase transition, 213
Subharmonic bifurcation, 84
Superconductor, 252, 388
    high T_c, 213
Superfluid, 213
Superposition principle, 3, 61
Surface exponent, 196, 197
Surface of section, 64
Surface tension, 202
Symbolic dynamic, 306
Symmetry breaking, 9
Symplectic relation, 67

T

Tangent bifurcation, 335
Taylor-Couette flow, 81
θ^4 equation, 236, 248
Thin film, 181, 189, 197
Time series, multivariate, 113
Time series, univariate, 113
Torus, 74, 80
Traffic flow, 267, 359
Trajectory, 58
Traveling wave, 213, 244
Turbulence, 308, 309
    fully developed, 308

predictability, 348
strong, 308, 341
transition to, 308
two-dimensional, 345
weak, 321
Turkevich-Scher law, 45, 46

U

Uncertainty exponent, 138
Unfolding, 88
Universality, 192, 196, 199, 329
Universe, 252
Urban growth, 395

V

Van der Pol oscillator, 79
Vapor deposition, 181
Variance, 122
Viscous fingering, 3, 6, 188
Vorticity equation, 345

W

Wagner number, 25
Weather, 5, 361
Wiener-Khinchine theorem, 130
Witten-Sander model, 179